BIOTECHNOLOGY-V

Including

Animal Cell Biotechnology, Immunology and Plant Biotechnology

(SECOND EDITION)

M.K. Sateesh
Faculty, P.G. Department of
Biotechnology and Microbiology
Bangalore University
Bangalore

PUBLISHING FOR ONE WORLD

NEW AGE INTERNATIONAL (P) LIMITED, PUBLISHERS

New Delhi • Bangalore • Chennai • Cochin • Guwahati • Hyderabad
Jalandhar • Kolkata • Lucknow • Mumbai • Ranchi
Visit us at www.newagepublishers.com

Published by New Age International (P) Ltd., Publishers
First Edition: 2003
Second Edition: 2010
Reprint: 2010

Branches:
- No. 37/10, 8th Cross (Near Hanuman Temple), Azad Nagar, Chamrajpet, **Bangalore** - 560 018. Tel.: (080) 26756823, Telefax: 26756820, E-mail: bangalore@newagepublishers.com
- 26, Damodaran Street, T. Nagar, **Chennai** - 600 017. Tel.: (044) 24353401, Telefax: 24351463 E-mail: chennai@newagepublishers.com
- CC-39/1016, Carrier Station Road, Ernakulam South, **Cochin** - 682 016. Tel.: (0484) 2377004, Telefax: 4051303 E-mail: cochin@newagepublishers.com
- Hemsen Complex, Mohd. Shah Road, Paltan Bazar, Near Starline Hotel, **Guwahati** - 781 008.Tel.: (0361) 2513881 Telefax: 2543669, E-mail: guwahati@newagepublishers.com
- No. 105, 1st Floor, Madhiray Kaveri Tower, 3-2-19, Azam Jahi Road, Nimboliadda, **Hyderabad** - 500 027. Tel.: (040) 24652456, Telefax: 24652457, E-mail:hyderabad@newagepublishers.com
- RDB Chambers (Formerly Lotus Cinema)106A, 1st Floor, S.N. Banerjee Road, **Kolkata** - 700 014. Tel.: (033) 22273773, Telefax: 22275247, E-mail:kolkata@newagepublishers.com
- 16-A, Jopling Road, **Lucknow** - 226 001. Tel.: (0522) 2209578, 4045297, Telefax: 2204098 E-mail: lucknow@newagepublishers.com
- 142C, Victor House, Ground Floor, N.M. Joshi Marg, Lower Parel, **Mumbai** - 400 013. Tel.: (022) 24927869 Telefax: 24915415, E-mail: mumbai@newagepublishers.com
- 22, Golden House, Daryaganj, **New Delhi** - 110 002. Tel.: (011) 23262370, 23262368, Telefax: 43551305 E-mail: sales@newagepublishers.com

---

ISBN : 978-81-224-3063-9

Rs. 275.00

C-10-09-5067

Printed in India at Ajit Printers, Delhi.
Typeset at Pagitek Graphics, Delhi.

PUBLISHING FOR ONE WORLD
NEW AGE INTERNATIONAL (P) LIMITED, PUBLISHERS
4835/24, Ansari Road, Daryaganj, New Delhi-110002
Visit us at **www.newagepublishers.com**

BIOTECHNOLOGY-V

Dedicated to
my beloved parents
Mr. B. Krishnaiah
and
Mrs. Susheela Krishnaiah

Preface to the Second Edition

The word "**Biotechnology**" is first used by a Hungarian agricultural engineer Prof. Kark Erkey in 1919 and from that time to to-date a tremendous progress has been made in multifarious offshoots. It uses the informations from basic sciences to advanced technical fields. Biotechnology is one of the most exciting areas of natural sciences and its applications have to reach the needy people. For better understanding of the biotechnology, basic knowledge in the microbiology, biochemistry, genetics, physiology, animal cell culture, immunology, plant cell culture and molecular biology is very essential. Biotechnology is a frontier area of science and technology having a great potential for welfare of humanity. Biotechnology is becoming a dream area for the present day science students. Large number of universities, deemed institutions started various undergraduate, postgraduate and diploma courses in the field of biotechnology.

Biotechnology has been a centre of attention worldwide, after the successful publications of the entire human genome (February 2001) by National Institute of Health and Celera Genomics. This branch is considered a field, which can solve the world's hunger and health problems in lesser duration. It is no longer a specialised segment of science confined to high tech laboratories. The topics of biotechnology are now entering into the texts of degree and pre-university classes. The common people have started talking about biotechnology in the same breaths they talk about information technology. Newspapers are daily giving informations on all fields of biotechnology to educate normal public. The Government of India framed draft of Biotechnology Policy in 2005.

Globally too, biotechnology is slowly entering the hallowed portals of the 'Knowledge economy'. Biotechnology companies, especially in the areas of pharma, agriculture, genomics, proteomics and bioinformatics, are grabbing the public attention. In India too, the similar trend is noticed in the business field and in academics. India is becoming an emerging leader in the biotechnology field.

Within a span of 10 years more than 265 Biotechnology Companies are established in India working on different fields. "These companies targeting for the revenue generation of US $ 5 billion by the end of 2010. The major companies include Biocon India, Astrazenaca, Metahelix, Mansanto, Bharat Biotech, Shantha Biotech, Bangalore Genei, Reddies Laboratories. At present in biotechnology field the focus should be made on research, development and intellectual property rights generation in the area of vaccines, diagnostics, and new medicines derived from ethanobotany, traditional knowledge, rural practice, value added products through employing food biotechnology, bioethics, increased productivity through pest and disease resistant transgenic animals and crops with ecofriendly technologies.

Biotechnology is essentially an interdisciplinary field, students should get exposure to wide varieties of basic and applied sciences. Owing to its interdisciplinary nature, biotechnology field appeals to both science students as well as engineering students equally. A huge number of biotechnology books are published within the country and out of the countries. These books generally deal with a few specialized areas in depth and are expensive in nature. Thus, basic understanding of the subject becomes difficult with respect to these books.

This book has been written as a compiled text to the requirements of biotechnology students for understanding major branches of biotechnology for undergraduate and postgraduate courses and engineering courses of various Indian Universities, for competitive entrance examinations, and serves as an insight into the basics of the biotechnology.

The book presents all integrated approach to 'Animal Cell Biotechnology', 'Immunology' and 'Plant Biotechnology' encompassing information on the basics of the subject. The subject matter has been written in a stepwise systematic manner to meet the student's immediate requirements. Simple, clearly understandable illustrations, correct and up to date informations are the main features of this book. "Thus, it can serve as good reference material for students, policy makers and teachers.

'Animal Cell Biotechnology' part deals with the general metabolism of cultured animal cells, animal cell culture media, growth factors for culture system, serum-free media, chemically defined media, protein-free media, production of commercially important pharmaceuticals and various types of bioreactors and media, expression of cloned proteins in animal cells, over production of cloned proteins, purification from animal cells, downstream processing of expressed proteins, large scale cell culture in biotechnology. Culture of animal cells is a primary requirement for successful isolation and culture of animal cell lines. These cell lines are used for genetic modification and production of transgenic products free from all types of contaminants with higher yield. Thus, emphasis is given to various aspects of animal cell culture experiments, applications of animal cell culture and transgenic animal production techniques. Vaccine production from animal cell lines and specific application of transgenic animals topics are added as new addition.

In 'Immunology' topics such as history of immunology, immune system, its types, antigen and antibody structures, effector mechanisms, immunity to infectious diseases and vaccines are dealt in detail. Hypersensitivity reactions and its types, blood groups, blood typing and blood transfusions, erythroblastosis fetalis, antigen-antibody reactions, immunoassays chapters are new additions to the book. In each chapter the clear definition and illustrations is expanded for better understanding of the topics. These topics are covered from the basics which help for better understanding of cellular mechanisms in protection. This paves the way for designing an appropriate biotechnology product useful in the health care systems.

'Plant Biotechnology' part gives introduction to plant tissue culture, different types of media, culture methods and types, embryo culture and rescue, haploid production, somaclonal variations, somatic hybridization and *Agrobacterium* genetic modification systems, Arabidopsis thaliana genomics, plants and industry are given from the basics. In all the chapters, their respective biotechnological applications are dealt in thorough details. At present large number of transgenic plants are produced worldwide to solve multifarious problems in agriculture, health care and environment. Hence, basic concept of plant culture system, genetic manipulation, plant genomics and production of transgenics with its applications are given insight in this part of the book.

For purposes of readability, references are generally omitted from the text. Comments and any type of suggestions, criticism from the readers will be deeply appreciated and carefully considered in the preparation of the next edition.

M.K. SATEESH

Preface to the First Edition

This book has been written as a compiled text to meet the requirements of 'Animal Cell Biotechnology', 'Immunology' and 'Plant Biotechnology'—Paper V for the final year B.Sc. Biotechnology students of Bangalore University and others.

Biotechnology is one of the most exciting areas of natural sciences and its applications have to reach the needy people. After the successful publication of the entire human genome (2002), biotechnology has been a centre of attention worldwide. The claims about the birth of Baby Eve, claimed to be the first cloned human being in December 2002 have only reinforced global attention on biotechnology. It is no longer a specialized segment of science confined to high tech laboratories. The common people have started talking about biotechnology in the same breaths they talk about information technology.

Globally too, biotechnology is slowly entering the hallowed portals of the 'Knowledge economy'. Biotechnology companies, especially in the areas of pharma, agriculture, bioinformatics and genomics, are grabbing the public attention. In India too, the same trend is seen.

This book is intended to be an introduction to the subject for students. The book presents an integrated approach to Animal Cell Biotechnology, Immunology and Plant Biotechnology encompassing information on the basics of the subject. The subject matter has been written in a stepwise systematic manner to meet the student's requirements. Clearly understandable illustrations, correct and latest information are the main features of this book. Thus it can serve good reference material for students and teachers.

I am extremely grateful to Prof. H.A. Ranganath, FNAS, Director, NAAC, Bangalore, Dr. V. Prabhu, Chief of Laboratories, Medinova Diagnostic Centre, Bangalore, my teachers, my colleagues, students, friends and all my well wishers for their constant support and encouragement.

Special thanks are extended to my parents, Mrs. Sowmya Sateesh, in-law and other family members who have shown great forbearance and encouraged me during the compilation of this book.

Finally, I am also grateful to M/s New Age International (P) Ltd., Publishers, for bringing out this book within a short period.

Comments and suggestions from the readers will be deeply appreciated and carefully considered in the preparation of the next edition.

M.K. Sateesh

Acknowledgements

This book could not have been completed without help and contributions from many individuals. Indeed, this venture would not have begun were it not for the initial encouragement I received from our colleagues in Microbiology and Biotechnology fields.

I am extremely grateful to Dr. T. Prabhu, former Principal, Bangalore Medical College, former Chief of Laboratories, Medinova Diagnostic Centre, Bangalore, for his creative energy and push that had a substantial impact on the initial planning of this book. Dr. N. Prabhu Dev, Vice-Chancellor, Bangalore University, Prof. H.A. Ranganath, Director, NAAC, Bangalore, former Vice-Chancellor, Bangalore University, for their enthusiastic support and sound advice for shaping the book into final form. I also thank the following individuals for their valuable comments and constructive criticism raised during writing of this book: The Director, Forensic Science Laboratories, Bangalore, Prof. Geetha Bali, Prof. K. Manjunath, Prof. S. Shankara Bhat, Prof. K.A. Raveesha, Prof. S. Shishupala, Dr. C. Srinivas, Dr. K.M. Srinivas Murthy, Dr. S.T. Girish, Prof. S.K. Sarangi, Prof. S. Chidananda Sharma, Dr. D.C. Mohana, Dr. J. Savitha, Dr. D. Manjulakumari, Dr. Tharasaraswathi, Dr. T. Ravikiran, Dr. K.S. Devaraj, Dr. M. Sreepriya, Dr. T. Sharmila, Dr. Lakshminarayana, Dr. P.H. Ramanjini Gowda, Dr. K.M. Harini Kumar, Dr. S. Ramesh, Dr. Mohan Ram, Dr. S. Shyamalamma, Dr. N. Eranna, Dr. Anitha Peter, Dr. N.G. Ravichandra, UAS, GKVK, Dr. A.B. Vedamuthy, Prof. V.S. Ranganathan, Dr. Jayaprakash, Dr. H. Nagabhushana, Dr. H.N. Ramesh Babu, Dr. Rajeshwari Ramesh Babu, Dr. Sumithra K. Urs, U.S.A., Dr. K.P. Guruprasad, Dr. Chandrashekara G. Joshi, Dr. S.I. Umesh, Dr. S. Puttaraju, Dr. L. Rajanna, Dr. Y.L. Krishnamurthy, Dr. Y.L. Ramachandra, Dr. H. Ramakrishnaiah, Mrs. Sowmya Sateesh, Mr. M.K. Girish, Mrs. M.K. Sandhya Sateesh Kumar, Mr. S. Satish Kumar, Ku. Samrudhi, Ku. Samskruthi, Mrs. Gayathri Girish, Mr. R. Devaraj, Super book service, Mr. B. Umamahesh, Sigma-Aldrich, Mr. C.R. Naveen Kumar, Biocon India, Dr. K.S. Mohan, Lead Biotech product support & Product launch Liaison, Monsanto Research Centre, my teachers, my colleagues, my students, friends and all my well wishers for their constant support and encouragement. I also express my gratitude to Mr. V.N. Narasimha Murthy, for his suggestions and critical reading of the manuscript.

This often consuming enterprise was made possible by the emotional, intellectual, and logistical support of my family numbers; parents Mr. B. Krishnaiah, Mrs. Susheela Krishnaiah, my wife Mrs. Sowmya Sateesh, my son M.S. Yajat Sateesh Gowda, my brother Mr. M.K. Girish, my sister

Mrs. M.K. Sandhya Sateesh Kumar, my in-laws and other family members who have shown great forbearance and encouraged me during the compilation of this book. Without the help of these people it was not possible to complete the book in time.

Finally, I am also grateful to Mr. Saumya Gupta, Managing Director for steering me through the production processes and Mr. R.C. Gupta, Mr. S.P. Sudharshan, Mr. G.N. Mahesh, Mr. C.K. Suryanarayana, Mr. M. Srinivas and other staff member of the New Age International (P) Ltd., Publishers, in bringing out this book.

M.K. SATEESH

Contents

Part A
Animal Cell Biotechnology

Chapter

1

Culture Media

1.1 INTRODUCTION

The artificial environment created in the laboratory is generally known as media. A media comprises an appropriate source of energy for the cells which they can easily utilize and compounds which regulate the cell cycle. The choice of media is cell type specific and often empirical and there is no "all purpose" medium. It should provide many nutrients, buffering capacity, isotonic, and should be sterile. Characteristics and compositions of the cell culture media vary depending on the particular cellular requirements. Important parameters include osmolarity, pH, and nutrient formulations.

A wider range of ingredients needed to support survival and proliferation or differentiation. *In vitro* animal cell cultivation requires a complex combination of nutrients, considering glucose and glutamine as main carbon, energy and nitrogen sources. Mineral salts, amino acids and vitamins are also required; other essential nutrients, like growth factors, hormones, and receptor and transport proteins are required in small quantities as well. The pH was maintained at 7.4 and often it includes pH indicator phenol red (red at 7.4, yellow at 6.5, purple at 7.8). A typical media may or may not comprise of serum. The latter is called a serum-free media. Some of the common sources of serum can be fetal bovine serum, equine serum, calf serum etc. both the types of media have their own set of advantages and disadvantages.

The culture media is prepared in such a way that it provides

1. The optimum conditions of factors like pH, osmotic pressure, etc.
2. It should contain chemical constituents which the cells or tissues are incapable of synthesizing. Generally the media is the mixture of inorganic salts and other nutrients capable of sustaining cells in culture such as amino acids, fatty acids, sugars, ions, trace elements, vitamins, cofactors, and ions. Glucose is added as energy source—its concentration varying depending on the requirement. Phenol Red is added as a pH indicator of the medium.

Basic Components in the Culture Media

Most animal cell culture media are generally having following 10 basic components and they are as follows:

1. Energy sources: Glucose, Fructose, Amino acids
2. Nitrogen sources: Amino acids

3. Vitamins: Generally water soluble vitamines B & C
4. Inorganic salts: Na^+, K^+, Ca^{2+}, Mg^{2+}
5. Fat and Fat soluble components: Fatty acids, cholesterols
6. Nucleic acid precursors
7. Antibiotics
8. Growth factors and hormones
9. pH and bufferig systems
10. Oxygen and CO_2 concentration.

Animal cell culture media vary in their complexity but most contain:

Amino acids	0.1-0.2 mM
Vitamins	ca. 1 μM
Salts	NaCl 150 mM
	KCl 4-6 mM
	CaCl 1 mM
Glucose	5-10 mM

1.2 CULTURE MEDIA

A cell culture medium is composed of a number of ingredients and these ingredients vary from one culture medium to another. The nutrient media used for culture of animal cells and tissues must be able to support their survival as well as growth, i.e., must provide nutritional, hormonal and stromal factors (See Chapter 4).

The various types of media used for tissue culture may be grouped into two broad categories:

1. Natural media
2. Artificial media.

The choice of medium depends mainly on the type of cells to be cultured (normal, immortalized or transformed), and the objective of culture (growth, survival, differentiation, production of desired proteins).

Nontransformed or normal cells (finite life span) and primary cultures from healthy tissues require defined quantities of proteins, growth factors and hormones even in the best media developed so far. But immortalized cells (spontaneously or by transfection with viral sequences) produce most of these factors, but may still need some of the growth factors present in the serum.

In contrast, transformed cells (autonomous growth control and malignant properties) synthesize their own growth factors; in fact, addition of growth factors may even be detrimental in such cases. But even these cultures may require factors like insulin, transferrin, silenite, lipids, etc.

Natural Media

These media consist solely of naturally occurring biological fluids and are of the following three types:

1. Cagula or clots
2. Biological fluids
3. Tissue extracts.

The natural biological fluids are generally used for organ culture. For cell cultures, artificial media with or without serum are used.

Clots

The most commonly used clots are plasma clots, which have been in use for a long time. Plasma is now commercially available either in liquid or lyophilized state. It may also be prepared in the laboratory, usually from the blood of male fowl, but blood clotting must be avoided during the preparation.

Biological Fluids

Of the various biological fluids used as culture medium, serum is the most widely used. Serum is one of the very important components of animal cell culture which is the source of various amino acids, hormones, lipids, vitamins, polyamines, and salts containing ions such as calcium, ferrous, ferric, potassium etc. It also contains the growth factors which promotes cell proliferation, cell attachment and adhesion factors. Serum may be obtained from adult human blood, placental cord blood, horse blood or calf blood (foetal calf serum, newborn calf serum, and calf serum); of these foetal calf serum is the most commonly used. Serum is the liquid exuded from coagulating blood. Different preparations of serum differ in their properties; they have to be tested for sterility and toxicity before use.

Tissue Extracts

Tissue or organ extracts and/or hydrolysates (e.g., bovine pituitary extract (BPE), bovine brain extract, chick embryo extract and bovine embryo extract), and animal-derived lipids and fatty acids, peptones, Excyte, sterols (e.g., cholesterol) and lipoproteins (e.g., high-density and low-density lipoproteins (HDLs and LDLs, respectively) are used in culturing of animal cells. Tissue extracts for example, Embryo extracts—Other biological fluids used as natural media include amniotic fluids, ascetic and pleural fluids, aqueous humour (from eye), serum ultra filtrate, insect haemolymph etc. Chick embryo extract is the most commonly used tissue extract, but bovine embryo extract is also used. Other tissue extracts that have been used are spleen, liver, bone marrow, etc. extracts. Tissue extracts can often be substituted by a mixture of amino acids and certain other organic compounds.

Embryo extract is the most commonly used tissue extract. It is a crude homogenate of a 10-day chick embryo clarified by centrifugation. Cohn (1966) fractionated crude extract to give high and low molecular weight fractions.

The low molecular weight fraction promoted cell proliferation, while the high molecular weight fraction promoted differentiation of pigment and cartilage cells. These fractions were not fully characterized, but recent evidences suggest that the low molecular weight fraction probably contained peptide growth factors and high molecular weight fraction contained proteoglycans and other matrix constituents. Smith and Schroedl (1992) showed that the embryo extract could be replaced by hemin in the induction of skeletal muscle differentiation.

Artificial Media

Different artificial media have been devised to serve one of the following purposes:

1. Immediate survival (a balanced salt solution, with specified pH and osmotic pressure is adequate),
2. Prolonged survival (a balanced salt solution supplemented with serum, or with suitable formulation of organic compounds),
3. Indefinite growth
4. Specialized functions.

The various artificial media developed for cell cultures may be grouped into the following four classes:

(*i*) Serum containing media
(*ii*) Serum free media
(*iii*) Chemically defined media
(*iv*) Protein free media.

1.3 SERUM

Liquid yellowish, clear content left over after fibrin and cells are removed from the blood is known as serum. Calf (bovine), foetal bovine, or horse are used, in some cases human. Fetal bovine serum (FBS) (10-20% v/v) is the most commonly applied supplement in animal cell culture media. Normal growth media often contain 2-10% of **serum.** These supplements provide carriers or chelators for labile or water-insoluble nutrients; bind and neutralize toxic moieties; provide hormones and growth factors, protease inhibitors and essential, often unidentified or undefined low molecular weight nutrients; and protect cells from physical stress and damage. Thus, serum and/or animal extracts (Table 1.1) are commonly used as relatively low-cost supplements to provide an optimal culture medium for the cultivation of animal cells. The role for all constituents (more than 200) is not clear proteins, peptides, special factors released during platelet aggregation e.g., PDGF, TGF-β, lipids, lipid transport proteins, carbohydrates, micronutrients such as minerals, etc.

Table 1.1: Serum Contents

Contents	*Concentration Range*
Proteins and Polypeptides	**40-80 mg/ml**
Albumin	20-50 mg/ml
Fetuin	10-20 mg/ml
Fibroncetin	1-10 µg/ml
Globulins	1-15 µg/ml
Protease inhibitors (α1-anti-trypsin)	0.5-2.5 mg/ml
Transferrin	2-4 mg/ml
Amino Acids	**0.01-1.0 µM**
Lipids	**2-10 mg/ml**
Cholesterol	10 µM
Fatty acids	0.1–1.0 µM
Linoleic acids	0.01-0.1 µM
Phospholipids	10.7-3.0 mg/ml
Carbohydrate	**1.0-2.0 mg/ml**
Glucose	0.6-1.2 mg/ml
Hexosamine	0.6-1.2 mg/ml
Lactic acid	0.5-2.0 mg/m
Pyruvic acid	2-10 µg/ml

(Contd.)

Contents	*Concentration Range*
Polyamines	**0.1-1.0 μM**
Putrescine, Spermidine	
Urea	**170-300 μg/ml**
Inorganics	**0.14-0.16 M**
Ca, Cl, Fe, K, P, Se, Na, In	
Hormones	**0.1-200 nM**
Hydrocortisone	10-200 nM
Insulin	1-100 ng/ml
Triiodothyronine	20 nM
Thyroxine	100 nM
Vitamins	**10 ng-10 μg/ml**
Vitamin A	10-100 ng/ml
Folate	5-2 ng/ml

Functions of Serum in the Culture Medium

1. It provides the basic nutrients for cells; the nutrients are present both in the solution as well as are bound to the proteins.
2. It provides several hormones, e.g., insulin, which is essential for growth of nearly all cells in culture, cortisone, testosterone, prostaglandin, etc.
3. It contains several growth factors, e.g., platelet derived growth factor (PDGF), transforming growth factor β (TGF-β), epidermal growth factor, etc.; these are present in concentrations of μg/l.
   Both hormones and growth factors are involved in growth promotion and specialized cell function. A given hormone or growth factor may stimulate growth of one cell type, may have no effect on another and may even be inhibitory to some others. For example, PDGF induces proliferation in fibroblasts, but induces differentiation of some types of epithelia.
   Further, proliferation of a single cell type may be induced by more than one growth factor, e.g., fibroblasts respond to PDGF, epidermal growth factor, fibroblast growth factor and somatomidins.
4. A major role of serum is to supply proteins, e.g., fibronectin, which promote attachment of cells to the substrate. It also provides spreading factors that help the cells to spread out before they can begin to divide. Although cells do produce these factors, but trypsinized cells are usually unable to attach to the substrate.
5. It provides several binding proteins, e.g., albumin, transferrin, etc., which carry other molecules into the cell. For example, albumin carries into cells lipids, vitamins, hormones, etc. Transferrin usually carries Fe in a nonbasic form, but binding of transferrin to its receptor in cell membrane is believed to be mitogenic.
6. It increases the viscosity of medium and thereby, protects cells from mechanical damages, e.g., shear forces during agitation of suspension cultures.
7. Protease inhibitors present in the serum protect cells, especially trypsinised cells, from proteolysis.
8. The serum also provides minerals, like Na^+, K^+, Fe^{2+}, Zn^{2+}, etc.
9. It also acts as a buffer.

Unfortunately, the use of serum or animal extracts in tissue culture applications has several drawbacks. For example, the chemical composition of these supplements may vary between lots, even from a single manufacturer. The supplements of animal or human origin may also be contaminated with infectious agents (e.g., mycoplasma and viruses) which can seriously undermine the health of the cultured cells when these contaminated supplements are used in cell culture media formulations and may pose a health risk in cell therapy and other clinical applications. A major fear is the presence of prions causing spongiform encephalopathy in humans or animals. Cell surface chemistry, which is a critical portion of the *in vitro* microenvironment for many cell types, can be adversely modified via adsorption or incorporation of serum or extract proteins. The use of undefined components such as serum or animal extracts also prevents the true definition and elucidation of the nutritional and hormonal requirements of the cultured cells, thus eliminating the ability to study, in a controlled way, the effect of specific growth factors or nutrients on cell growth and differentiation in culture. Moreover, undefined supplements prevent the researcher from studying aberrant growth and differentiation and the disease-related changes in cultured cells. Using cell culture media in the industrial production of biological substances, serum and animal extract supplementation of culture media can also complicate and increase the costs of the purification of the desired substances from the culture media due to nonspecific co-purification of serum or extract proteins.

1.4 SERUM-FREE MEDIA

To overcome these drawbacks of the use of serum or animal extracts, a number of serum-free media have been developed. These media, which often are specifically formulated to support the culture of a single cell type, incorporate defined quantities of purified growth factors, lipoproteins and other proteins usually provided by the serum or extract supplement. Since the components (and concentrations thereof) in such culture media are precisely known, these media are generally referred to as "defined culture media" and often as "serum-free media" or "SFM." A number of SFM formulations are commercially available, such as those designed to support the culture of endothelial cells, keratinocytes, monocytes/macrophages, fibroblasts, neurons, lymphocytes, chondrocytes or hepatocytes which are available from Life Technologies, Inc. (Rockville, Md.).

SFM generally provide several distinct advantages to the user. For example, the use of SFM facilitates the investigation of the effects of a specific growth factor or other medium component on cellular physiology, which may be masked when the cells are cultivated in serum- or extract-containing media. In addition, SFM typically contain much lower quantities of protein (indeed, SFM are often termed "low protein media") than those containing serum or extracts, rendering purification of biological substances produced by cells cultured in SFM far simpler and more cost-effective.

Some extremely simple SFM, which consist essentially of vitamins, amino acids, organic and inorganic salts and buffers have been used for cell culture. Such media (often called "basal media"), however, are usually seriously deficient in the nutritional content required by most animal cells. Accordingly, most SFM incorporate into the basal media additional components to make the media more nutritionally complex, but to maintain the serum-free and low protein content of the media. Examples of such components include serum albumin from bovine (BSA) or human serum albumin (HSA), animal-derived lipids such as human Excyte, sterols, etc., and certain growth factors or hormones derived from natural (animal) or recombinant sources.

The use of such animal-derived supplements in cell culture media, however, also has certain drawbacks. For example, there is a risk that the culture medium and/or products purified from it may be immunogenic, particularly if the supplements are derived from an animal different from the source of the cells to be cultured. Thus, if biological substances to be used as therapeutics are purified from such culture media, certain amounts of these immunogenic proteins or peptides may be co-purified and may induce an immunological reaction, up to and including anaphylaxis, in an animal receiving such therapeutics.

To overcome this potential problem, supplements derived from the same species as the cells to be cultured may be used. For example, culture of human cells may be facilitated using HSA as a supplement, while media for the culture of bovine cells would instead use BSA. This approach, however, runs the risks of introducing contaminants and adventitious pathogens into the culture medium (such as HIV or Hepatitis B virus from HSA preparations, or Bovine Spongiform Encephalopathy virus from BSA preparations), which can obviously negatively impact the use of such media in the preparation of animal and human therapeutics. In fact, for such safety reasons, the biotechnology industry and government agencies are increasingly regulating, discouraging and even forbidding the use of cell culture media containing animal-derived products which may contain such pathogens.

Non-animal Peptide Supplements

To overcome the limitations of the use of animal proteins in SFM, several attempts have been made to construct animal cell culture media that are completely free of animal proteins. For example, some culture media have incorporated extracts of yeast cells into the basal medium to provide sources of nitrogen and other essential nutrients. In another approach, hydrolysates of wheat gluten have been used, with or without addition of yeast extract, to promote *in vitro* growth of animal cells. Still other media have been developed in which serum is replaced by enzymatic digests of meat, or of proteins such as α-lactalbumin or casein (e.g., peptone), which have been traditionally used in bacterial culture. None of these approaches, however, provided a culture medium optimal for the cultivation of a variety of animal cells. In fact, the approach using wheat peptides is likely to be quite unfavorable for culture of many animal cells and tissues, since wheat peptides are known to be toxic or to induce toxic effects *in vitro* and *in vivo*, particularly in the cells and tissues of the gastrointestinal systems of some mammals, including humans. Moreover, extracts from certain plants, including wheat, barley, rye and oats have been shown to inhibit protein synthesis in cell-free systems derived from animal cells, suggesting that the use of peptides derived from these plants in cell culture media may actually inhibit, rather than stimulate, the growth of animal cells *in vitro*.

Thus, there remains a need for a serum-free, low-protein culture medium suitable for cultivation of animal cells, which is completely devoid of animal or human proteins. Such a medium formulation will facilitate studies of the effects of growth factors and other stimuli on cellular physiology, will allow easier and more cost-effective purification of biological substances produced by cultured animal cells in the biotechnology industry, and most importantly will eliminate the risk of the introduction of adventitious animal and human pathogens.

Chemically Defined Media: These media contain contamination free ultra pure inorganic and organic constituents, and may contain pure protein additives, like insulin, epidermal growth factor, etc. that have been produced in bacteria or yeast by genetic engineering with the addition of vitamins, cholesterol, fatty acids and specific amino acids. The CHO cell lines are widely used for being highly stable expression systems for heterologous genes (those from a different organism), and for its relatively simple adaptation to adherence-independent growth in serum and protein free media.

Protein-Free Media: In contrast, protein free media do not contain any protein; they only contain non-protein constituents necessary for culture of the cells. The formulations MEM, DME, RPMI-1640, etc. are protein free; where required, protein supplementation is provided. Compared with serum-supplemented media, use of protein-free media promotes superior cell growth and protein expression and facilitates downstream purification of the expressed product. The protein-free, chemically defined formulation allows the growth of many cell lines but at the same time this becomes lipid dependence, such as NSO myeloma cells without further lipid supplementation does not grow in protein-free, chemically defined medium. Through the use of cyclodextrin-based additives to medium, it is possible to solubilize significant quantities of sterols and other lipids and to maintain a protein-free, chemically defined cultivation environment for NS0 cells. The culture system can be kept entirely free of animal-derived components if the supplement is made with plant-derived or synthetic lipids.

A chemically-defined protein-free medium called FMX-Turbodoma has been improved for the production of monoclonal IgA antibodies by hybridoma cells, using a systematic method by Stoll et al. in 1996 for IgA production and the price of the protein-free medium is about 20% of the serum-containing medium. This makes such a protein-free medium very convenient for laboratory and large-scale production. A recombinant CHO cell clone which has been cultivated in serum- and protein-free medium for at least 40 generations and which expresses a recombinant product. When adapting cells initially grown under serum-containing conditions to protein-free medium, it has repeatedly been found that the yield of expressed protein and the productivity of recombinant CHO cells greatly drops after adaptation in protein-free medium as compared to serum-containing conditions. This is the consequence of an instability or reduced growth of the recombinant clones due to the changed culturing conditions. Despite the use of a stable original clone, on account of the altered fermentation conditions, repeatedly a large portion of the cells become cells with reduced expression or also non-producers, which overgrow product producers during the production process, whereby the fermenter culture finally largely consists of non-producers or of such cells having a low expression.

A chemically defined, protein-free medium (designated CFBI 1000, where CFBI = Clayton Foundation Biochemical Institute) that supports human peripheral lymphocyte proliferation has been recently developed. This medium allows exploration of individual metabolic differences by varying the medium composition as well as providing a base to explore further the mechanisms of lymphocyte activation in a system initially free of added macromolecular species other than mitogen. The peripheral blood lymphocyte is an ideal system for metabolic studies because it is easily obtained, is a primary resting cell that can be activated to proliferate, and presumably reflects both the genetic make-up and biochemical environmental history of the individual at the time the cells were formed. Examination of the role of various factors in lymphocyte activation and subsequent events may be simplified by the utilization of a medium that is protein-free and chemically defined. The CFBI 1000 medium supports the growth response of human peripheral lymphocytes to mitogen as measured by [^{3}H]thymidine incorporation to an extent comparable to other media used widely in assessment of lymphocyte proliferation.

Many researchers are now produced stable recombinant cell clones which contain the encoding sequence for a recombinant blood factor, such as factor II, factor V, factor VII, factor VIII, factor IX, factor X, factor XI, protein S, protein C, an activated form of any one of these factors, and which are capable of stable expression of this factor over several generations.

1.5 APPLICATIONS OF ANIMAL CELL CULTURE

The animal cell cultures are used for a diverse range of research and development. These areas are:

(*a*) Production of antiviral vaccines, which requires the standardiznion of cell lines for the multiplication and assay of viruses.

(*b*) Cancer research, which requires the study of uncontrolled cell division in cultures.
(*c*) Cell fusion techniques.
(*d*) Genetic manipulation, which is easy to carry out in cells or organ cultures.
(*e*) Production of monoclonal antibodies requires cell lines in culture.
(*f*) Production of pharmaceutical drugs using cell lines.
(*g*) Chromosome analysis of cells derived from womb.
(*h*) Study of the effects of toxins and pollutants using cell lines.
(*i*) Use of artificial skin.
(*j*) Study the function of the nerve cells.
(*k*) Many commercial proteins have been produced by animal cell culture and there medical application is being evaluated. Tissue Plasminogen activator (t-PA) was the first drug that was produced by the mammalian cell culture by using rDNA technology. The recombinant t-PA is safe and effective for dissolving blood clots in patients with heart diseases and thrombotic disorders.

EXERCISE

1. Define culture media.
2. Explain main components of animal cell culture medium.
3. Differentiate between natural media and artificial media.
4. Write a note on different natural components used in culture media.
5. What do you mean by serum and mention its advantages in using for culture media?
6. Describe different functions played by serum components in animal cell culture media.
7. What is meant by serum free media?
8. Mention advantages of using serum free media in animal cell culture.
9. Explain the need of protein-free media in animal cell culture.
10. Outline the applications of animal cell culture system.

Chapter

2

Animal Cell Culture and Metabolism

2.1 INTRODUCTION

During the last fifty years, the prerequisites for the maintenance and propagation of animal cells in culture have been worked out systematically. The present state of development is characterized by the fact that the cultivation of animal cells has been established in many laboratories and clinics in order to deal with biochemical, physiological and morphological questions.

The development of animal cell biotechnology owed much to the needs of two major branches of medical research; the production of antiviral vaccines and the understanding of cancer biology. The standardization of conditions and cell lines for the production and assay of viruses undoubtedly provided much impetus to the development of modern animal cell biotechnology, particularly the production of large numbers of cells suitable for biochemical analysis. This and other technical improvements made possible by the commercial supply of reliable media, sera and by the greater control of contamination with antibiotics and clean-air equipment, have made animal biotechnology accessible to a wide range of interests.

In addition to cancer biology and virology, other areas of research have come to depend heavily on animal cell culture systems. The introduction of cell fusion, genetic manipulation, animal cloning, gene expression analysis, monoclonal antibodies production, cell interactions, intracellular control mechanisms in cell differentiation, nerve function analysis, hormone industries and transplantation studies are dependent upon the animal culture systems for understanding and production technologies.

The cultivation of animal cells on a large technical scale started with BHK (baby hamster kidney) cells which were adapted to growth in suspension in 1962 and have been used industrially since 1967 in the United Kingdom, Italy, Germany, Spain, Denmark and France, particularly for the production of foot and mouth disease vaccines. Girard (1977) reported the construction of a factory in which every year 500,000 liters of cell suspension are produced in 3000 liter fermenters. More advanced processes are already based on fermenters with a capacity up to 10,000 liters.

As majority of animal cells are suitable for *in vitro* cultivation, the present annual demand of around 300 millions of experimental animals worldwide will be reduced as further developments become available.

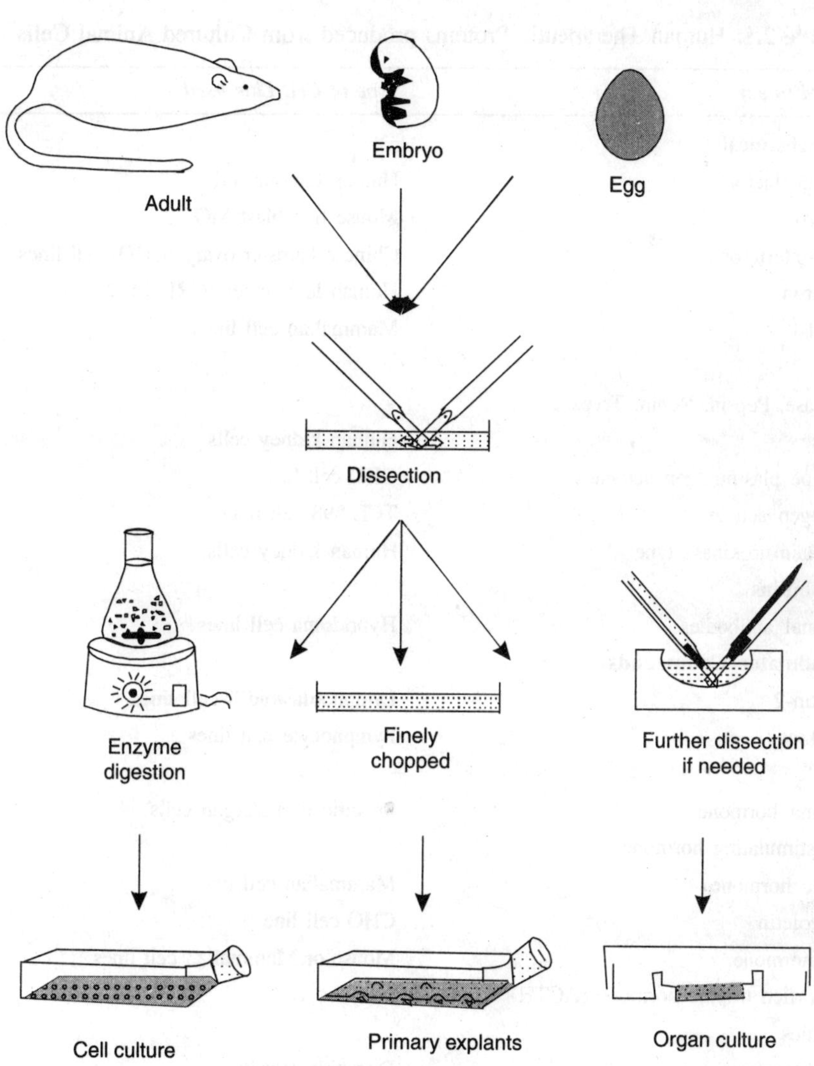

Fig. 2.1. Different types of animal tissue culture.

Animal cell culture deals with culturing the parts of organs, tissues or individual cell *in vitro*. The starting material for a culture initiation is known as **explant**. As long as the material used maintains its intact structure is called **tissue** or **organ culture** (Fig. 2.1). If the organization of a tissue is destroyed by mechanical, chemical or enzymatic action that forms individual cells is known as **cell culture**. Cells or tissues taken from an organism for culture form the **primary culture**. The generations of cells obtained after the first subcultivation and all subsequent ones are referred as **cell lines**.

At present in animal biotechnology varieties of normal and transformed cell types and cell lines are established and maintained for utilization in different purposes from simple study of cell culture characteristics to large scale production of vaccines and other therapeutic and commercial proteins (Table 2.1).

Table 2.1: Human Therapeutic Proteins produced from Cultured Animal Cells

Therapeutic Protein	*Type of Cell Line used*
Cellular biochemicals	
Angiogenic factor	Human tumour cells
β-Interferon	Mouse fibroblast MG 63
β- and γ-Interferons	Chinese hamster ovary (CHO) cell lines
α-Interferon	Human leukocytes (CRL 1432)
Factor VIII	Mammalian cell lines
Enzymes	
Collagenase, Pepsin, Renin, Trypsin	–
Urokinase	Human kidney cells
Tissue-type plasminogen activator	CHO cell lines
Plasminogen activator	TCL 598 cell lines
Single chain urokinase type	Human kidney cells
Immunoglobulins	
Monoclonal antibodies	Hybridoma cell lines
Immunomodulator Compounds	
Interleukin-2	Lymphoblastoid T-cell lines
Lymphokines	Lymphocyte cell lines
Hormones	
Luteinizing hormone	Specific tissue/organ cells
Follicle stimulating hormone	
Chorionic hormone	Mammalian cell line
Erythropoietin	CHO cell line
Growth hormone	Mouse or Mammalian cell lines
Adrenocortico trophic hormone (ACTH)	
Viral Vaccines	
Canine distemper	Dog kidney cells
Foot and Mouth disease	Cow kidney cells
Hepatitis-B	Human plasma-desired
Influenza	Chick embryo
Measles	Chick embryo
Mumps	Chick embryo
Polio	Primate kidney
Rabies	Chick embryo/human diploid cells

2.2 GENERAL METABOLISM OF ANIMAL CELLS IN CULTURAL CONDITION

To date, the function of many components of the growth medium essential for cellular proliferation and biological production has not been precisely defined at the molecular level. Animal or human

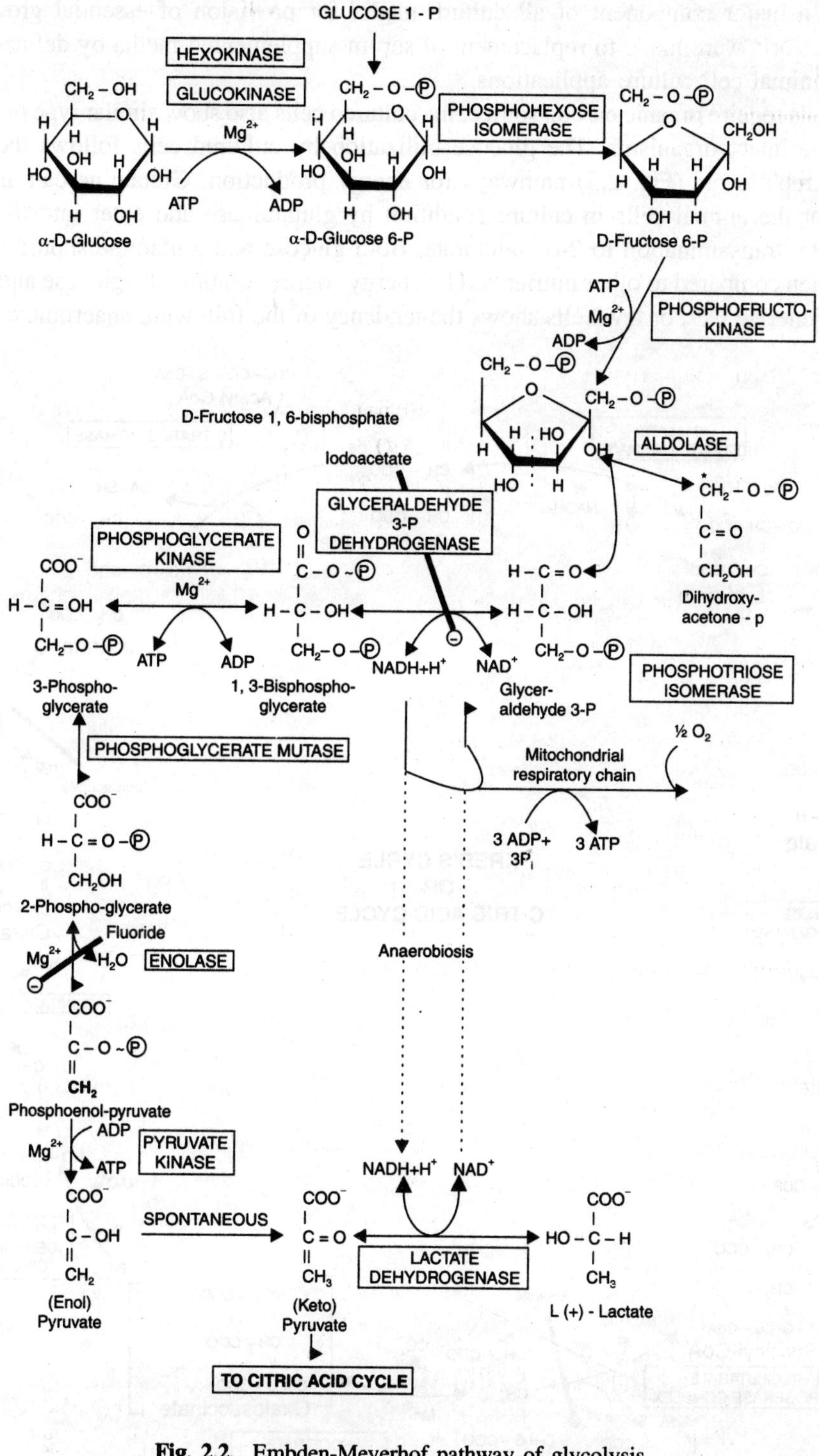

Fig. 2.2. Embden-Meyerhof pathway of glycolysis. (Ⓟ, $-PO_3^{2-}$; P_i, $HOPO_3^{2-}$; ⊖, inhibition)

serum has been a major component of all culture media for provision of essential growth factors. However, some efforts were made to replacement of serum-supplemented media by defined media for most industrial animal cell culture applications.

As animal cells require organic carbon source, the cultured cells also show similar type of metabolism that occurs in the intact organisms. The glucose utilization by cultured cells follows the glycolysis (Fig. 2.2) and Kreb's cycle (Fig. 2.3) pathways for energy production. Glutamine can also serve as energy source for the animal cells in culture condition by glutaminase and enter into the citric acid cycle (Fig. 2.4) by transamination to 2-oxoglutarate. Both glucose and glutamine should be at higher concentration when compared to other nutrients. The energy source compounds (glucose and glutamine) are oxidized to generate CO_2 or few cells shows the tendency of the following anaerobic conversion to

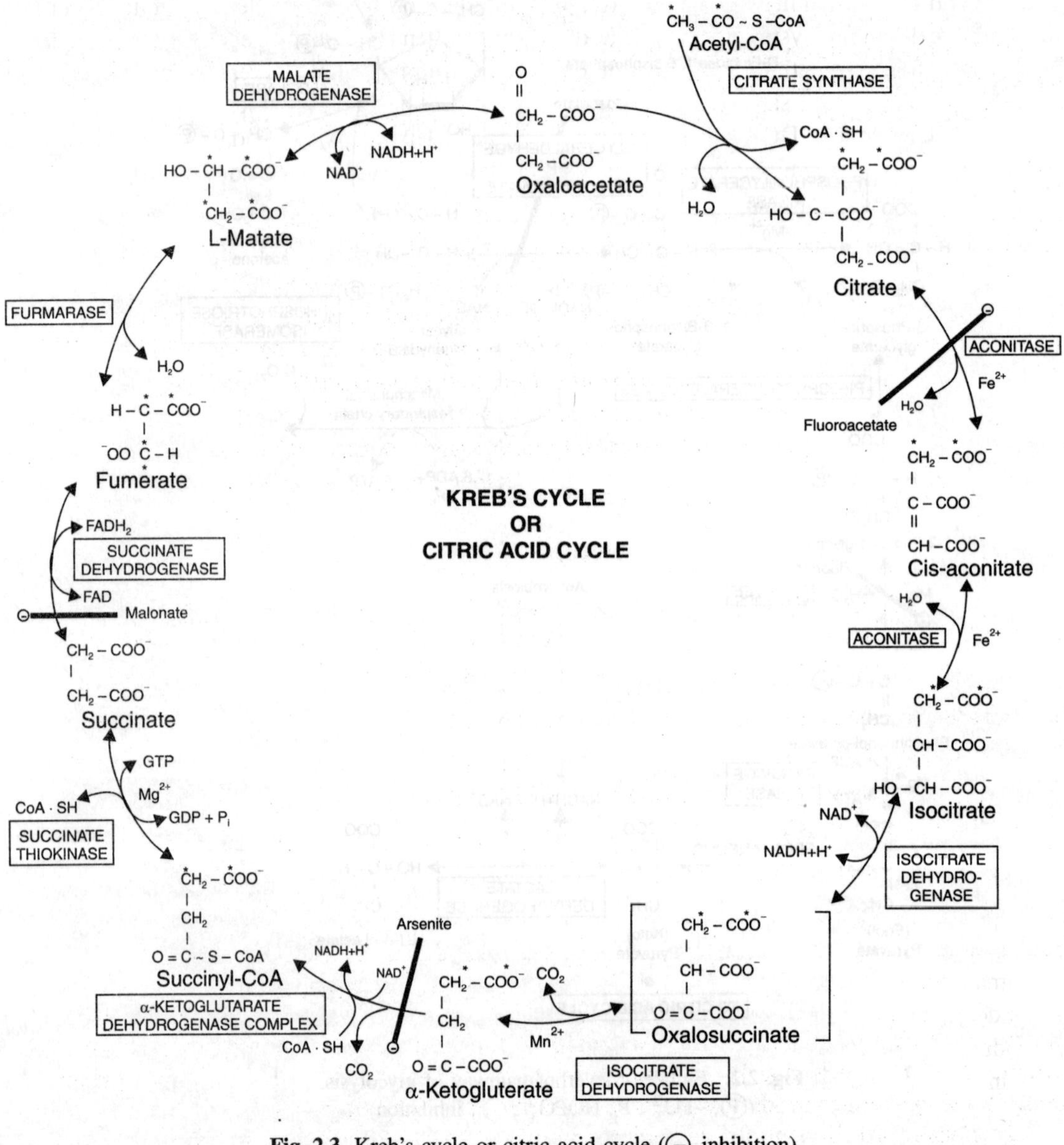

Fig. 2.3. Kreb's cycle or citric acid cycle (⊖ inhibition).

accumulate lactic acid, ammonium and keto acids in the medium. Accumulation of lactic acid leads to a reduction in the pH of the medium, resulting in the loss of cell viability. At low glucose concentrations, a large amount of glucose and glutamine is utilized by oxidative method. Thus, it is desirable to maintain a low glucose concentration in the bioreactor to prevent the accumulation of lactic acid. The toxic ammonia (limits the cell growth) can be formed by utilization of glutamine or by spontaneous degradation of glutamine. This can be avoided by using comparatively low concentration or substitution by glutamate or glutamyl-alanine or glutamyl-glycine (dipeptides which are more stable also) or removal by ion-exchange resins, ion-exchange membranes, gas-permeable membranes or electrodialysis.

Generally cultured cells have high variability for oxygen requirements. The proper oxygen and carbon dioxide concentration is very much essential as these two gases influence the pH and HCO_3^- ion concentration of the culture medium. Oxygen is required for energy metabolism of glucose and glutamine and for the synthesis of cholesterol. Oxygen transport can be a potential limitation in large-scale animal cell culture system. The optimal dissolved oxygen (DO) is not generally the same for cell growth and product formation. Typically, the oxygen uptake rate (OUR) increases with an increase in DO up to about 10 to 30% saturation with air. Increase in DO beyond 30% does not lead to any further increase in OUR. Higher DO concentrations are toxic and lead to oxidative damage, including DNA degradation, lipid peroxidation, polysaccharide depolymerization and metabolism of nutrients in culture medium at rates greater than that required for consumption.

The range of optimal concentrations of amino acids in the medium for cell growth is wide. Amino acid consumption is often cell line dependent and affects cell growth but not protein synthesis. The balance of amino acids that is critical for the cell growth.

Cultured cells and tissues known to synthesize their lipid requirements from simple chemical substances such as acetate. It is not yet clear that fat compounds are directly taken up from the medium and stored. But numbers of cells are known to biosynthesize cholesterol.

Animal cells in culture condition known to readily metabolize nucleic acids from simple compounds in a medium such as formate, glycine and bicarbonate. They contain all the functional active enzymes for the nucleic acids biosynthesis.

2.3 SPECIAL SECONDARY METABOLITES

All living organisms require some chemical compounds that are essential to the structure and function of the cell. These compounds are referred as **primary metabolites**. Some of the primary metabolites act as building blocks of macromolecules and some act as coenzyme. The example for primary metabolites include carbohydrates, amino acids, organic acids, purines, pyramidines and vitamins. Metabolism of these compounds is known as primary metabolism. The **secondary metabolites** are compounds which are not essential for the structure and function, but necessary for maintaining the proper health and state of the cell or body. The best known example of secondary metabolites are antibiotics from fungi and bacteria. In case of animals also secondary metabolites are produced which are much essential to maintain proper health and metabolic activity. Hence they are considered special secondary metabolites. These include hormones, interferons, polypeptide growth factors and proteins of therapeutic value (Table 2.1).

Human cell contains more than 29,000 genes, most of them code essential proteins. Sometime a gene is defective due to mutation resulted in production of defective protein. In some cases protein is not produced at all. A deficiency of the protein may not be noticed, but sometimes the deficiency results in disease. The deficiency is generally inherited and thus the disease is genetically linked.

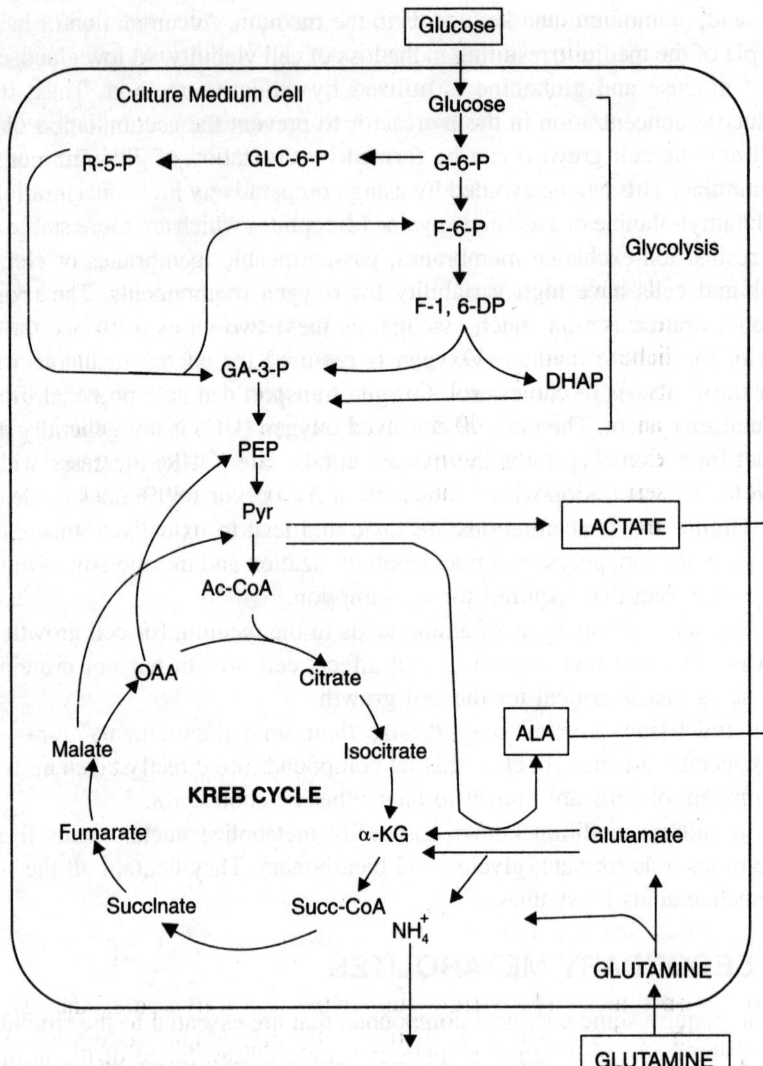

Fig. 2.4. Model of glucose and glutamine catabolism in mammalian cells in culture condition. Compounds in bold letters are consumed from or excreted into the medium. Abbreviations used: G-6-P, glucose-6-phosphate: GLC-6-P, 6-phosphategluconate; R-5-P, ribose-5-phosphate; DHAP, dihydroxyacetons phosphate; PEP, phosphoenolpyruvate; Pyr, pyruvate; Lac, lactate; Ac-CoA, acetyl-CoA; α-KG, α-CoA; OAA, oxaloacetate; Ala, alanine; Succ-CoA, Succinyl CoA.

Now it is possible to treat certain genetically associated diseases provided the protein deficiency is identified. There are many human therapeutic proteins which are produced by the recombinant DNA technology in large scale to meet the demand of the needy people.

2.4 INSULIN: PRODUCTION OF HUMAN INSULIN

An example of genetically-linked disease is diabetes mellitus. It is the most common and widespread disease. The beta cells of the pancreas are responsible for the production of hormone Insulin. If these cells fail to produce the required quantities of the hormone causes diabetes mellitus. In 1961, Sir Edard Sharpy-Schafer discovered insulin, while studying the Islets of Langherans of the pancreas. Frederick Banting and Howard Best in Canada reported antidiabetic role of insulin in 1921. It actually facilitates the cellular uptake of glucose. Without insulin, glucose remains in the bloodstream resulting in increase of blood sugar as well as lowering the production of energy rich ATP. The patient with diabetes mellitus is very weak and tired because the cells are very low in ATP. The kidney removes excessive glucose from the bloodstream and excretes it in the urine (mellitus = sweet). To keep the glucose diluted, the kidney also puts out an abnormally large amount of water (diabetes = siphon = large volume of water siphoned from the body). Diabetics are prone to eye damage and blindness, as well as kidney disorders, nerve damage, and circulatory diseases, including gangrene and stroke. They have to be careful to avoid the diabetic coma, a loss of consciousness caused by pH imbalance arising from the body's breakdown of fats. Approximately 150 million people in the world are believed to be diabetic and it is estimated that an increase to 300 million in the year 2025. In India at present there are 35 million diabetics and is expected to rise 3 times in another 20 years. They must take regular injections of insulin to avoid the serious effects of the disease. According to a recent estimate, by year 2025, the global cost of treating diabetes and its complications could reach US $1 trillion (1000 billion) annually.

Insulin was for the first time applied to a diabetic nine-year-old boy in 1922 with success. Eli Lilly immediately set it on for production in 1923 and animal insulin (from Pancreas) was marketed. Frederick Sanger and his coworkers (1957) at Cambridge University finally succeeded in achieving the first complete description of the structure of protein molecule insulin. Sanger's work was not only to discover the overall chain configuration of the insulin molecule but also to learn the sequence of all the amino acids in the chain. Insulin is a protein made up of 51 amino acids in two interconnected chains, an A chain of 21 amino acids and a B chain of 30 amino acids. Insulin has been purified from the pancreas of pigs and cows obtained at slaughter houses. Animal insulin is similar to human insulin. However, there is a difference of one (pig) or three (cow) amino acids. Porcine (pig) insulin is similar to human insulin (Humulin) in that the C-terminal threonine on the A chain is replaced by alanine, while bovine (cow) is different at three positions (Table 2.2). This difference in chemical structure causes allergic responses in diabetic patients after taking the animal insulin. All patients treated with injections of bovine insulin develop circulating anti-insulin antibodies that are responsible for subcutaneous fat atrophy and also neutralize insulin in the circulation. Consequently the patient needs a higher dose in due course. Moreover, some diabetics are allergic to animal hormones. Besides bovine insulin, shows a greater immunogenicity as compared to porcine. Therefore there is a requirement of human insulin. Another important factor is that to treat one diabetic patient for one year, 7-10 pounds of pancreas tissues (70 pigs) are needed. About 800-1000 kilograms of pancreas produce 100 gm of insulin. Since then it has been produced commercially by several pharmaceutical concerns.

In 1970s biochemists used the recombinant DNA technology for the production of insulin. They separated the DNA segment (gene) coding for insulin. Then they attached the structural gene for insulin to the bacterial plasmid of *E. coli*. They also inserted promoter sequence of lac operon before insulin gene to produce insulin. In July 1980, for the first time 17 patients received injections of genetically engineered insulin at Guy' Hospital in London, England and this insulin worked as the animal insulin. Later on they inserted a short segment of DNA between the promoter site and the structural genes. The new segment directs the production of 24 amino acids peptide chain that is continued with the insulin. This encourages the bacterium to secrete the synthesized insulin into the surrounding media (outside the cell) which in turn saved the trouble of retrieving insulin from the cell body. Further, the bacterium continues its metabolism as the food is supplied from one end of the fermentation tank and the insulin is removed from the other end.

Table 2.2: Amino acid sequence of Human, Porcine and Bovine Insulins

Insulin	*A8*	*A10*	*B30*
Human	Threonine	Isoleucine	Threonine
Porcine	Threonine	Isoleucine	Alanine
Bovine	Alanine	Valine	Alanine

In the body pancreatic cells first synthesize a large chemical molecule called pre-proinsulin, which is 105 amino acids in length. This molecule includes a signal sequence (24 amino acids) and three sections of amino acids (A, B and C chains). During natural insulin production the signal sequence is removed to form proinsulin and later a section of 33 amino acids is cut free at the Golgi apparatus and the remaining two chains (A and B) are joined to form a molecule of insulin (Fig. 2.5).

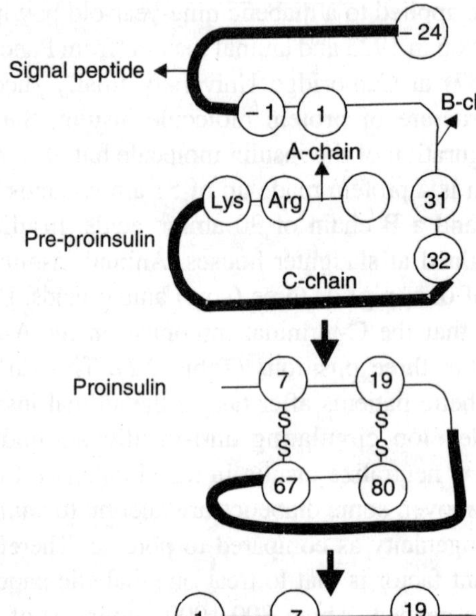

Fig. 2.5. Processing of pre-proinsulin to produce insulin.

Synthesizing of pre-proinsulin is too difficult in bacteria. However, the biotechnologists used two separate bacteria for synthesis of A and B chain of insulin molecule. These chains are extracted after the synthesis from the media and purified. Later they were joined chemically by disulfide linkages resulted in synthetic insulin. In 1986, the Eli Lilly company received approval to market Humulin, the first genetically engineered insulin for general use. In addition to this, biotechnologists discovered a way to alter the porcine insulin to insulin, which is identical to human insulin. Thus today there are two biotechnological methods to manufacture the insulin (Fig. 2.5).

2.5 PRODUCTION OF HUMAN GROWTH HORMONE (HGH)

HGH is a protein that stimulates overall body growth by increasing the absorption of amino acids by the cells which in turn helps in protein synthesis. It also promotes the use of fat as body fuel for energy production. This hormone is produced by pea-shaped gland present at the base of the pitutary gland. This pitutary human growth hormone is also known as somatotropin. Under production of this protein leads to pitutary dwarfism, which is another metabolic disorder. Giving the same protein controls this disease. When the pitutary gland produces insufficient amount of HGH in childhood, the growth is retarded. The child develops a chubby face; abundant body fat at the waist, height of about 4 feet and intelligence is normal. However, the adult has unusual body proportions and is a dwarf. If dwarfism is diagnosed at an early age and while bone growth is still occurring, injections of HGH can be given. This is very expensive and pitutary glands from nearly 80 cadavers must be used to obtain enough HGH for a year's therapy (8-10 year individual). Further, it is dangerous to give the HGH of infected brain (e.g., CJ syndrome).

DNA technologists can now produce HGH exclusively through the mechanisms of genetic engineering. The process is similar to that for insulin. Natural HGH comprises 191 amino acids and this structural detail was given by Genetech group consisting of Goeddel and coworkers. During production of HGH in the body the additional signal peptide of 26 amino acids is cut free to produce mature HGH during secretion. However, the complementary DNA molecules that biotechnologists isolate also contain this signal sequence. This is first removed by using restriction enzyme EcoRI. Then the gene is inserted to the plasmid (Fig. 2.6).

The synthetic HGH when injected to the undersized children, it stimulates body growth and helps a child to reach his or her normal height and size. The hormone is species specific and its main source has been human cadavers. It has been demonstrated that an intramuscular dose of 6-10 mg per week can increase the stature by more than 6 cm during the first year of treatment. However, the hormone has to be used while bone formation is actively occurring which means that adults would not benefit from HGH therapy.

Genetically engineered HGH is used after 1985 and marketed as Protropin by the Genentech Company and Humatrope by Eli Lilly & Company. The synthetic hormone has tripled the growth rate of hormone deficient babies in the first year and doubled the growth rate in the second and third years of therapy. Currently it is prescribed for children described as very short. To reach normal heights, daily intramuscular injections of the drug may be required for many years at a cost of about 20,000 dollars per year. As of 1998, 30,000 short children were taking hormone therapy each year in the US.

Recently Somatotropin (HGH) gene was synthetically produced by a combination of chemical synthesis and isolation of natural molecule. This is a polypeptide of 191 amino acids. The segment of gene that codes for the first 24 amino acids part of the peptide chain was constructed chemically from different blocks of nucleotide. The recombinant human growth hormone plasmid was constructed by usual genetic engineering techniques. Reverse transcriptase enzyme was used to copy the gene for the hormone from mRNA obtained from human pituitary tissues. Restriction endonucleases which cut out the needed part of the DNA. Then DNA ligase was then used to join the natural and synthetic fragments. The complete gene was inserted into a modified version of plasmid vector pBR 322 with the lac operon. The synthetic part of the growth-hormone gene had been constructed with its own initiation codon (ATG) required for the protein synthesis. The hormone could, therefore, be produced

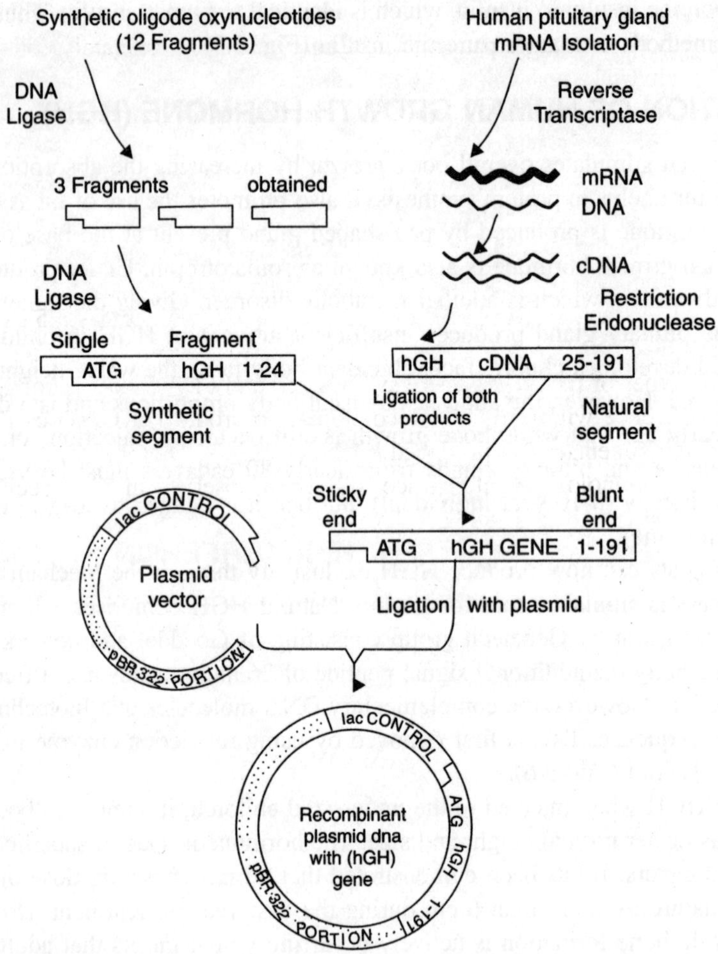

Fig. 2.6. Synthesis of recombinant molecule of human growth hormone.

independently in prokaryotic bacterial cells, without the need for attachment to a bacterial protein. Thus purification becomes very easy. This growth hormone may have many clinical uses, but the extremely limited availability of the human growth hormones is a serious limitation. Microbiological production may not only increase the commercial availability of the drug but also make it possible to investigate in detail its potential applications. Now rat tumor cell lines and mammalian cell lines are used for production of human growth hormone. Currently several industries including Genentech, Pasteur Institute, Kabi Vitrum AB involved in large scale production of somatotropin. Biosynthesis of somatotropin might reduce the cost of treatment.

2.6 FACTOR VIII: PRODUCTION OF FACTOR VIII

It is a blood protein that contributes to the formation of a blood clot and in doing so prevents hemophilia. In US the most common inherited blood clotting disorder is hemophilia A. This disorder affects about 1 in 10,000 males in the US. and is caused by a genetic defect for normal blood clotting. The defect is traced to be the inability of the body to produce an important factor responsible for blood clotting process called Factor VIII.

Previously this was obtained from units of whole blood (8000 pints of blood are needed to yield enough clotting factor for a single patient for a single year). Unfortunately contamination by unknown agents is possible and this possibility was observed during 1980s by the HIV infection. More than 10,000s of patients have got AIDS due to contaminated blood. Afterwards, the DNA technologists worked to produce synthetic factor VIII using genetic engineering. They noticed that the gene for factor VIII is located on the X-chromosome and it comprises 186,000 bp within 26 exons. They isolated mRNA that contains 9000 bases and encodes the final factor VIII, a protein consists of 2332 amino acids. Scientists synthesized a cDNA molecule using this RNA as a template. Then inserted it into hamster kidney cells where it encoded the factor VIII protein. Mammalian cell lines were used for the production because factor VIII is very large and complex, with at least 25 sites where carbohydrate molecules are attached.

By 1992, two types of factor VIII were synthesized and FDA (Federal Food and Drug Administration) was offered licence within a year. Recombinate, a product of Genetics Institute in Cambridge, Massachusettes and Kogenate was produced by Miles Lab.

Symptoms of immunological intolerance have been observed in some recipients.

2.7 INTERFERONS: PRODUCTION OF INTERFERON

Alick Isaacs and Jean Lindemann (1957) reported the production of some antiviral substances in cells of many vertebrates after viral infections. Isaacs (1963) called such substances as interferons because these substances play an important role in virus interference. Similarly such proteins were also observed in plants such as tomatoes and tobacco (Wallace, 1944). Thus the virus-infected cells produce a protein that protects uninfected cells against the infection of virus or inhibits the productive multiplication cycle of the virus called interferons.

Interferon is not a single substance but a group of more than twenty substances with molecular weights between 20,000-30,000 Daltons, and are broadly classified into three types on the basis of physicochemical and antigenic properties: alpha (IFN-α), beta (IFN-β) and gamma (IFN-γ) interferons. All interferons are protein molecules and many are also glycoproteins. They have carbohydrate molecules attached to the amino acids at various points along the chain. The interferons trigger a nonspecific reaction that protects cells against numerous different viruses, including the ones that stimulated their production. Interferons are produced by cells after a virus releases its DNA or RNA into the cell's cytoplasm. The interferons secreted by the cells then bind to receptor sites on nearby cells. Here they trigger the production of protective proteins. These proteins inhibit viral replication by methods, which are not fully understood. However, recently it has been shown that one protein binds to mRNA encoded by virus. In early days it was very difficult to get one gram of interferon to work with (90,000 pints of blood were required to produce a single gram of interferon). However, after deciphering the genetic code of interferon by Swiss and Japanese biotechnologists (1980), it was possible to synthesize the interferon using genetic engineering techniques. It was found that the genetically engineered (bacteria derived) interferons reduced the symptoms of hepatitis, diminished the spread of herpes zoster (Shingles) and shrank certain tumors.

In 1984, Swiss biotechnology firm started marketing alpha interferon as 'Intron' and in 1986, FDA approved the use of this against leukemia; in 1988 approved the use of this against genital warts, sarcoma, malignant melanoma, multiple myeloma and certain kidney cancers. Interferon beta licensed in 1993 for use against multiple sclerosis. Interferons can be used to treat common cold and broad spectrum of viral diseases. It is produced by separating WBCs from human blood and exposing them to Sendai Virus. The viral capsid stimulates a human cell chromosome (presumably number 5) to produce interferon. However, these interferons are unable to affect viral replication in the cell in which they were produced. These interferons are absorbed by adjacent cells where they induce chromosome number 21 to produce an unknown protein that binds to mRNA molecules specified by the virus (Fig. 2.7). Interferon is finally released in the culture medium and is purified using monoclonal antibodies by employing affinity chromatography technique. Negative aspect of this is that the interferons production is costly because bacteria cannot synthesize glycoproteins such as interferons as they lack such modification mechanisms.

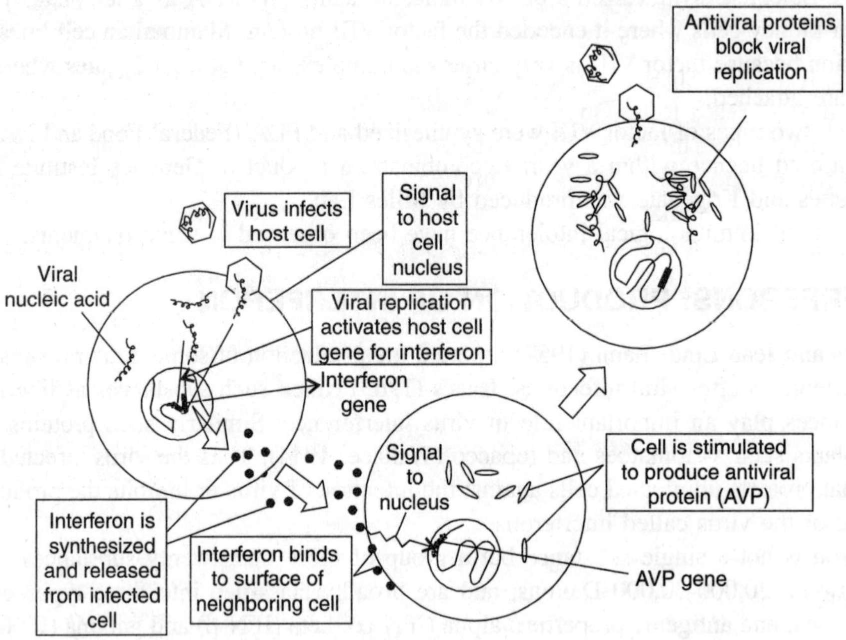

Fig. 2.7. The mechanism of interferons α and β action on viral replication.

Throughout the world a large number of firms are manufacturing leukocyte interferon in the United States including Genetech, Interferon Sciences, Schering Plough as well as Roche Institute of Molecular Biology. In Japan also, Yamanouchi Pharmaceutical Company, Tokyo, has manufactured interferon. In Europe there are a large number of public and private concerns which are attempting/having produced interferon such as Searle Laboratories and Wellcome Laboratories in U.K., University of Louvain in Belgium, Snafu Company in France, Hoechst in Federal Republic of Germany and Kabi Vitrum AB in Sweden, Israel is also on the world market for successful production of Interferon. However, the major difficulty is how to reduce the cost of production to produce interferon to be utilized by an average customer.

When compared to earlier methods which used leukocytes in serum-free medium for α-interferon production, the recent methods employing mammalian cell lines. The lymphoblastoid cell line called Namalva (ATCC No. CRL 1432) obtained from tumor is used for α-interferon production. This cell line produced up to 20,000 units of α-interferons per 10^6 cells. The Wellcome Foundation (Beckenhan, U.K.) and Sumitomo Chemical Company (Japan) are major producers of α-interferon.

The β-interferon are produced by diploid fibroblasts, particularly cell line the FS-4. This cell line has a limited life span corresponding to about 30 passages. Now the MG63 (ATCC No. CRL 1427) obtained from osteosarcoma patient is considered as a superior cell line for β-interferon production.

This cell line produces up to 10^8 units per litre of culture medium. The Bioferon (Germany) company and Rega Institute (University of Louvain, Belgium) are major producers of β-interferon in the market.

The γ-interferon was produced from B-lymphocytes or T-lymphocytes obtained from fresh blood collections. The mammalian cell lines such as monkey tumor cell lines are used for the large scale production of γ-inteferons. Mizrali in 1983 listed more than ten firms producing γ-interferons in the large scale bioreactors.

2.8 TISSUE PLASMINOGEN ACTIVATOR (TPA): PRODUCTION OF TPA

Biotechnology has also extended to disease therapy as researchers and physicians seek to develop new drugs and approaches to treating disease using DNA technology. The two examples are: Tissue plasminogen activator and Interferon. Both drugs once in very short supply, are now produced in large quantities for therapeutic use. In addition entirely new class of compounds called antisense molecules has been developed to stimulate the specific molecules associated with disease.

The blockage of blood vessels by clots is the commonest cause of death in the developed world. Blood clots in the vessels, which supply the heart (coronary heart disease), brain (stroke) or lungs (pulmonary embolism) account for more than half of all deaths in the West. The processes involved in the formation and subsequent destruction of a blood clot (thrombus) are complex and finely balanced. An array of enzymes and other substances ensure that clots normally form only when and where they are required to prevent blood loss, and that they are broken down once their task is completed. The basis of a blood clot is a network of fibres of the protein, fibrin. This network can be destroyed by the action of plasmin, a fibrinolytic enzyme, which is formed in the blood from an inactive precursor, plasminogen. The role of plasminogen activators (Pas) is to remove part of the plasminogen molecule to yield plasmin (Fig. 2.8). The Pas are found in blood, tears, saliva, urine, semen and cerebrospinal and other body fluids. They are also found in many tissues and cells, including cancer cells. In

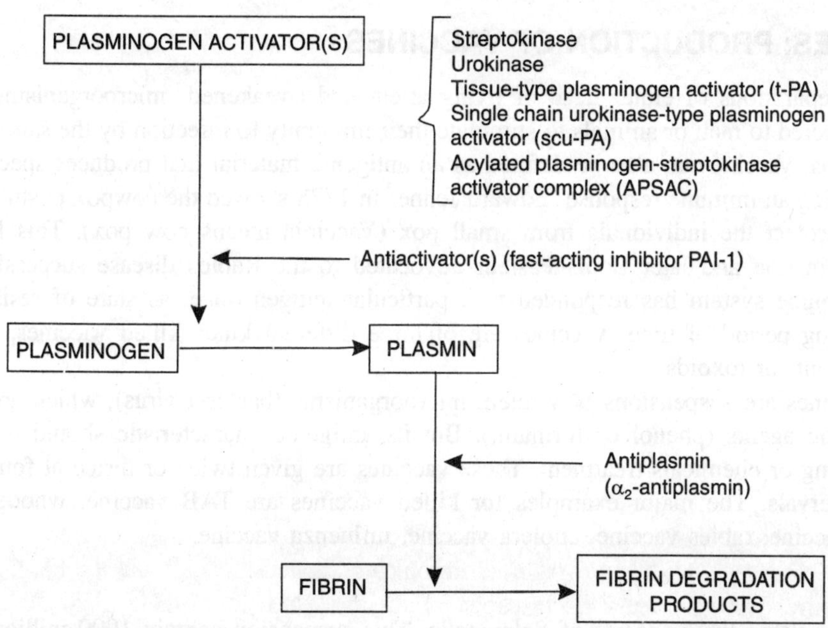

Fig. 2.8. A diagram of the fibrinolytic system stimulated by tissue plasminogen activator.

addition, *Streptomyces* bacteria manufacture an enzyme, streptokinase, which is also able to activate plasminogen.

Since many years, coronary thrombosis has been treated with streptokinase and other enzymes that destroy blood clots in the blood vessels. A recent addition to the list of therapies is a protein called tissue plasminogen activator (TPA). It is a naturally occurring protease expressed in mammalian cells. TPA exhibits its protease activity by degrading an inactive precursor enzyme, namely, plasminogen, changing it into plasmin. Plasmin is also a protease that degrades fibrin, the protein of which blood clots are formed. TPA attaches itself to blood clots and stimulates other blood borne components to breakdown the clot without inhibiting the blood clotting power in other parts of the body (wounds). Whereas the another clot dissolving enzyme used in the therapy is urokinase decreases the clotting ability of the blood throughout the body resulted in risk of internal bleeding. TPA can be administrated intravenously whereas other clot dissolver must be administrated directly into the blocked vessel. TPA appears to work quickly than other dissolvers of clots. This is another advantage of TPA.

TPA was the first drug produced by a mammalian cell culture. The production of tissue plasminogen activator from a continuous line of procine kidney cells (LCC-PK_1; ATCC No. CL 101) has been patented by Lilly Research Laboratories (US patent No. 3904480). During 1980s the DNA technologists first synthesized complimentary DNA molecule for TPA. Then they biochemically attached it to a synthetic plasmid containing a TPA signal sequences, promoter and terminator sites. Then plasmid was introduced to mammalian cells, and cells secreting high levels of TPA were obtained from the mixture. Cultivated in large fermenters, the cells supplied sufficient TPA for pharmaceutical use.

In 1987, the biotechnology company Genentech received a licence from the FDA to market TPA under the trade name Activase. TPA is not used as a heart attack preventive but it is used as therapy for those having heart attacks. It is a protein product of DNA technology that attaches to blood clots and stimulates blood components to breakdown clot. The protein works locally rather than throughout the body where it could cause hemorrhaging.

2.9 VACCINES: PRODUCTION OF VACCINES

Vaccines are preparations of either dead or living attenuated (weakened) microorganisms. Vaccines may be administered to man or animals to stimulate their immunity to infection by the same or closely related organisms. Vaccine can also be defined as an antigenic material that produces specific immunity by stimulating an immune response. Edward Jenner in 1778 showed the cowpox pustule materials are known to protect the individuals from small pox (Vaccinia means cow pox). This lead to the process of vaccination and later Louis Pasteur advocated to the Rabies disease successfully. If an individual's immune system has responded to a particular antigen once, as state of resistance will remain for a long period of time. Vaccines are of three different kinds killed vaccines, Attenuated vaccines (avirulent) or toxoids.

Killed vaccines are suspensions of virulent microorganisms (bacteria. virus), which are killed by heat or antiseptic agents (phenol or formalin). But its antigenic characteristic should not denature during the heating or chemicals treatment. These vaccines are given twice or thrice at four weeks or longer time intervals. The major examples for killed vaccines are TAB vaccine, whooping cough vaccine, salk vaccine, rabies vaccine, cholera vaccine, influenza vaccine.

TAB vaccine

It is prepared by using three species of *Salmonella*. This suspension contain 1000 million bacilli of *Salmonella typhi,* 750 million bacilli each of *S. paratyphi* A and *S. paratyphi* B per ml. Then suspension was sterilized by heating at 60°C for 30 minutes and preserved in 0.5% phenol. This vaccine is used against enteric fever and known to provide immunity for about two years. Booster dose of 0.5 ml is given after every two years to give a long lasting immunity.

Whooping cough vaccine

Bacterium *Bordetella pertusis* cells are killed by formalin or formalin merthiolate and suspension made at the rate of 200,000 million cells per ml. Intramuscularly three doses of one ml each was given at four week interval to all the children.

Poliomyelitis salk vaccine

Three polio viruses (type 1, type 2 and type 3) are killed by formalin and two doses of intramuscular injection containing 1 ml each was given at an interval of three weeks. Then two booster doses are required for the long term immunity.

Rabies vaccine

This was prepared from sheep brain infected with poliomeylitis virus. The 4% suspension of sheep brain in which the virus is killed by 0.5% phenol at 37°C is used. Nowadays a new rabies vaccine is used from virus grown in human diploid cells.

Influenza vaccine

Influenza vaccine consists of formalin killed viral strains. Influenza, on the other hand may be prevented by the parenteral injection of vaccines of influenza viruses that have been grown in chick embryos and rendered noninfectious by formalin or ultraviolet irradiation. With such a dead vaccine, the duration of immunity rarely exceeds a year and since the causative agent (virus) of influenza is capable of mutating frequently, different vaccines usually must be prepared to combat each mutant strain.

The French company, Transgene, has been aiming at producing a vaccine against rabies; in 1982 the workers strived to synthesized surface epitope of the rabies virus. They isolated a clone of *E. coli* that produced a 58,000-dalton viral epitope of which the encoding DNA had been introduced into the bacterial host. Rabies is a zoonois that occurs endemically in animals and humans in several countries of Africa, South America and Asia and is increasing in European countries also. Preventive vaccination is impeded by the high cost of the currently used vaccines. This is primarily due to difficulties countered in virus propagation in human cells and viral purification. Generally human cell culture requires high cost. Therefore, the production of cheap vaccine from a viral protein with the required immunogenic properties would be useful from the industrial production and public health viewpoints.

Both antibiotics and vaccines are used in modern medicine to fight infectious diseases. Vaccines are used as prophylactic agents to increase immunity and to prevent subsequently infection with more virulent strains. Antibiotics are employed most often as therapeutic agents to control disease after its onset, although they may also be used prophylactically against some organisms.

Now, vaccine for Rickettsial Rocky Mountain spotted fever and typhus fever are also available. However, the major problem is allergic reactions often arise from residual egg protein in which the rickettsiae have been cultured. This has not been achieved successfully for the causative microbe of syphilis (*Treponema pallidum*) nor has it been possible to make vaccines with conventional techniques relating to the extraction and purification of antigens. The DNA of this microorganism was isolated from infected rabbit testicles and cloned in *E. coli* host cells with the help of a coliphage used as a vector. This has enabled the development of more specific diagnostic screening tests for syphilis causative agent and the manufacture of an effective vaccine against disease. The need for such an effective vaccine is particularly important, as the strains of *Treponema* tend to become resistant to the antibiotics currently in use for the treatment of syphilis.

Foot and mouth disease vaccine

In veterinary medicine efforts are also being made to produce effective vaccine in genetically modified microbes, which could be effective against most of the animal diseases such as food and mouth disease and swine fever. Foot and mouth disease is highly contagious and threatening for cattle, sheep and swine farming. The virus causing this disease belongs to the Aphthovirus (single stranded RNA) genus of the family Picornaviridae. Recently German scientists have been successful in cloning double-stranded DNA copies of the viral RNA into plasmid pBR 322 and grown in *E. coli* host. It was found that more than 1,000 molecules of viral protein were synthesized per bacterial cell with a positive serological test.

The Aphtho virus is surrounded by a capsid made of sixty copies of four polypeptides, VP_1, VP_2, VP_3 and VP_4. Treatment of the virus particles with trypsin enzyme results in the cleavage of VP_1 and a considerable decrease in infectivity and a loss of immunizing activity. It has, therefore, been advantageous to synthesize the polypeptide precursor of VP_1 and to use it as the protein conferring long-lasting immunity. A lot of work is currently going on to improve the immunizing ability of VP_1 protein. A vaccine made up of VP_3 has also been synthesized by genetically engineered bacteria by workers from Plum Island Animal disease Center, New York. Intervet International, pharmaceutical company from the Netherlands has produced two vaccines, cloned in bacteria, to prevent infectious diarrhoea of swine and calves caused by pathogenic strains of *E. coli*.

The development of synthetic vaccines produced from genetic recombination and gene cloning, is likely to increase their medical utility and availability. This offers a safe, less cost and more potent vaccines than the conventional vaccines. The development of synthetic vaccines has long term prospects as far as replacement of current bacterial and viral vaccines is concerned. Producing vaccines using peptides with specific immunogenic properties is also being investigated. This includes polio, cholera, tuberculosis and others.

Nowadays a new approach for producing vaccines is the cell culture technique. The new vaccines based on cell culture are purer and safer than some of the existing vaccines. In the existing vaccines all the antigens are obtained after treatment of animals with particular bacteria or virus causing the disease are used. This causes "overloading". In the new method, cells are cultured with the specific bacteria or virus and only those antigens or antibodies, that are specifically required against that disease are injected into the human body.

2.10 HYBRIDOMA: PRODUCTION OF MONOCLONAL ANTIBODIES

Monoclonal antibodies are ultrapure antibodies synthesized by hybridoma cells. Hybridoma cells are hybrid cells formed by fusion of B-lymphocytes (antibody producing cell) and malignant myeloma (cancer) cells. B-lymphocytes produces antibodies specific for a particular antigens. Myeloma cells proliferate rapidly and can be cultured indefinitely (immortal) from a single cell. They cannot, however, be induced to yield antibodies to a specific antigen. That difficulty was overcome in 1975 when Cesar Milstein, G. Kohler and their colleagues at the Medical Research Council laboratory of Molecular Biology in Cambridge, England conceived the idea of fusing mouse myeloma cells with B-lymphocytes from the spleen of a mouse immunized with a specific antigen. The resulting hybrid myeloma or hybridoma cells had the properties of both parent cells: immortality and the ability to secrete large

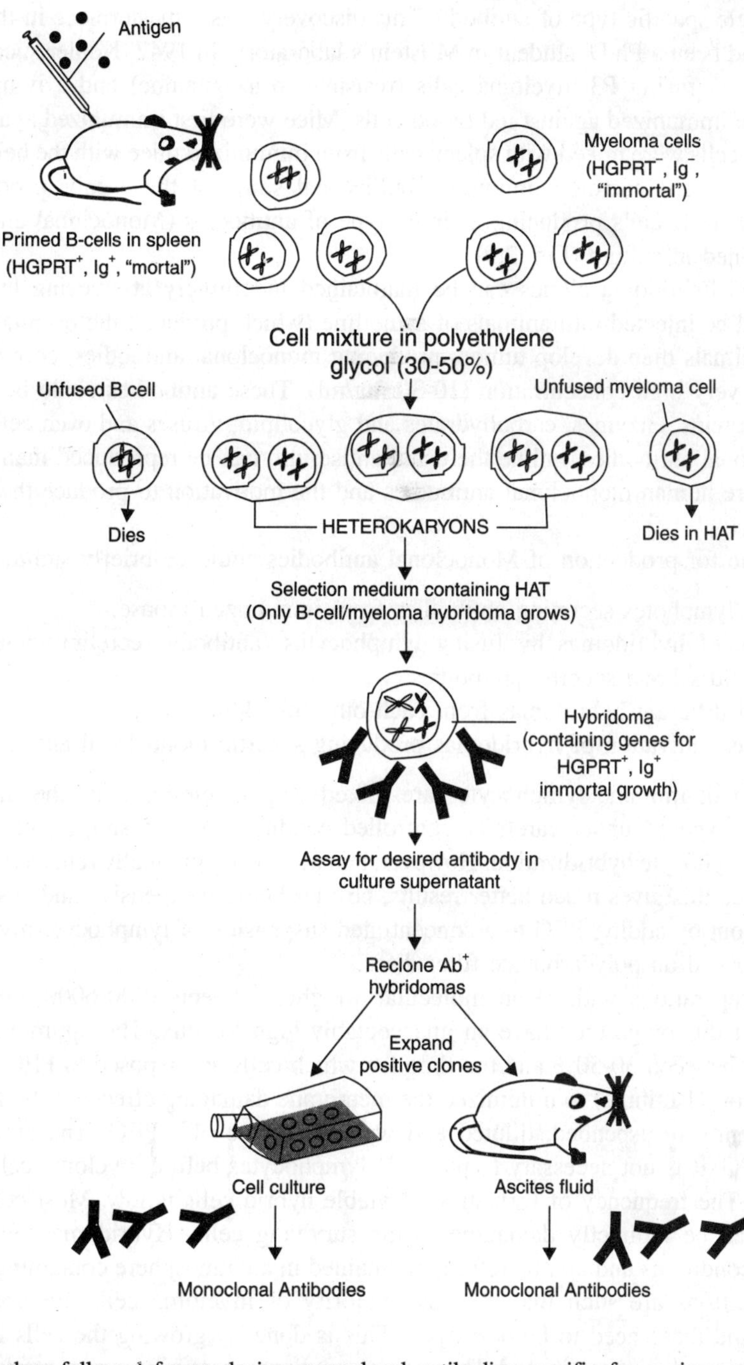

Fig. 2.9. The procedure followed for producing monoclonal antibodies specific for a given antigen developed by G. Kohler and C. Milstein. Spleen cells ($HGPRT^+$ and lg^+) from an antigen-immunized mouse are fused with mouse myeloma cells ($HGPRT^-$ and lg^-). The spleen cell provides the necessary enzymes for growth on HAT (Hypoxanthine, Aminopterin and Thymidine) medium, while the myeloma cell provides immortal-growth properties. Unfused myeloma cells or myeloma/myeloma fusions fail to grow because of their lack of HGPRT. Unfused spleen cells have limited growth capabilities *in vitro* and will die within a few days.

amounts of a single specific type of antibody. This discovery was a masterpiece in the research career of Kohler who had been a Ph.D. student in Milstein's laboratory. In 1947, Kohler succeeded in producing a hybridoma by fusing P3 myeloma cells (resistant to azaguanine) and a lymphocyte from the spleen of a mouse immunized against red blood cells. Mice were first immunized against these RBC's; then P3 myeloma cells were mixed with spleen cells from immunized mice with the help of polyethylene glycol. The hybrid cells secreted immunoglobulins and some of these cells produced against the RBC's. Clones of these cells producing a single type of antibodies (Monoclonal antibody) were isolated and maintained in culture (Fig. 2.9).

The separated hybridoma clones can be maintained indefinitely at freezing temperature. These cells/clones could be injected into animals of same line (which produced the original cell required for fusion). These animals then develop tumors producing monoclonal antibodies, corresponding to those cells/clones, at a very high concentration (10-30 mg/ml). These antibodies could be produced against small haptens, proteins, enzymes, carbohydrates and glycolipid, viruses and even cell membrane components. Monoclonal antibodies, unlike the usual antiserum, can be reproduced many times. It is now possible to prepare human monoclonal antibodies and the motivation to produce these for therapeutic use is tremendous.

The technique for production of Monoclonal antibodies could be briefly summarized as follows:

1. Obtaining lymphotes secreting antibodies from immunized mouse.
2. Production of hybridomas by fusing lymphocytes (antibody secreting) with myeloma cells which would secrete specific antibodies.
3. Separating different hybridomas from each other by cloning.
4. Continuous cultivation of hybridomas producing specific monoclonal antibody.

A suspension of immune lymphocytes are mixed with myeloma cells, the mixture is washed, concentrated and exposed under carefully controlled conditions to a fusing agent. Sendai virus was used earlier on lymphocyte hybridization. However, it is now being gradually replaced with polyethylene glycol (PEG) since this gives much better results, besides being inexpensive and easily available. The fusion is carried out by adding PEG to a concentrated suspension of lymphocyte-myeloma cells either in a tube or deposited on polycarbonate filter discs.

The PEG preparations with mean molecular weights between 1000-6000 are suitable for cell fusion; outside of this range they have an unacceptably high toxicity. The optimum concentration of PEG is normally between 30-50% and the time for which cells are exposed to PEG at this concentration is usually short (1 minute) to minimized the membrane damaging effects of the fusing agents. The cells are then gently resuspended, diluted and washed to free off PEG. The entire operations are conducted at 37°C. It is not necessary to purify B-lymphocytes before myeloma cells selectively fuse with these cells. The frequency of formation of viable hybrid cells is low. Most cells die at an early stage and this can be indirectly damaging to the surviving cells. Hybridoma cells are sensitive to alkaline culture conditions and are, therefore, maintained in an atmosphere containing at least 5% CO_2. The fusion conditions are such that the vast majority of myeloma cells are not fused to spleen B-lymphocytes and these need to be destroyed. This is done by growing the cells in the presence of HAT (Hypoxanthine, Aminopterin and Thymidine) medium containing aminopterin or amethopterin drugs which inhibit folic acid metabolism essential for purine synthesis (*de novo* pathway) and, therefore, kill unfused the myeloma cell deficient in $HGPRT^-$ (enzyme hypoxanthine guanine phosphoribosyl transferase). These myeloma cells ($HGPRT^-$) are mutational variants of parent myeloma and produced for purposes of selecting the hybrid myeloma cells from that of HGPRT minus cells that cannot carry out salvage pathway of DNA synthesis. The unfused B-cells are able to survive in HAT medium, but these cells do not live for longer periods *in vitro* and thus die out within few days. Hybrid cell lines secreting specific monoclonal antibody are isolated and cloned by routine laboratory procedure. It is technically simpler to clone cells in fluid phase media.

Many monoclonal antibodies are now available for detecting pregnancy, diagnosing numerous pathogens, detecting tumor antigens, toxins, drugs in the blood and matching histocompatibility antigens. Several public and private enterprises have commercialized production of monoclonal antibodies all over the world since 1981. The price ranges between Rs. 1500-4000 per milli gram. Many pharmaceutical firms have been interested in large-scale production of ultrapure monoclonal antibodies because of enormous applications of these molecules.

Procedure followed for Production of Hybridoma and Monoclonal Antibodies

1. Inject an appropriate antigen into a mouse first by subcutaneously or in intraperitoneal cavity with successive doses and finally by intravenously.
2. Three days after the last intravenous dose animal was killed.
3. Spleen was asceptically removed and gently disrupted to release lymphocytes and red blood cells.
4. Lymphocytes are separated from RBC and spleenic fluid by density gradient centrifugation.
5. Lymphocytes are washed to remove all contaminating materials.
6. These lymphocytes are mixed with HGPRT – negative myeloma cell line.
7. Expose to Polyethylene glycol (PEG) at 30-50% concentration for one minute.
8. Cells are then washed to make free from PEG.
9. Washed cells comprise mixture of hybridomas and unfused myeloma cells and lymphocytes.
10. Hybridoma are selected on HAT medium. After 10-14 days the only surviving cells on HAT medium are the hybridoma cells.
11. Dilute hybridoma (different types of antibody producing cells) to a level so that aliquot contains single hybridoma cell (growth not easy on dilution).
12. Each clone is then examined (mass screening).
13. Store it in refrigerated conditions.
14. Culture whenever required on large scale.

Applications of Monoclonal Antibodies

MCA has multifarious applications because of their specificities and thus gives a clear-cut results. A large number of hybridoma are produced and stored by cryopreservation technique. Hybridoma banks have been set up in a number of Institutes and Research centres.

(*i*) Monoclonal antibodies help in classification of blood groups into ABO types and Rh types.

(*ii*) By using monoclonal antibodies of an animal immunised against a particular medicine, dose determination of the medicine can be carried out. The reliability of such an immunodosage and the savings it entails are of substantial advantages.

(*iii*) Specific tools based on hybridoma technology could be produced to carry out hormone tests to diagnose viral diseases, to detect certain types of cancers, to monitor the presence or appearance of malignant cells after surgical or radio-therapeutic treatments and to detect allergies.

In 1982, Farrands *et al.*, succeeded in isolating an antibody reacting specifically to malignant tumors of the colon and rectum. Using same technique Mach detected cancer of thyroid and gastrointestinal tract and Tublana detected epithelial cancer.

(*iv*) Monoclonal antibodies could also be used for the purification of complex mixtures of substances which are biologically important (proteins, hormones, etc.)

(*v*) Monoclonal antibodies are also being used for the labelling and precise identification of specialized cells, such as neurons in order to gain better knowledge of the ways in which these cells associate and operate (localisation of chemical neutro-transmitters).

(*vi*) Cell membranes proteins are difficult to purify, since they are present in small amount and their biological activity is not easily measured. These difficulties are overcome by characterising membrane proteins by immunological method such as monoclonal antibody technique.

(*vii*) Monoclonal antibodies could be used as standard reagents to identify molecules present in cell membranes and also to distinguish between cell populations. Promising results are obtained with several blood forming and lymphoid cells.

(*viii*) The differential diagnosis of various leukaemias and related disorders is also possible by these technologies.

(*ix*) Serotherapy, a direct therapy, can be made more effective with the administration of a purified or monoclonal antibodies.

(*x*) Monoclonal antibodies could also be used in the preparation of very specific vaccines, particularly against certain viral strains and parasites.
Croce and his co-workers (1980) obtained a human hybridoma by fusing B lymphocytes from a patient suffering from multiple myeloma with lymphocytes from another patient with subacute panencephalitis. These hybridomas produced human M immunological molecules, specific for certain components of the measles virus. Such a result indicated the possibility of obtaining hybridomas from human B lymphocytes capable of secreting antibodies against pathogenic viruses.

(*xi*) Monoclonal antibodies could also neutralise the action of lymphocytes responsible for the rejection of grafts and destroy the auto-antibodies produced in auto-immune diseases (certain forms of diabetes, multiple sclerosis, rheumatic diseases).

(*xii*) Monoclonal antibodies could considerably increase the effectiveness of medicinal substances on the target cells while avoiding the serious side effects of cancer therapies.

2.11 NEED TO EXPRESS IN ANIMAL CELLS

Animal cell culture system has become an important aspect of biotechnology for the production of proteins, which are of medicinal importance. Now the major compounds that are in wide usage by animal cell culture system include varieties of vaccines, monoclonal antibodies, follicle stimulating hormone (FSH), IFN-α, IFN-β, IFN-γ, tissue plasminogen activating factors, factor VII, factor VIII, factor XI, factor X and erthryopoetins. Because efforts made to produce the required proteins (specifically glycoproteins) especially proteins which are of medicinal importance in the prokaryotic microorganisms and plants are known to having improper function, activity or antigenic characters.

Sometime the required proteins may not be obtained in large quantities or complex modification of final protein synthesized is not possible from the prokaryotic organisms. Generally prokaryotic organisms are not able to synthesize the foreign proteins in large quantities required for fulfilling the demand of the market. The complex modifications of protein include intricate tertiary structure formed by disulfide bond formation or "post-translation modifications" (changes in protein structure after the protein is made from m RNA). The post-translation modifications (Table 2.5) consist of addition of small chemical groups (e.g., an acetyl, methyl or phosphate group), glycosylation, lipid addition and gamma carboxylation.

Table 2.3: Post-translation Modification of Proteins

Modification	*Amino Acids that are Modified*
Addition of small chemical groups	
Acetylation	Lysine
Methylation	Lysine
Phosphorylation	Serine, Threonine, Tyrosine
Hydroxylation	Proline, Lysine
N-formylation	N-terminal glycine
Addition of sugar side chains	
O-linked glycosylation	Serine, Threonine
N-linked glycosylation	Asparagine
Addition of lipid side chains	
Acylation	Serine, Threonine, Cysteine
N-myristoylation	N-terminal glycine
Addition of biotin	
Biotinylation	Lysine

2.12 POST-TRANSLATION MODIFICATIONS

Addition of chemical groups was carried out in a highly specific manner, the same amino acids being modified in the same way in every copy of the protein in a cell. This chemical modification often plays an important role in determining the precise biochemical activity of the target protein.

Glycosylation of a protein means addition of carbohydrate moiety to the protein molecules to form glycoproteins. The above mentioned therapeutic proteins are good example for glycoproteins. This occurs in the endoplasmic reticulum and Golgi complex of eukaryotic cell and depends on the presence of specific enzymes; glycosyltransferases and glycosidases. Glycosylation are generally two types i.e., O-linked glycosylation; is the attachment of a sugar side chain through hydroxyl group of a serine or threonine amino acid and N-linked glycosylation involves attachment through the amino group on the side chain of asparagnine (Fig. 2.10).

Glycosylation can result in attachment to the protein grand structures having networks of 10-12 sugar units of various types.

The prokaryotic organisms neither contain these organelles nor the enzymes and therefore not able to perform this post-translational modification. Eukaryotic yeast and filamentous fungi are able to glycosylate proteins but they do so differently from that of the animal cells.

The presence and conformation of the sugar moiety of a glycoprotein is very essential for a functional product in many cases. This functional properties usually prolongs the half-life in the blood stream, and diverts the glycoproteins to its specific location in the cell. Only the eukaryotic animal cells can do these changes appropriately. Also in many cases the production can be done by plants (especially secondary metabolites), but it involves several enzymatic functions most of which are not known. In some cases, such as immunoglobulins, progress has been made to assemble these molecules in bacterial cell or plant cell systems. However, these processes are not able to work efficiently at large scale. Hence these products can be obtained/expressed only from the cultivation of eukaryotic animal

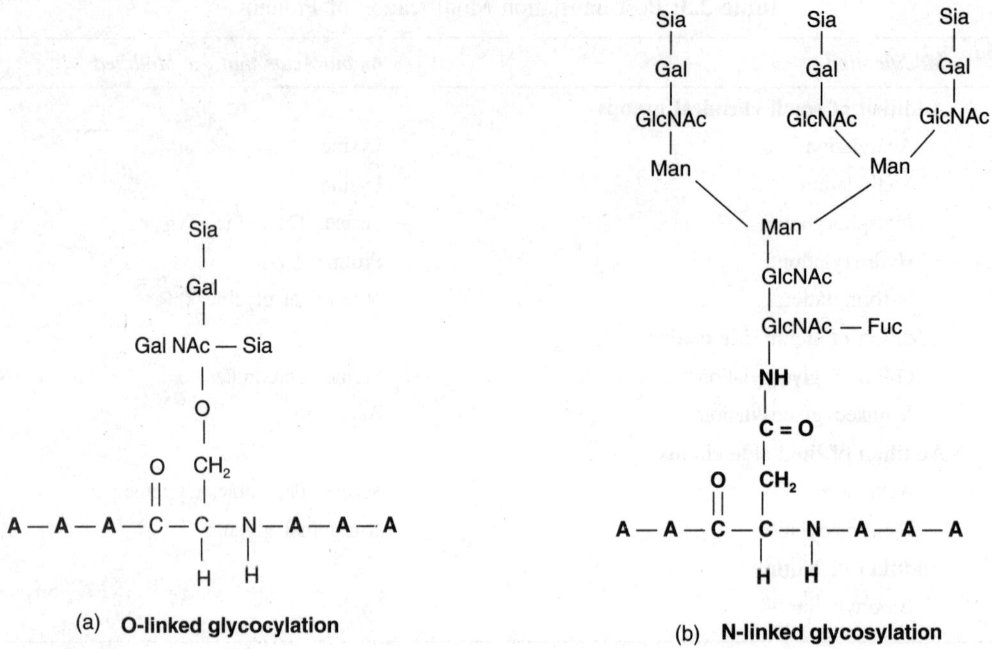

Fig. 2.10. Glycosylation of proteins (a) O-linked glycosylation. Sugar units added to serine amino acid (b) N-linked glycosylation. Larger sugar structure added to asparagine amino acid. Abbreviations used: A, different amino acids; Fuc-fucose; Gal-galactose; GalNAc-N-acetyl galactosamine; GlcNAc-N-acetyl glucosamine; Man-mannose; Sia-Sialic acid.

cells especially mammalian cell culture system. In recent years, the advance of insect cell culture system has allowed some of these molecules to be processed with post-translation modifications similar to higher eukaryotic cells. Pharmaceutical proteins that are not glycosylated or need not be glycosylated for proper function such as insulin, human growth hormone, human serum albumin and hemoglobin may be produced more cost-effectively with bacteria, yeasts and filamentous fungi.

EXERCISE

1. Define cell lines and primary culture.
2. What is meant by cell culture and tissue culture?
3. Explain importance of insulin.
4. Write a note on interferons. (BU Oct. 2003)
5. Define vaccine and give an account on various types of vaccines produced. (BU Oct. 2001)
6. What are monoclonoal antibodies? Explain in detail their production. (BU Oct. 2002)
7. Given an account on metabolism of cultured cells.
8. Why some proteins have to produce only in animal cells?
9. Write a note on glycolysis. (BU Oct. 2001)
10. Write a note on vaccines. (BU April 2001)
11. Give an account of glycosylation.
12. What is meant by post-translation modification? Add a note on different types of post-translation modifications.
13. Define secondary metabolites. Explain production of recombinant insulin and growth hormone. (BU April 2001)

14. Write a note on HAT medium. (BU April 2001, Oct. 2001)
15. Elucidate the applications of monoclonal antibodies. (BU Oct. 2002)
16. Give an account of types and properties of polio vaccines. (BU Oct. 2001)
17. Write a short note on monoclonal antibodies. (BU April 2003)
18. Discuss aminoacid catabolism with examples. (BU April 2003)
19. Explain the production of recombinant factor VIII. Add a note on interferons. (BU April 2003)
20. How tissue plasminogen activator is produced by animal cell biotechnology?

Chapter 3

Growth Factors Promoting Proliferation of Animal Cell Culture

3.1 INTRODUCTION

In animal cell culture system it is now very clear that *in vitro* condition cultured cell undergo divisions only when blood serum is added to the medium. Serum means liquid portion of the blood (plasma) without fibrinogen and blood cells. First des Ligners in 1928 reported the successful cultivation of different mammalian tissues in dilute serum medium. Then in 1933 Parker showed cultivation of chick tissues in serum medium. Fischer *et al.* in 1948 gave a clue of low molecular weight growth factors present in serum and their importance.

Recently these serum proteins were identified and are used instead of whole serum in the culture medium. They include epidermal growth factor (EGF), endothelial cell growth factor (ECGF), fibroblast growth factor (FGF), platelet derived growth factor (PDGF), Interleukin 1(IL-1), Interleukin 2 (IL-2), nerve growth factor (NGF), transforming growth factor (TGF), eye derived growth factor (EDGF), granulocyte-macrophage colony stimulating factor (GM-CSF), granulocyte colony stimulating factor (G-CSF), macrophage colony stimulating factor (M-CSF), stem cell factor (SCF) and erythropoietin (EPO). Thus serum provides some of the growth factors or stimuli necessary for the growth and maintenance.

Growth factors are signal substances released by certain cells and acting on other cells in growth and differentiation. The first such signal substances to be identified were NGF and EGF by Rita Levi-Montalcini and Starley Cohen respectively during 1950s-1960s. This led to the active research in this field and several other growth factors were identified that are released by different core type (Table 3.1).

3.2 COMMON FACTORS AFFECTING THE CELL CULTURE

Commonly in cultural condition addition of epidermal growth factor (EGF), platelet derived growth factor (PDGF) and transforming growth factor (TGF) can stimulate the division in most of the cell types. Whereas for specific cells such as nerve cells require nerve growth factor (NGF) and endothelial

Table 3.1: Growth Factors their Principal Source and Primary Activities

Factor	*Principal Source*	*Primary Activity*	*Comments*
PDGF	Platelets, endothelial cells, placenta	Promotes proliferation of connective tissue, glial and smooth muscle cells	Two different protein chains form 3 distinct dimer forms; AA, AB and BB
EGF	Submaxillary gland, Brunners gland	Promotes proliferation of mesenchymal, glial and epithelial cells	
TGF-a	Common in transformed cells	May be important for normal wound healing	Related to EGF
FGF	Wide range of cells; protein is associated with the ECM	Promotes proliferation of many cells; inhibits some stem cells; induces mesoderm to form in early embryos	At least 19 family members, 4 distinct receptors
NGF	Submaxillary gland	Promotes neurite outgrowth and neural cell survival	Several related proteins first identified as proto-oncogenes; trkA (***trackA***), trkB, trkC
Erythropoietin	Kidney	promotes proliferation and differentiation of erythrocytes	
TGF-β	Activated TH_1 cells (T-helper) and natural killer (NK) cells	Anti-inflammatory (suppresses cytokine production and class II MHC expression), promotes wound healing, inhibits macrophage and lymphocyte proliferation	At least 100 different family members
IGF-I	Primarily liver	Promotes proliferation of many cell types	Related to IGF-II and proinsulin, also called somatomedin C
IGF-II	Variety of cells	Promotes proliferation of many cell types primarily of fetal origin	Related to IGF-I and proinsulin
IL1-α and -β	Macrophages and other antigen presenting cells (APCs)	Costimulation of APCs and T cells, inflammation and fever, acute phase response	Plays a role in hematopoiesis
IL-2	Activated TH_1 cells, NK cells	Proliferation of B cells and activation of T cells	Also stimulates NK functions
IL-3	Activated T cells	Hematopoietic progenitor cell stimulation	Necessary for many cell types

cells require endothelial cell growth factor (ECGF). Now more specific type of growth factors for a particular cell type has been discovered. These factors are supplemented to the medium instead of whole serum. Thus it will decrease the non-defined medium usage with different lots, animals, strains and so on. Sometime the serum extracted may contain the infective viruses, which may be discovered or undiscovered type.

Generally calf, bovine, porcine, horse or goat serum are used for the different cell culture. Culturing of human lymphocytes autologous AB serum was used. The serum from calf foetus is preferable because there was no gammaglobulins when compared to the adult animal serum, which are known to suppress the growth of many animal cell cultures. Serum factors generally act as mitogenic agents

(mitogens). Any factor which stimulates the cell division in culture conditions are referred to as mitogenic agents. These mitogenic agents are very much essential for the induction or division and maintenance of cells in culture condition.

3.3 EPIDERMAL GROWTH FACTOR (EGF)

Stanley Cohen in 1965 discovered this growth factor during the experiments of animal cell culture and EGF is a potent nitrogen for epithelial and mesenchymal cells. This consists of 53 amino acid polypeptide chain and submaxillary gland and kidney produces this growth factor. The EGF known to stimulate the epidermal cell proliferation and nerve cell regeneration. EGF is having major action on stem cells, leukocytes, macrophages and tissue cells. More number of animal cells shows response to the epidermal growth factor in the medium. This is also required for the embryo development. It stimulates synthesis of DNA, collagen, hyaluronic acid and fibronectin.

3.4 ENDOTHELIAL CELL GROWTH FACTOR (ECGF)

This growth promoting factor is isolated from eye and brain. It is also produced in the developing embryo and having molecular weight of 16-17 kD. A large number of mesoderm and endoderm derived cells including vascular endothelial cells, glial cells, smooth muscle cells and fibroblasts are induced by this factors in the culture condition.

3.5 FIBROBLAST GROWTH FACTOR (FGF)

It is a 18 kD protein produced in many adult tissues and also in the embryo. These two forms of FGF namely acidic FGF and basic FGF. It acts as a mitogenic substance, which induces cell proliferation in a number of cells derived from mesoderm and ectoderm such as fibroblasts, smooth muscle cells, vascular endothelial cells and glial cells. Heparin potentiate FGF activity by protecting it against denaturation and degradation. It also stimulates the production of collagenase & plasminogen activator.

3.6 PLATELET DERIVED GROWTH FACTOR (PDGF)

This is an endogenous growth promoting protein known to induce cell division in epidermal cells, leukocytes and T cells. The PDGF is the major mitogen found in serum. It stimulates the proliferation and chemo taxis of fibroblasts and smooth muscle cells. It also stimulates collagen, glycosaminoglycan is collagenase production by fibroblasts. There are two types of PDGF, namely, platelet derived growth factor-A, which is 125 amino acid polypeptide chain and platelet derived growth factor-B, which is a 109 amino acid polypeptide chain. These forms either occurs as homodimers (AA) or heterodimer (AB). PDGF is mainly involved in wound healing process.

3.7 INTERLEUKIN 1(IL-1)

This cytokine is earlier known as endogenous pyrogen, leucocyte activating factor (LAF), B cell activating factor (BAF), lymphocyte activating factor or catabolin which later named as interleukin-1 in 1979. This is known to produce by macrophages, monocytes B cells, endothelial cells and fibroblasts but most abundantly by macrophages. In culture condition or *in vivo* they have effect on varieties of cell types. They help in proliferation of stem cells, lymphocytes, firbroblasts, endothelial cells, B cells, T_H cells and epithelial cells.

Interleukin 1 was known to induce protein synthesis (collagenase, plasminogen activator, fibrinogen, collagen), adhesion molecules production, extracellular matrix production, prostaglandin synthesis in various cell types and alters the permeability of cells. The receptor for IL-1 is present on all the cells of the animal body, hence they can act on all the cell types in culture system. IL-1 is a stable polypeptide retaining its activity upto 56°C and between pH 3-11. It occurs in two molecular forms IL-1 alpha (IL-1α) and IL-1 beta (IL-1β).

3.8 INTERLEUKIN 2 (IL-2)

This cytokine when discovered in 1976 is known as T cell growth factor (TCGF), later it was renamed as interleukin-2. It is 133 amino acid polypeptide chain. These are produced by T cells (majority of the time by CD4+ cells) and generally help in proliferation and differentiation of all types of T cells. It also supports long term growth/maintenance of the cultures. Sometimes they also show the activation and division of B cells.

3.9 NERVE GROWTH FACTOR (NGF)

This was the first "growth factor" to be isolated and characterised. Rita Levi-Montalcini and Hamburger discovered this in the 1951 and then purified by Cohen in 1960. This growth factor was also known as neurotrophins. It is 118 amino acid polypeptide chain with molecular weight of 26 kD and produced by submaxillary gland. The nerve growth factor regulates the development, differentiation and survival of neurons in cultural condition. The NGF is not a good mitogen (induces mitotic division) for the cells in culture.

3.10 TRANSFORMING GROWTH FACTOR (TGF)

It is produced in most adult tissues and many cell types in culture. The platelets, macrophages, lymphocytes and mast cells are major cells that secretes TGF. The TGF is a 25 kD protein and also produced by embryo cells. This factor generally makes normal cells anchorage independent. The anchorage independent character is a feature of transformed animal cells (e.g. tumor cells). Hence this factor is known as transforming growth factor. This elicits multiple cellular responses such as stimulation of proliferation, reversible growth inhibitory activity against epithelial, endothelial, fibroblast, neuronal, lymphoid and hematopoietic cell types.

3.11 ERYTHROPOIETIN (EPO)

It is a glycoprotein and stimulates the division and differentiation of myeloid stem cells to red blood cells. This was produced by kidney and fetal liver cells. The molecular weight is approximately 30,400 daltons and it is 166 amino acid polypeptide chain. The protein contains two disulfide bonds and four sites of glycosylation. The disulfide bonds and glycosylation are very much essential for the biological activity. It acts as mitogenic substance and enhances survival of cells in culture condition and prevents programmed cell death.

3.12 EYE DERIVED GROWTH FACTOR (EDGF)

This is having 14,000 daltons molecular weight. The exact action on the culture cells is not yet discovered but it is known to have positive effect on the cells in cultural condition.

EXERCISE

1. Name different types of growth factors for animal culture. (BU April 2001)
2. Explain application of interleukins and erythropoietin.
3. Write a note on transforming growth factor.
4. What are the applications of growth factors in culture?
5. Explain the role of growth factors in animal cell proliferation. (BU Oct. 2001)
6. Write a short note on erythropoietin. (BU April 2001)
7. What is meant by mitogen?
8. Give the composition of serum and its application in animal cell culture.
9. Name types of cells induced by platelet derived growth factor.
10. Describe effects of interleukin 1 on target cells.

Chapter 4

Bioreactors for Large-Scale Culture of Cells

4.1 INTRODUCTION

Bioreactor can be defined as a vessel or container used to culture the cells at controlled environmental condition for the production of some commercially valuable product. The bioreactor is usually of large volume vessels varies from one litre to 10,000 litres. These bioreactors has provisions for maintaining the aeration, stirring to achieve medium and cell mixing, contamination control and replacement of used medium or used medium with cells.

Animal cells produce macromolecules of medical interest that are not available by any other chemical methods or microbial source. The only available source of such substances previously was urine, animal tissue or even human tissue when human-specific molecules such as growth hormones were required. Later various cell lines have been identified for such purposes. In 1925, on the recommendation of the Oncological Council and the National Cancer Research Institute a cell bank was founded in the USA (American Type Culture Collection; (ATCC), Rockville, Maryland). In 1979 ATCC lists 400 cell lines and 400 species, including about 200 human fibroblast lines with special genetic features. Thus, now it is possible to establish animal culture system for the production of varieties of compound both in laboratory scale and large-scale.

Before starting large-scale cell culture system, the conditions such as cell type, source of cells (embryonic or adult), subculture, serum free growth, medium adoptability, gas phase requirement, culture system and capacity to grow in suspension cultures are to be properly determined and standardized. Otherwise the large scale production cannot be achieved with greater success.

Large-scale cell cultures are used for the production of various biomedical compounds such as interferons, interleukins, hormones, antibodies, enzymes and vaccines (Table 4.1). The bioreactors up to 10,000 litres are generally used for the production purposes. The large-scale culture of cells can be done by three methods such as (1) monolayer cultures, (2) suspension cultures and (3) immobilized cell cultures.

4.2 MONOLAYER CULTURES

Monolayer cultures system are used for anchorage dependent cells. Increase in cell concentration in culture is based on the principle of increasing the surface area available for the growth by using plates, ceramics, spirals and microcarrier particles. Some of the monolayer cultures are as follows.

Table 4.1: Various Biomedical Compounds produced in Animal Cell Culture System

Biomedical Compound	*Type of Cells*	*Used for*
Currently in Therapeutic Use		
Human growth hormone (HGH)	Mouse mammary tumour cell line	Dwarfism
Tissue plasminogen activator (tPA)	Chenase hamster ovary (CHO) cell line	Thrombolysis
Erythropoietin*	Chenase hamster ovary (CHO) cell line	Anemia
Blood clotting factor VIII*	CHO cell line	Hemophilia
In Advanced Stages of Development		
Antithrombin VIII	CHO cell line	Anticoagulant
Blood factor VIII	Baby hamster kidney (BHK)-21 cell line	Hemophilia
Blood factor IX	CHO cell line	Hemophilia-B
Interleukin-2 (II-2)	Lymphocytes	Cancer therapy
Human recombinant antibodies	CHO cell line and Mouse myeloma lines	Infections, various cancers
Hepatitis-B surface antigen cell line	CHO cell line and C-127 against	As vaccine (HBSAg) Hepatitis-B
Interferons (IFN)	Lymphocytes	Tumours
Plasminogen activator (including chimeric and modified)	CHO cell line and *Spodoptera frugiperda*	Thrombolysis
CD4 protein	Mammalian cell line	AIDS vaccine
gp160 protein	Mammalian cell line	AIDS vaccine
gp24 protein	Mammalian cell line	AIDS vaccine
Interleukin-1 receptor antagonist	Mammalian cell line	Severe sepsis
Atrial natriuretic	Mammalian cell line	Kidney failure
Ciliary neurotropic factor	Mammalian cell line	Amyotrophic sclerosis
DNase	Mammalian cell line	Cystic fibrosis

* Approved for marketing in India.

Roux Bottle

It is a flat glass bottle having more internal surface area. This was more often used in the laboratory for the production purpose (Fig. 4.1). Roux bottle is kept stationary after pouring the medium, thus it provides space approximately 175-200 cm^2 internal surface area for the cell attachment on the medium and it occupies around 750-1000 cm^3 area.

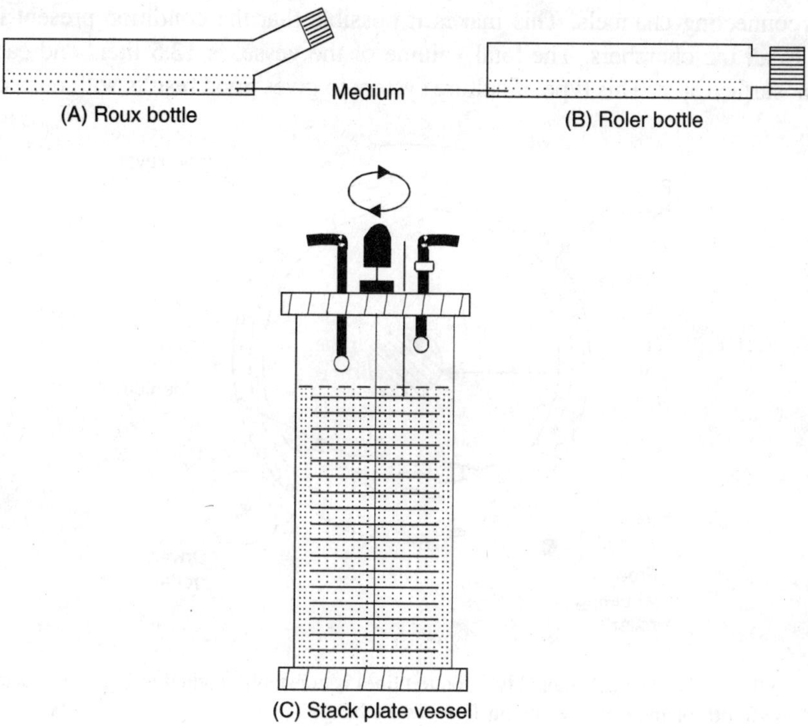

Fig. 4.1. Figures showing three types of culture vessels used for large scale animal cell cultures. (A) Roux bottle, (B) Roller bottle, (C) Stack plate vessels

Roller Bottle

The glass tube/bottle is used which is rotated in special incubators at 2 to 4 rpm. If the cells are nonadherent, they will be agitated by the rolling action, but will remain in the medium. If the cells are adhesive, they will gradually attach to the inner surface of the bottle and grow to form monolayer (Fig. 4.2). This provides an inner surface area of 600-1500 cm^3 area. A further development to this is made by semiautomatic exchange of medium. This helped in an inoculation with 4×10^5 cells/ ml in 100 ml of medium, a yield of 8×10^8 cells per flask can be obtained within six days.

The roller bottle culture has four advantages over static Roux bottle culture: (1) the increase in surface area, (2) the constant but gentle, agitation of the medium, (3) the increased ratio of the medium's surface area to its volume, which allows gas exchange to take place at an increased rate through the thin film of medium over cells not actually submerged in the deep part of the medium and (4) large numbers of cells or for collecting supernatant medium this method is most economical.

Later several modifications of roller bottle further manifold increase in the useful surface area. Such changes include introduction of glass tubes, introduction of polyester spirals or corrugated (ribbed) bottle surface or mechanics of rotations on roller racks or roller drum. These modified surfaces are known to offer the equivalent of up to 100 bottles.

Stacked Plate Vessel

In this vessel number of chambers are stacked one above the other and each chamber is interconnected by the channels (Fig. 4.1C). The standard unit has around 10 chambers stacked one above the other and have interconnecting channels. This makes it possible that the condition present in one chamber will be going to all the chambers. The total volume of the vessel is 12.5 liters and each chamber has around 600 cm^3 surface area. This type of culture system to gives good results with respect to production.

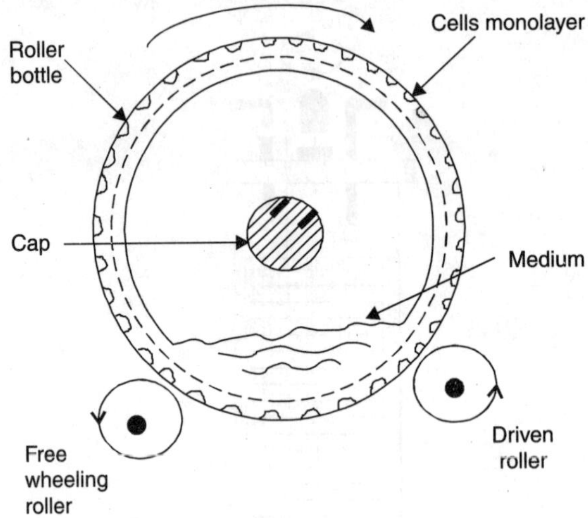

Fig. 4.2. Roller bottle culture. The cell monolayer (dotted line) is constantly bathed in liquid, but is submerged only for about one-fourth of the cycle, enabling free gas exchange

In the monolayer culture system by one or the other reason it is difficult to get good production when compared to the other described methods.

4.3 SUSPENSION CULTURES

In this method cells are grown in the liquid medium without anchorage. Here medium must be continuously stirred to keep the cells in suspension and to make culture homogeneous. The various types of stirrers range from simple magnetic stirrers, flat blade turbine impellers to that complex airlift stirrers or perfusion of medium (taking medium out by hydraulic energy and giving back to the culture vessel). The proper mixing can be done by making alteration in the design of stirrer paddle or by using multiple impellers. The suspension cultures are monitored by using pH, oxygen, CO_2 and glucose electrodes that read from the culture *in situ* and by assaying the utilisation of nutrients, such as glucose and amino acids or the buildup of metabolism such as lactate or ammonia (Fig. 4.3). The stirring operation should be done without causing damage to the cells. Because animal cells are highly fragile as these cells are not like plant cells or microbial cells having rigid cell wall.

Another factor, which influences the growth and production, is the sufficient amount of oxygen supply without affecting the cells. The mean oxygen consumption by cultured cells is about 6 mg/10^6 cell/hour. The oxygen is slightly soluble in medium at the rate of 17 mg/cm^2/hour. Bubbling the air through spargers into the medium can do the proper aeration. The air bubble should be large (as small bubbles damages plasma membrane by its high surface energy), lower gassing rates and by the addition of methyl cellulose, sodium carboxymethyl cellulose, polyvinylpyrrolidone, Dextarn 50000, Ficol 400 and polyglycol (non-nutritional supplements) at 0.1 to 2% which protects from shear forces and bubbles.

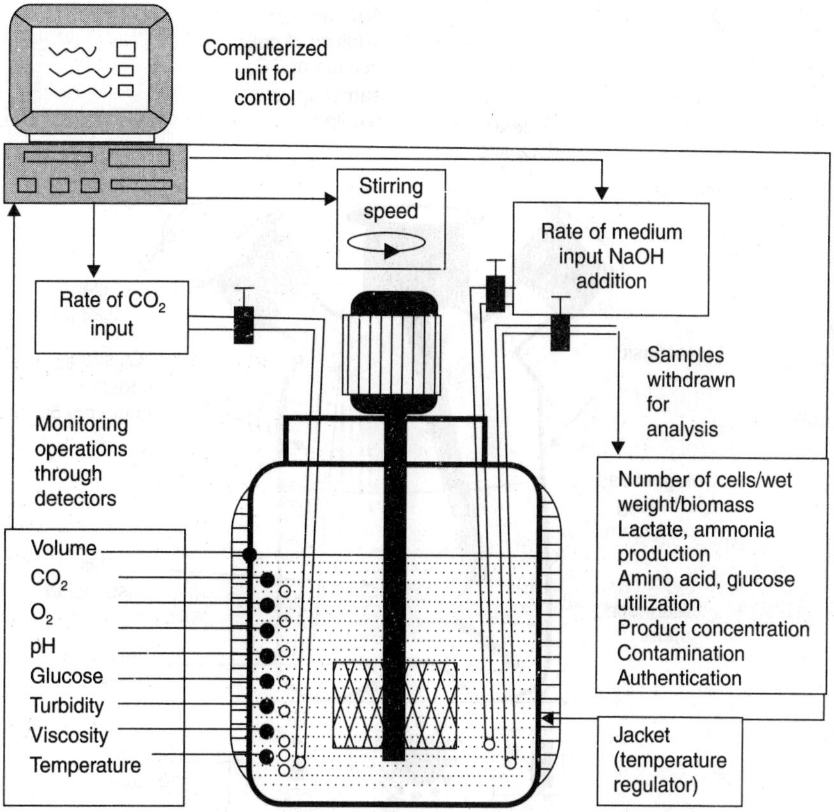

Fig. 4.3. Bioreactor process control. Schematic representation of a paddle-stirred bioreactor with direct-reading probes on the left, feeding to a control unit (top left) that stores the data and also regulates conditions within the bioreactor. A sampling port on the right withdraws the cell suspension from the bioreactor for analysis.

There are three types of suspension cultures (a) stirred bioreactors, (b) continuous flow reactors and (c) airlift bioreactors.

Stirred Bioreactors

These types of bioreactors are closed vessels with fixed culture volume. The help of motor-driven stirrers along with variation in design to stir the medium. These bioreactors are glass (smaller) or stainless steel (larger) culture vessels of 1 to 1000 or even 8000 liter capacity (Fig. 4.4). But normally 20 liter capacity culture vessels are appropriate because it is easy to handle, autoclave and agitate the culture medium without any problem. Stirring of such cultures is done slowly with a magnet encased in a glass pendulum or with a large surface area paddle. The stirring speed should be between 30-100 rpm, sufficient to prevent cell sedimentation. The stirring speed should not create shear forces that would damage the cells. These types of bioreactors are used for the production of interferons, interleukin, hormones, monoclonal antibodies and vaccines.

Continuous Flow Bioreactors

In these types of bioreactors, culture was maintained continuously by adding fresh medium after a specific period of growth. They may be of chemostat or turbidostat method of culture system. Initially both the methods of cultures started as batch culture but later became continuous culture (Fig. 4.5).

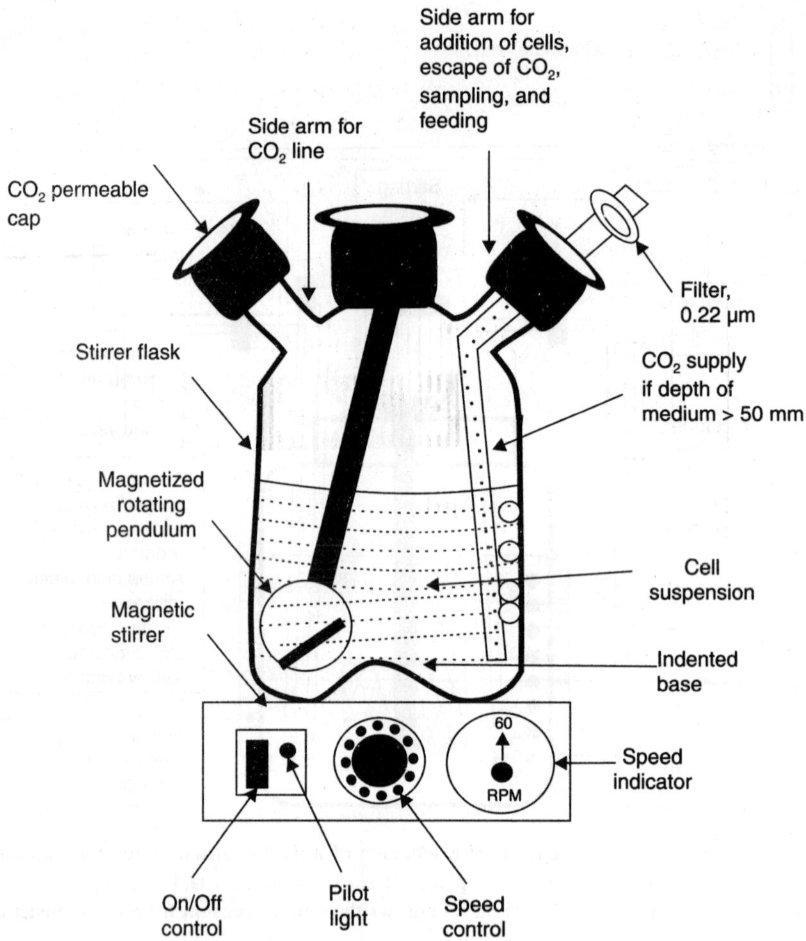

Fig. 4.4. Stirred animal cell culture. Diagrammatic representation of a large stirred flask suitable for volumes up to 8 liters

In the **chemostat** method the inoculated cells grow to the maximum density in the medium when some nutrient (e.g. vitamin) becomes growth limiting. At a constant rate fresh medium is added after 24-48 hours of growth and at an equal rate the culture content is withdrawn. When rate of growth equals the rate of cells withdrawn, the culture is in a steady state and both the cell density and medium composition remain constant. In this condition one of the growth component/nutrient is used at a lower concentration to make it growth limiting. But chemostat is the less efficient or controllable at the cell's maximum growth rate, hence the steady state growth rates in them are much lower than the maximum.

In the **turbidostat** method cells are allowed to grow up to known density which will be measured as turbidity by using photoelectric cell. Once the required turbidity is achieved a fixed volume of culture is withdrawn from the vessel and at the same time equal volume of fresh normal medium is added. This brings down the turbidity of the culture or lowers the cell density. Then cells again start to grow and if known density is reached, the fixed volume of culture is replaced by fresh medium. This rbidostat system works in appropriate manner when the growth rate of the culture is close to the maximum for the cell line. The continuous flow bioreactors give continuous sources of cells and are tu most suitable for production formation (interferons, hormones and vaccines). In animal culture

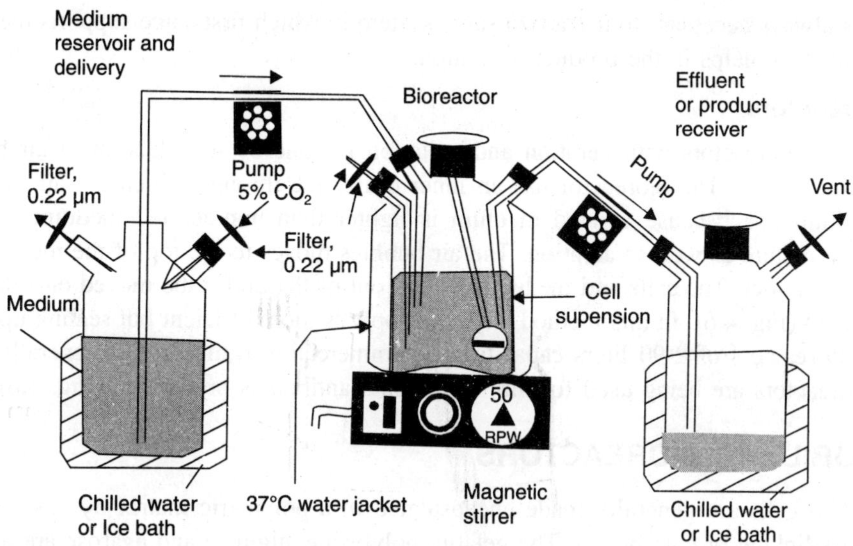

Fig. 4.5. Chemostat/turbidostat. A modification of the suspension culture vessel in Fig. 4.4, with continuous matched input of fresh medium and output of cell suspension. The objective is to keep the culture conditions constant rather than to produce large number of cells.

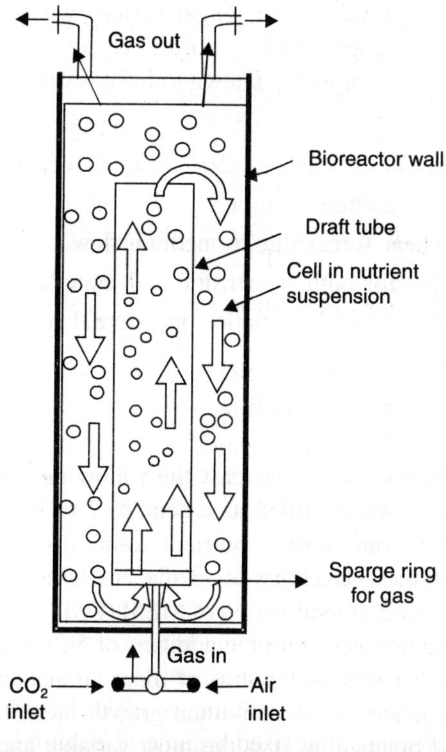

Fig. 4.6. Airlift fermentor. In an airlift fermentor, the bioreactor consists of two concentric cylinders, with the inner cylinder being shorter at both ends than the outer, thereby creating an outer and inner chamber. The bottom of the inner chamber carries a sintered steel ring through which 5% CO_2 in air is bubbled. The bubbles rise, carrying the cell suspension with them. CO_2 is vented from the top, and displacement ensures the return of the cell suspension with them

systems it is always necessary to use a two-stage system in which first stage supports the cells growth, while second stage helps in the product formation.

Airlift Bioreactors

In this type of bioreactors both aeration and agitation is done by introduction of air bubbles at the bottom of the reactor. The bioreactor has an inner draft tube through which the air bubbles and the aerated medium rise because aerated medium is lighter than non-aerated medium. This results in mixing of the culture as well as aeration. The air bubbles moves to the top of the medium and passes out through an outlet. The cells and medium that lift out of the draft tube moved outside the tube and are recirculated (Fig. 4.6). In this method oxygen supply is more efficient but scaling up poses certain problems. Bioreactors of 2-90 liters capacity are commercially available, but generally 20,000 liters capacity bioreactors are being used for the monoclonal antibodies production (Fig. 4.6).

4.4 IMMOBILIZED BIOREACTORS

In this method cells are generally made encapsulated in a polymeric matrix by adsorption, covalent linkage, cross-linking or entrapment. The gelatin, polylysine, alginate and agarose are used as matrix. This makes cells fixed in one position or immobilized. Immobilization effectively protects cells from mechanical damage in large bioreactors, allows medium to diffuse freely into the matrix and to the cells, the products freely comes out of the cells into the medium, the product formation (hormones, immunochemicals and enzymes) prolonged for longer periods when compared to the suspension cultures. The bioreactors of capacity ranges from 1-3 litres are commercially available.

Immobilization of cells in the cultures are known to have several advantages over other methods in the following ways.

1. Higher cell densities can be maintained (50-200 × 10^6 cells/ml).
2. Stability and longevity of cultures is more.
3. Protect the cells from shear forces due to medium flow.
4. This system is applicable for both suspension and monolayer cultures.
5. Less dependence of cells at higher densities on external supply of growth factors, which saves culture cost.

4.5 MEDIA

All media for animal cell cultures contain at least the following components: a hexose (glucose or galactose), the 13 amino acids to be regarded as essential, 7 to 8 vitamins of the B series, choline, inositol and inorganic salts. Such an isotonic nutrient solution can be used for the maintenance of many types of cells. Some of these category of media are basal medium Eagle (BME), minimum essential medium (MEM), Dulbecco's modification of MEM (DME), McCoy 5A, Leibovitz's 15 medium (L 15) (Table 4.2). These media are used with the addition of 5-20% of a homologous or heterologous serum (preferably fetal serum, because γ-globulins of older animals serum inhibits the growth). Even though great efforts are being made to use individual growth factors of the serum, still whole serum usage is continued as standard practice, as it is the most variable and expensive additive. Along with this addition of some hormones such as glucagon, hydrocortison, estradiol, prostaglandins, etc. is having stimuli effect on cultured cells.

Table 4.2: Nutritional Composition of Commonly Used Media (with Serum), for Animal Cell and Tissue Culture System

Nutrient	*mg/lt*								
	Eagle's MEM	*Dulbecco's Modification*	*Ham's F12*	*CMRL 1066*	*RPMI 1640*	*Medium 199*	*L15*	*Fischer's*	*Waymouth MB 752*
Amino acids									
L-alanine	–	–	8.90	25.0	–	25.0	225	–	–
L-arginine (free base)	–	–	–	–	200	–	550	–	–
L-arginine-HCl	126	84.0	211	70.0	–	70.0	–	15.0	75.0
L-asparagine	–	–	–	–	50.0	–	–	–	–
L-asparagine H_2O	–	–15.0	–	–	–	260	11.4	–	
L-aspartic acid	–	–13.3	30.0	20.0	30.0	–	–	60.0	
L-cysteine (free base)	–	–	–	–	–	–	120	–	61.0
L-cysteine	24	48.0	–	20.0	50.0	–	–	–	15.0
L-cysteine 2Na	–	–	–	–	–	23.7	–	23.7	–
L-cysteine $HCl.H_2O$	–	–	35.1	260	–	0.09 87	–	–	–
L-glutamic acid	–	–14.7	75.0	20.0	66.8	–	–	150	–
L-glutamine 292	584	146	100	300	100	300	200	350	–
Glycine	–	30.0	7.50	50.0	10.0	50.0	200	–	50.0
L-histidine (free base)	–	–	–	–	15.0	–	250	–	128
L-histidine $HCl.H_2O$	42.0	42.0	21.0	20.0	–	21.9	–	81.1	–
L-hydroxyproline	–	–	–	10.0	20.0	10.0	–	–	–
L-isoleucine	52.0	105	3.94	20.0	50.0	20.0	125	75.0	25.0
L-leucine	52.0	105	13.1	60.0	50.0	60.0	125	30.0	50.0
L-lysine. HCl	73.1	146	36.5	70.0	40.0	70.0	93	50.0	240
L-methionine	15.0	30.0	4.48	15.0	15.0	15.0	75.0	100	50.0
L-phenylalanine	33.0	66.0	4.96	25.0	15.0	25.0	125	67.0	50.0
L-proline	–	–	34.5	40.0	20.0	40.0	–	–	50.0
L-serine	–	42.0	10.5	25.0	30.0	25.0	200	15.0	–
L-threonine	48.0	95.0	11.9	30.0	20.0	30.0	300	40.0	75.0
L-tryptophan	10.0	16.0	2.04	10.0	5.0	10.0	20.0	10.0	40.0
L-tyrosine	36.0	72.0	5.40	40.0	20.0	–	–	–	40.0
L-tyrosine. 2Na	–	–	–	–	–	49.7	373	74.6	–
L-valine	47.0	94.0	11.7	25.0	20.0	25.0	100	70.0	65.0

(Contd.)

Nutrient	*mg/lt*								
	Eagle's MEM	*Dulbecco's Modification*	*Ham's F12*	*CMRL 1066*	*RPMI 1640*	*Medium 199*	*L15*	*Fischer's*	*Waymouth MB 752*
Vitamins									
L-ascorbic acid	–	–	–	50.0	–	0.050	–	–	17.5
Biotin	–	–	0.0073	0.010	0.200	0.010	–	0.010	0.02
D-Ca-pantothenate	1.00	4.00	0.480	0.010	0.250	0.010	1.00	0.500	1.00
Calciferol	–	–	–	–	–	0.100	–	–	–
Choline chloride	1.00	4.00	14.0	0.500	3.00	0.500	1.00	1.50	250
Folic acid	1.00	4.00	1.30	0.010	1.00	0.010	1.00	10.0	0.40
i-inositol	2.00	7.20	18.0	0.050	35.0	0.050	2.00	1.50	1.00
Nicotinamide	1.00	4.00	0.04	0.025	1.00	0.025	1.00	0.50	1.00
Pyridoxal. HCl	1.00	4.00	0.04	0.025	–	0.025	–	0.50	–
Riboflavin	0.10	0.40	0.038	0.010	0.20	0.010	–	0.50	1.00
Thiamin. HCl	1.00	4.00	0.34	0.010	1.00	0.010	–	1.00	10.0
Vitamin B_{12}	–	–	1.36	–	0.005	–	–	–	0.20
Pyridoxine HCl	–	–	0.062	0.025	1.00	0.025	–	–	10.0
Cholesterol	–	–	–	0.200	–	–	–	–	–
Para-amino benzoic acid	–	–	–	0.050	1.00	0.050	–	–	–
Nicotinic acid	–	–	–	–	–	0.025	–	–	–
Menaphthone Sodium bisulphite $3H_2O$	–	–	–	–	–	0.019	–	–	–
Dl-Â tocopherol PO_4.2Na	–	–	–	–	–	0.01	–	–	–
Vitamin A acetate	–	–	–	–	–	0.115	–	–	–

(Contd.)

Riboflavin	–	–	–	–	–	–	0.10	–	–
$PO_4.2Na$	–	–	–	–	–	–	1.00	–	–
Thiamine mono PO_4, $2H_2O$									
Inorganic salts									
$CaCl_2$ (anhyd.)	200	200	–	200	–	–	–	–	–
$CaCl_2.2H_2O$	–	–	44.0	–	–	186	186	91.0	120
$Fe(NO_3)_3.9H_2O$	–	0.10	–	–	–	0.10	–	–	–
KCl	400	400	224	400	400	400	400	400	150
KH_2PO_4	–	–	–	–	–	60.0	60.0	–	80.0
$MgCl_2.6H_2O$	–	–	122	–	–	–	–	–	240
$MgSO_4.7H_2O$	200	200	–	200	100	200	400	121	200
NaCl	6,800	6,400	7,599	6,799	6,000	800 0	800 0	8,000	6,000
$NaHCO_3$	2,200	3,700	1,176	2,200	2,200	350	–	1,125	2,240
$Na_2H_2PO_4.H_2O$	140	125	–	140	–	–	–	78.0	–
Na_2HPO_4 (anhyd)	–	–	–	–	–	47.5	190	60.0	–
$Na_2HPO_4.7H_2O$	–	–	268	–	1,512	–	–	–	566
$CuSO_4.5H_2O$	–	–	0.0024 9	–	–	–	–	–	–
$FeSO_4.7H_2O$	–	–	0.834	–	–	–	–	–	–
$ZnSO_4.7H_2O$	–	–	0.863	–	–	–	–	–	–
$CaNO_3.4H_2O$	–	–	–	–	100	–	–	–	–
Other components									
D-glucose	1,000	4,500	1,802	1,000	2,000	1,000	–	1,000	5,000
D-galactose	–	–	–	–	–	–	9.00 0	–	–
Lipoic acid	–	–	0.21	–	–	–	–	–	–
Phenol red	10.0	15.0	12.0	20.0	5.00	17.0	10.0	5.00	10.0
Sodium pyruvate	–	110	110	–	–	–	550	–	–
Hypoxanthine	–	–	4.10	–	–	0.30	–	–	–
Linoleic acid	–	0.084	–	–	–	–	–	–	25.0
Putrescine 2HCl	–	–	0.161	–	–	–	–	–	–
Thymidine	–	–	0.73	10.0	–	–	–	–	–
Cocarboxylase	–	–	–	1.00	–	–	–	–	–
Coenzyme A	–	–	–	2.50	–	–	–	–	–
Deoxyadenosine	–	–	–	10.0	–	–	–	–	–
Deoxycytidine HCl	–	–	–	10.0	–	–	–	–	–
Deoxyguanosine	–	–	–	10.0	–	–	–	–	–
Diphosphopyridine nucleotide									

The pH optimum (6.9-7.8) should be known for each cell line. This may be helpful in saving the half the quantity of serum used in the cultural condition. Earlier for pH control use of 5% CO_2 in the air space with 0.22% sodium hydrogen carbonate ($NaHCO_3$) are followed. Recently, use has been made of non-toxic organic buffers e.g., HEPES (N-2-hydroxyethylpiperazine-N-2-ethanesulfonic acid) in combination of up to 850 mg $NaHCO_3$/liter and 2% of CO_2 in the external air space. Because, the traditional concentration of CO_2 is above the optimum requirement by the cell culture.

The use of antibiotics in media is disputed but a mixture of 20 to 200 units of penicillin/ml and 10 to 100 μg of streptomycin/ml is added almost as routine and helps against the majority of bacterial contamination. A proper routine test for Mycoplasmas must be carried out. According to one estimate these organisms contaminate almost the 60% of cell lines. Mycoplasmas do not lead to the immediate death of the cells but changes the nutritional requirements and cell properties.

EXERCISE

1. What is meant by monolayer cultures?
2. Explain different types of monolayer culture system.
3. Write a note on stirred bioreactors.
4. Briefly explain Airlift bioreactor.
5. Mention advantages of immobilized bioreactors.
6. Write a short note on bioreactors. (BU Oct. 2002)
7. Differentiate between chemostat and turbidostat.
8. With neat diagram explain roller bottle culture system.
9. Describe suspension culture systems.
10. Define immobilization and outline the advantages of immobilized bioreactor.

Chapter 5

Expression of Cloned Proteins in Animal Cells

5.1 INTRODUCTION

Eukaryotic cells are about 10 times the size of a typical prokaryote and can be as much as 1000 times greater in volume. The major difference between prokaryotes and eukaryotes is that eukaryotic cells contain membrane-bound compartments in which specific metabolic activities take place. Most important among these is the presence of a cell nucleus, a membrane-delineated compartment that houses the eukaryotic cell's DNA. It is this nucleus that gives the eukaryote its name, which means "true nucleus." The eukaryotic DNA is organized in one or more linear molecules, called chromosomes, which are associated with histone proteins. All chromosomal DNA is stored in the *cell nucleus*, separated from the cytoplasm by a membrane. Some eukaryotic organelles such as mitochondria also contain some DNA.

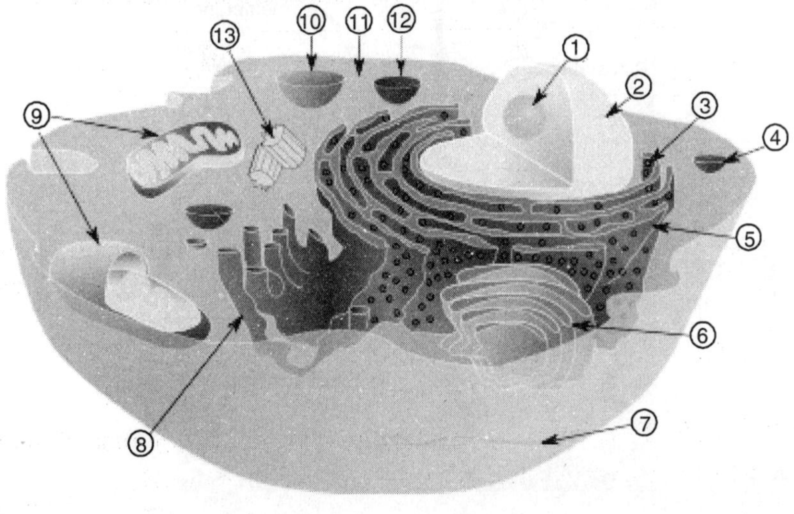

Fig. 5.1. Diagram of a typical animal (eukaryotic) cell, showing subcellular components.

Organelles:

(1) Nucleolus
(2) Nucleus
(3) Ribosome
(4) Vesicle
(5) Rough endoplasmic reticulum (ER)
(6) Golgi apparatus
(7) Cytoskeleton
(8) Smooth endoplasmic reticulum
(9) Mitochondria
(10) Vacuole
(11) Cytoplasm
(12) Lysosome
(13) Centrioles within centrosome.

5.2 PROTEIN SYNTHESIS IN EUKARYOTIC ANIMAL CELL

Within the nucleus of the cell (*gray area*), genes (DNA) are transcribed into RNA. This RNA is then subject to post-transcriptional modification and control, resulting in a mature mRNA that is then transported out of the nucleus and into the cytoplasm, where it undergoes translation into a protein. The mRNA is translated by ribosomes that match the three-base codons of the mRNA to the three-base anti-codons of the appropriate tRNA. Newly-synthesized proteins are often further modified, such as by binding to an effector molecule, to become fully active.

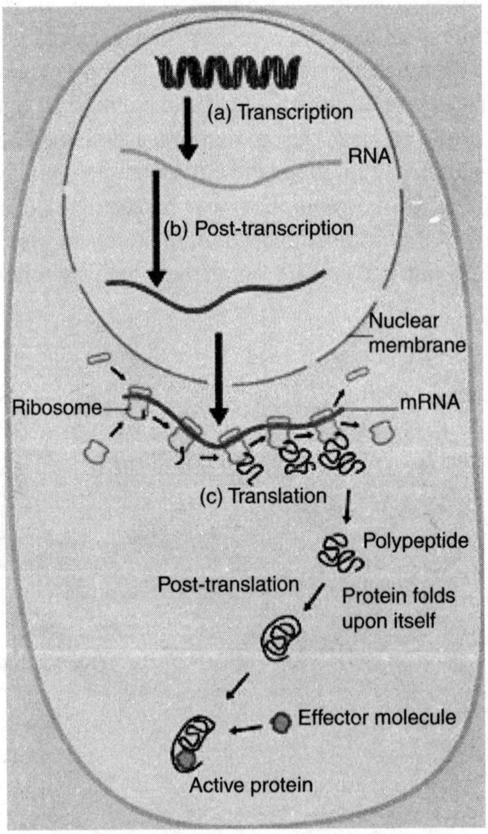

Fig. 5.2. Protein production pathway in the eukaryotic cells.

The need for cell culture, especially at large scales, became apparent with the need for viral vaccines. Major epidemics of polio in the 1940s and 1950s promoted a lot of effort to develop an effective vaccine. When it was shown in 1949 that polio virus could be grown in cultures of human cells, considerable interest was shown to develop large quantities of the polio vaccine using cell culture. The polio vaccine, produced from de-activated virus, became one of the first commercial products of cultured animal cells.

Recombinant DNA technology (also known as genetic engineering) was developed in the 1970s to express mammalian genes in bacteria. It soon became apparent that large complex proteins (and especially those having therapeutic value) couldn't be produced in bacteria, as they do not have the appropriate metabolism to add sugar chains to these proteins. Therefore, genetically engineered animal cells were developed for large-scale commercial production of such important proteins. Animal cells can be grown either in an unattached suspension culture or attached to a solid surface. Suspension cultures have been successfully developed to quite large bioreactor volumes, with successful production of viruses and therapeutic proteins.

The range of commercially available biologicals produced by cell culture technologies has increased rapidly over the last decade. This is particularly so for therapeutic proteins synthesized from selected or genetically engineered mammalian cells. They are needed in large quantities and hence require careful study of the underling biochemical, cell biological and engineering principles for control of production processes. Mammalian cells are used in industry as well as in research for a variety of applications. Examples are the production of monoclonal antibodies by hybridoma cells, proteins for diagnostic or therapeutic use with recombinant cells, production of retroviral vectors for gene therapy as well as the cultivation of tissue cells for artificial organs. In addition to the engineering aspects systems biology approach is increasingly used in our study. Cell culture has been established as a working horse of the modern pharmaceutical industry, the intracellular events and their regulation in the cell during protein production in a bioreactor.

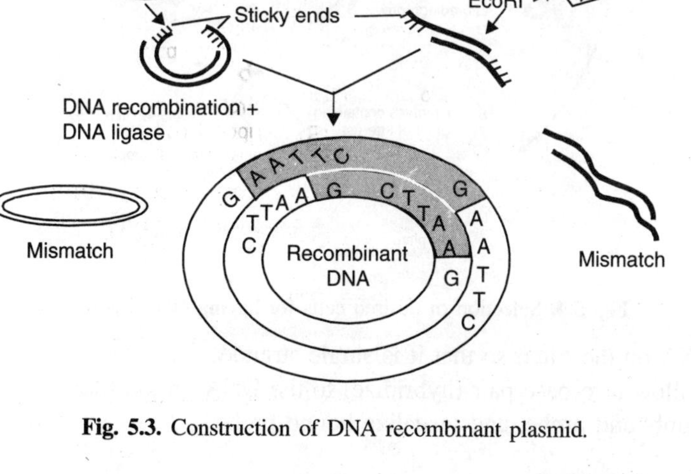

Fig. 5.3. Construction of DNA recombinant plasmid.

5.3 CLONING OF REQUIRED GENE INTO HOST CELLS

With standard techniques, the gene sequence information can be used to construct a gene clone (*Gene Synthesis*). Cloning is accomplished by inserting the synthesized DNA segment into a cloning vector, usually a specific microbe or animal cells or bacterial virus designed to over-express the protein of interest (*Cloning*).

Choice of vector will vary, since all DNA sequences cannot be cloned in the same vector, nor can all proteins be produced in the same vector. In some cases, specific DNA sequence modifications will be needed before insertion [e.g., to increase the resultant protein's solubility or to change the way it interacts with other proteins (*Modification*).

5.4 IDENTIFICATION OF CELL CLONES CARRYING THE GENE OF INTEREST

- Can identify the DNA of the desired gene by **nucleic acid hybridization** Fig. 5.4)
  - * Depends upon base pairing of the gene and a complementary sequence on another short nucleic acid, called the **probe**.
  - * Probe can be DNA or RNA, is labeled with a radioisotope or a fluorescent tag.
  - * First transfer the DNA to be tested to a nitrocellulose (or similar) filter.

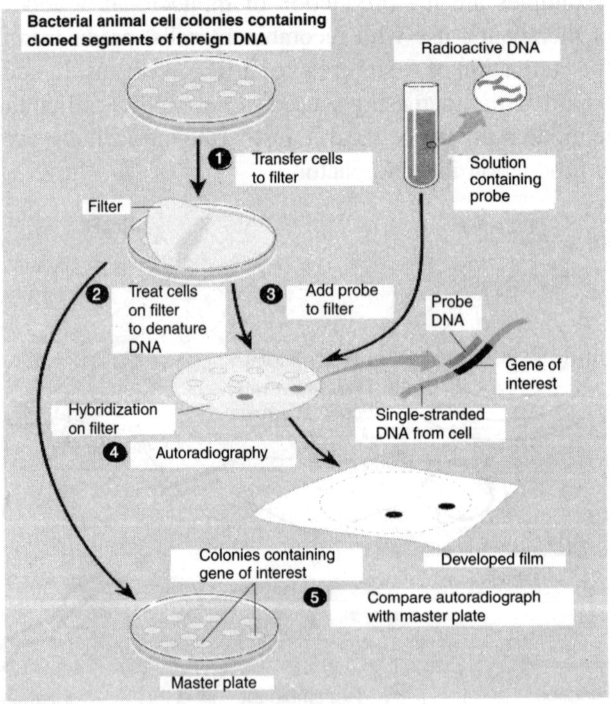

Fig. 5.4. Selection of desired cells for having cloned gene.

  - * Denature DNA on the filter, so that it is single stranded.
  - * Add probe, allow it to base pair (hybridize) to the DNA on the filter.
  - * Wash away unbound probe, and visualize bound probe.

• Can identify the protein product of the desired gene
  * Usually transfer protein to nitrocellulose filter.
  * Assay for enzymatic activity if enzyme, with substrate that yields colored or fluorescent product.
  * Assay using antibodies labeled with radioactive or enzymatic tag = immunoassay.

Cloning and modification can introduce errors into a given DNA sequence. A critical quality-control step, one of several in the protein-production process, is verification that the gene clone's DNA sequence is correct.

No single method will work equally well for all proteins, so several methods will be needed to produce difference proteins from gene clones or gene sequence information. Preproduction screening will optimize production and purification methods for each protein of interest. The presence, level, and purity of protein expression will be checked using microchannel separations of reaction products combined with molecular-weight markers and various detection techniques (e.g., mass spectrometry, ultraviolet absorbance, and light scattering). Data will be entered into a computer and analyzed, and the best conditions and methods for large-scale protein production will be identified automatically.

During cellular protein production, vectors carrying the gene clone of interest are inserted into a animal host whose cellular machinery is used to produce the specific protein of interest (*Cellular Production*). The protein is extracted from the host cells and purified. Alternatively, proteins can be produced by mixing a DNA template with a set of purified enzymes and chemicals normally used by the cell for protein production; only the protein of interest is produced without the need for a living cell (*Cell-Free Production*). Finally, well-established chemical-synthesis methods can be used to make short strings of amino acids that must then be hooked together to make complete proteins (*Chemical Synthesis*). These methods, especially chemical synthesis, can be used to introduce specific changes in a protein sequence such as modification of protein subunits or incorporation of radioactive isotopes needed in downstream analysis. All proteins will undergo purification using a variety of separation technologies (e.g., liquid chromatography, capillary electrophoresis, or affinity columns). Proteins also will need to be collected and maintained under specific conditions that enable them to fold into their natural, functionally active configurations.

The protein production process can be automated and run simultaneously on hundreds of samples to generate a vast array of normal or modified proteins ready for characterization.

5.5 BENEFITS OF EXPRESSION OF PROTEIN IN ANIMAL CELLS

- Eukaryotic post-translational modification
- Proper protein folding and function
- High expression levels
- Easy scale up and protein purification

Production of therapeutic recombinant proteins from mammalian systems is one of the fastest growing segments of the pharmaceutical market. Several analyses indicate that demand will eventually outpace production of these proteins. This imminent shortage has served to emphasize the need to improve on the yield that can be obtained from the systems currently in place. The majority of these proteins are produced in Chinese Hamster Ovary (CHO) cells, while the remainder are produced in a variety of other cell types including mouse myelomas NSO and SP2/0, baby hamster kidney (BHK) and a variety of others. CHO cells were first cultured in the late 1950's after their isolation from a Chinese hamster (*Cricetulus griseus*) ovary epithelial tumor. Produced in CHO, tPA was the first of these recombinant proteins to receive approval for therapeutic use in 1986. Many changes to the process have occurred since that time, including changes to the cells, the growth medium and the reactor conditions to improve from that first venture.

5.6 EXPRESSION VECTOR

An expression vector with protein producing DNA is imported into a cell through the use of plasmids or any other vector into a host cell. This causes the host cell to then produce said protein. This method can be used in medicine to form proteins on a wide-scale such as Insulin, Erythyroprotein, or Tissue Plasminogen Activator. An image of this can be seen below:

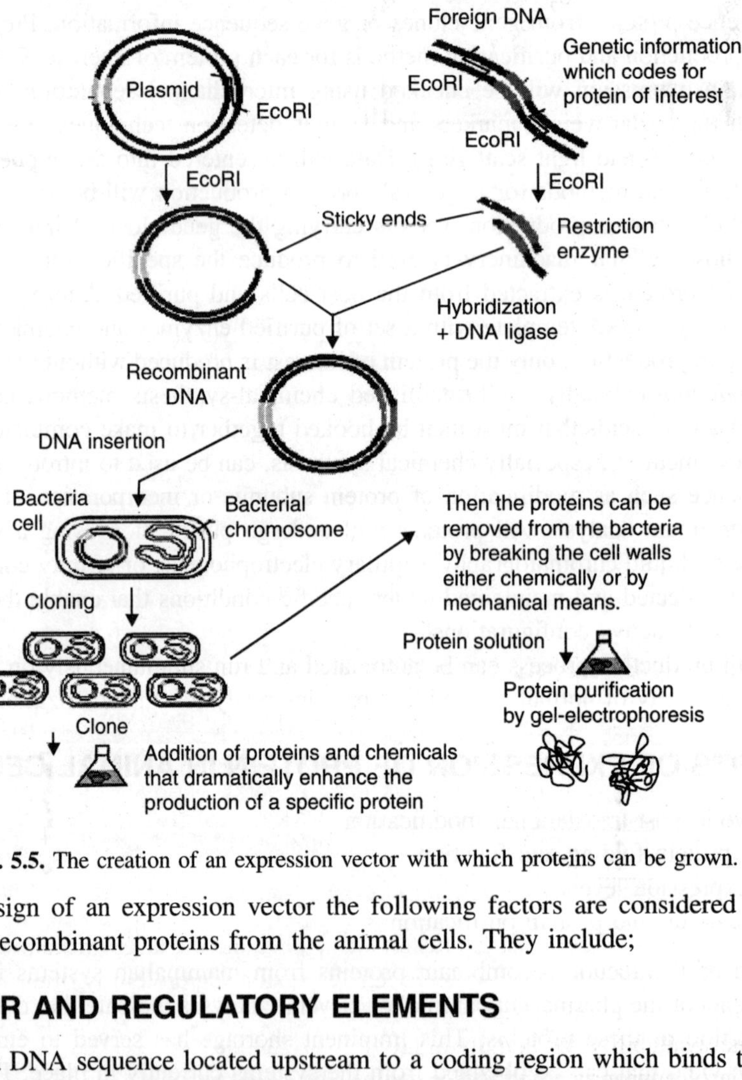

Fig. 5.5. The creation of an expression vector with which proteins can be grown.

During the design of an expression vector the following factors are considered for appropriate expression of the recombinant proteins from the animal cells. They include;

5.7 PROMOTER AND REGULATORY ELEMENTS

The promotor is a DNA sequence located upstream to a coding region which binds the transcription complex thus allowing transcription initiation. The promotor is positioned 10-100 nucleotides upstream of the ribosome binding site. The ideal promoter exhibits the following features:

- It is strong enough to allow product accumulation up to 50% of the total cellular protein.

- It exhibits a minimal level of **basal** transcription (a highly repressible promoter is especially important, when the protein of interest is toxic or detrimental to the growth of the host cell).
- It is easy to induce.

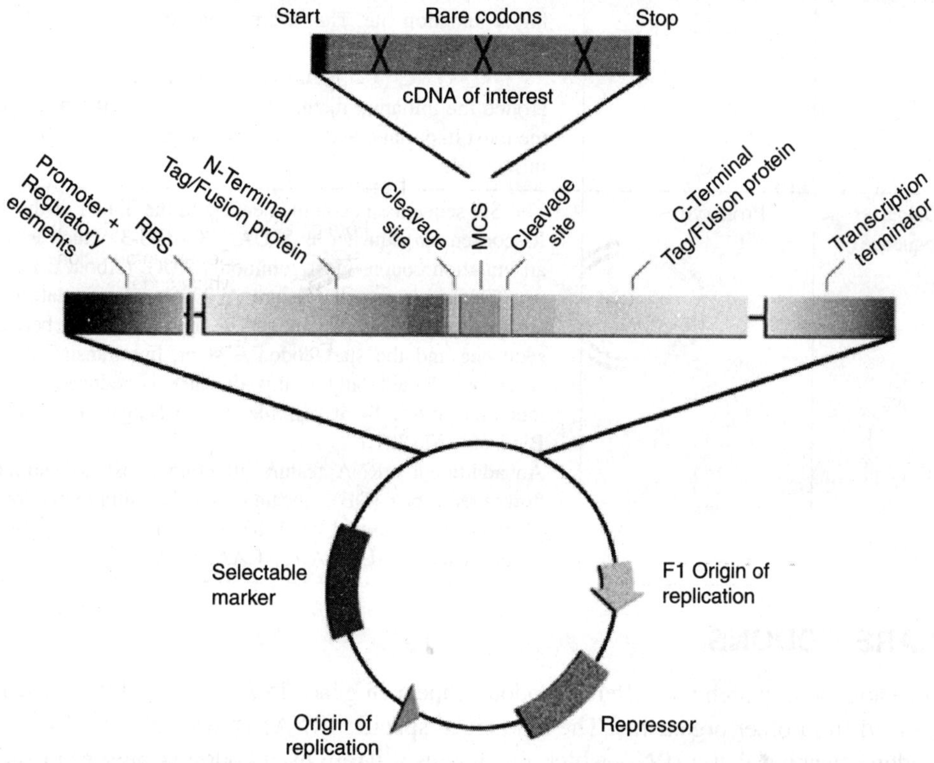

Fig. 5.6. Expression vector.

5.8 RIBOSOMAL BINDING SITE (RBS)

- This site, located at the 5' untranslated region (UTR) of the mRNA transcript, regulates protein synthesis based on its sequence and structure. There are two consensus sequences, named Kozak and Shine Dalgarno (after their discoverers), which are known to facilitate efficient translation in Eukaryotes and Prokaryotes respectively. See Table 5.1 for more details.

5.9 CLEAVAGE SITE

This sequence is positioned between the fusion tag/protein and the target protein where it is recognized by proteases or other cleaving agents. Although the fusion tag/protein is often important for protein production and/or purification, it may interfere with:

- Active site availability for activity assay
- Production of appropriate antibodies against the target protein
- Determination of 3D structure of the protein

Addition of proteases or cleaving agents which specifically recognize the cleaving site, allow for easy and efficient release of the target protein from the fusion tag/protein.

Table 5.1: Description of Organisms Consensus Sequences

Name	*Type of Organism*	*Description*
Kozak Sequence	Eukaryote	Preferred sequence context which allows effective ribosomal binding at initiation site. The favored context is: –6 –5 –4 **–3** –2 –1 1 2 3 **4** G C C **A** C C **A U G G** around the initiating methionine (**AUG**), with the **A** at the –3 being the most important, and a purine at +4, preferably **G** being next most influential.
Shine Dalgarno (SD) Sequence	Prokaryote	The SD sequence is complementary to the 3' end of the 16S rRNA. Its consensus sequence is 5'-UAAGGAGG-3' which is followed by an initiation codon, most commonly AUG. (About 8% of start sites use GUG, whereas UUG and AUU are rare initiators present in autogenously regulated genes). The optimal spacing between the SD sequence and the start codon is 8 nt, but translation initiation is severely affected only if this distance is reduced below 4 nt or increased above 14 nt. [Shine, J. and Dalgarno, L. (1975) Eur. J. Biochem. 57, 221.] An additional mRNA feature affecting translation initiation is the downstream box (DB), located after the initiation codon and complementary to bases 1469-1483 of the 16S rRNA. DBs consensus sequence is: 5'-AUGAAUCACAAAGUG-3'.

5.10 RARE CODONS

The frequencies with which the different codons appear in genes in *E. coli* are different from those in genes derived from other organisms. The amount of specific tRNAs is also reflected by the frequency of the codon, meaning that a tRNA which recognizes a rarely used codon is present in low amounts. Therefore, various genes that contain codons which are rare in *E. coli* may be inefficiently expressed by this organism. Rare codons can cause premature termintion of the synthesized protein or misincorporation of amino acids. *Clusters* of rare codons stand a higher chance to create translation errors and reduce the expression level. This problem can be solved either by exchanging codons in the target gene for codons which are more frequently used in *E. coli*.

5.11 MCS (MULTIPLE CLONING SITE) / POLYLINKER

This region contains several restriction enzyme recognition sites very close to each other. Each of these sites cuts the plasmid only once. Hence, cutting the plasmid with one of these restriction enzymes does not disrupt any of its essential features, and may serve to insert the foreign DNA. By having many different restriction enzyme recognition sites close together, polylinkers offer many choices in regard to which enzymes to use for cloning.

Fusion Tag /Protein

A protein or a peptide located either on the C- or N- terminal of the target protein, which facilitates one or several of the following characteristics:

- **Improved solubility (S):** Fusion of the N-terminus of the target protein to the C-terminus of a soluble fusion partner often improves the solubility of the target protein.

- **Improved detection** (D): Fusion of the target protein to either terminus of a short peptide (epitope tag) or protein which is recognized by an antibody (Western blot analysis) or by biophysical methods (e.g. GFP by fluorescence) facilitates the detection of the resulting protein during expression or purification.
- **Improved purification** (P): Simple purification schemes have been developed for proteins used at either terminus which bind specifically to affinity resins.
- **Localization (L)**: Tag, usually located on N-terminus of the target protein, which acts as address for sending protein to a specific cellular compartment.
- **Improved Expression (E)**: Fusion of the N-terminus of the target protein to the C-terminus of a highly expressed fusion partner results in high level expression of the target protein.

The fusion protein must be continuous with the target protein—the same open reading frame must be maintained. Stop codons between the target protein and the fusion partner must be omitted.

Transcription Terminator

The transcription terminator is placed at the 3' end of the target gene. It enhances plasmid stability by preventing transcription through the replication region and through other promoters located on the plasmid. Continued transcription from strong promoters into the replication region can destabilize plasmids due to overproduction of the ROP protein, which is involved in the control of plasmid copy number. The presence of transcription terminators also prevents formation of antisense RNAs from downstream promoters operating in reverse orientation with respect to the target gene. In addition, the transcription terminator enhances the stability of the mRNA transcript by a stem-loop formation at the 3' end.

Selectable Marker

This element is required for the maintenance of the plasmid in the cell. Due to the presence of the selective marker, the plasmid becomes useful for the cell. Under the selective conditions, only cells that contain plasmids with the appropriate selectable marker can survive. Commonly, genes that confer resistance to various antibiotics are used as selective markers in cloning vectors. The drawbacks of this aproach are: (1) loss of selective pressure as a result of antibiotics degradation and inactivation. (2) contamination of the product or biomass by antibiotics, which may be unacceptable from medical or regulatory considerations.

Origin of Replication

An origin of replication is a sequence of DNA at which replication is initiated on a chromosome, plasmid or virus. For small DNAs, including bacterial plasmids and small viruses, a single origin is sufficient. Larger DNAs have many origins, and DNA replication is initiated at all of them; otherwise, if all replication had to proceed from a single origin, it would take too long to replicate the entire DNA mass.

The origin of replication determines the **vector copy number**, which could typically be in the range of 25–50 copies/cell if the expression vector is derived from the low-copy-number plasmid pBR322, or between 150 and 200 copies/cell if derived from the high-copy-number plasmid pUC. The copy number influences the plasmid stability, i.e. the maintenance of the plasmid within the cells during cell division. A positive effect of a high copy number is the greater stability of the plasmid when the random partitioning occurs at cell division. On the other hand, a high number of plasmids generally decreases the growth rate, thus possibly allowing for cells with few plasmids to dominate the culture, since they grow faster. There appears to be no significant advantage of using higher-copy-number plasmids over pBR322-based vectors in terms of production yields.

The origin of replication also determines the plasmid's **compatibility**: its ability to replicate in conjunction with another plasmid within the same bacterial cell. Plasmids that utilize the same replication system cannot co-exist in the same bacterial cell. They are said to belong to the same compatibility group. The introduction of a new origin, in the form of a second plasmid from the same compatibility group, mimics the result of replication of the resident plasmid. Thus any further replication is prevented until after the two plasmids have been segregated to different cells to create the correct prereplication copy number.

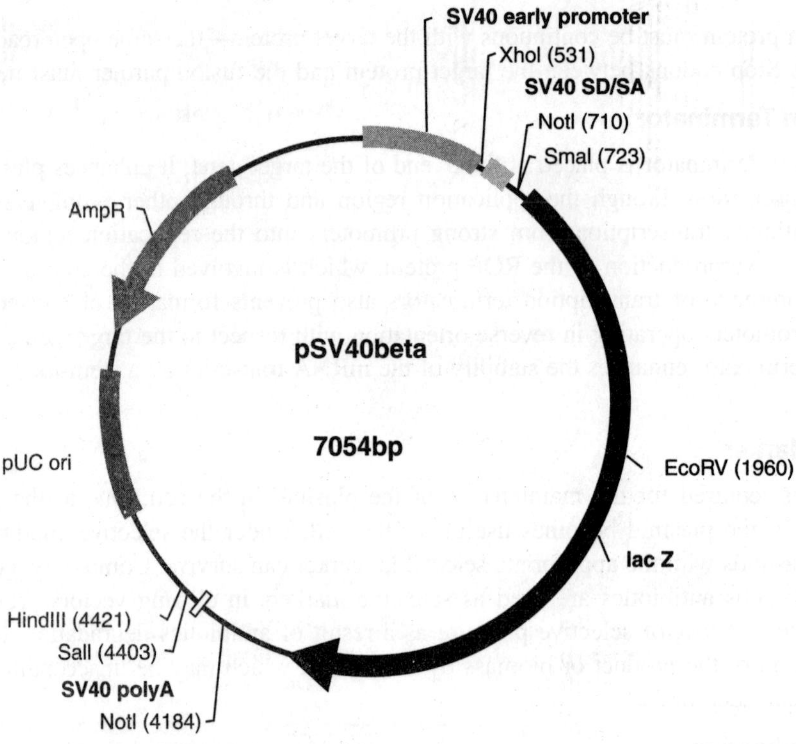

Fig. 5.7. pSV40β Mammalian *lacZ* Vector.

pSV40β Vector: This common eukaryotic expression vector, pSV40β expresses the full-length β-galactosidase gene under the control of simian virus 40 (SV40) early promoter. This vector is very useful for transfection of mammalian cells in culture and for use in other species. The β-galactosidase enzyme expression is enhanced by elements including: SD/SA-RNA splice donor and acceptor sequence, and SV40 late polyadenylylation signal. pSV40β expression vector also contains β-lactamase, which acts as a selection marker (100 μg/ml ampicillin resistance) in *E. coli* host.

The four base letters can be used to compose "words" inside of cells. We know that this language has exactly twenty words plus a couple of "punctuation marks" (*i.e.*, DNA segments that mean "this is the end of a sentence.") A word is specified by three letters and corresponds to one of twenty amino acids. Extending the analogy, since a protein consists of many words, we can think of a protein as a long sentence or a paragraph. The mapping from base triplets to amino acids is the gist of the genetic code. The Fig. 5.8A illustrates the process by which protein is fabricated inside of a cell. It begins inside of a cell nucleus, where the double helix of DNA is pulled apart like a zipper. Messenger RNA (denoted mRNA) can then match up to a strand of the DNA so that a copy of the information can be safely transported outside of the cell. Once the mRNA travels through the cell's cytoplasm, it can then hookup with a ribosome, which takes care of the remainder of the protein making process.

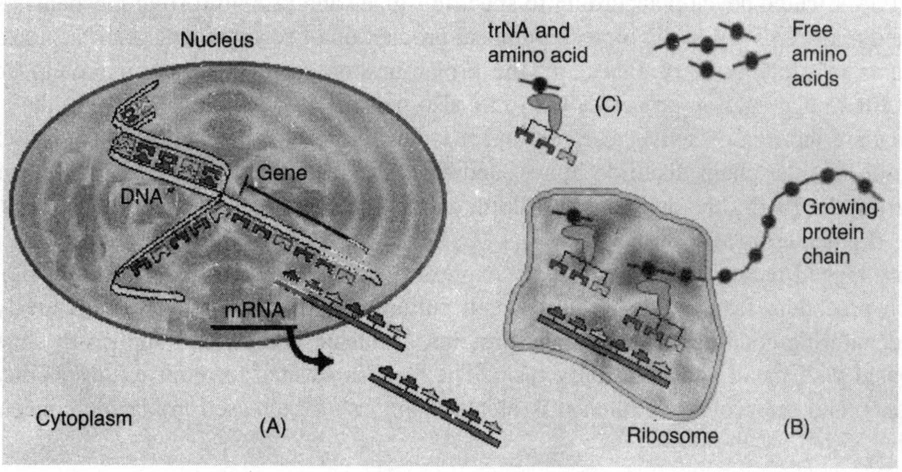

Fig. 5.8. Gene expression in three stages: (A) the information encoded in DNA is copied to a strand of messenger RNA (mRNA) which (B) moves outside of the nucleus and eventually to a ribosome where (C) transfer RNA (tRNA) combines with the mRNA to assemble a protein.

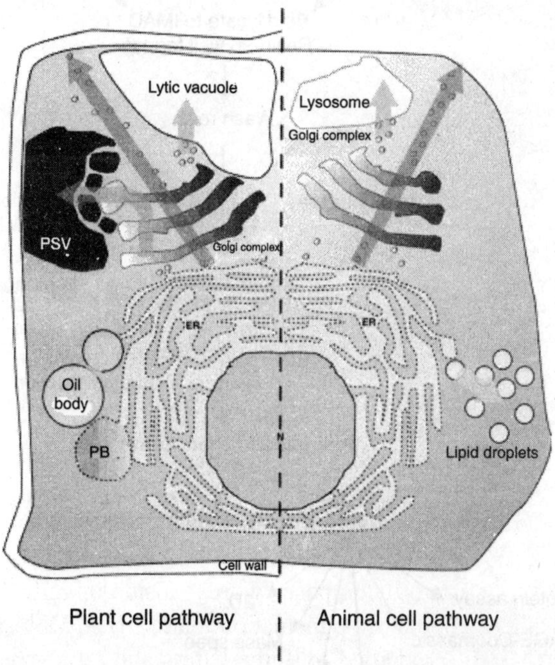

Fig. 5.9. The secretory pathways of plant cells versus animal cells: Opportunities to exploit animal cell as bioreactors.

Secretory proteins enter the ER (delimited here by the broken line, indicating membrane-bound ribosomes) in both cell types and travel to the Golgi complex via vesicular trafficking. In the absence of ER-retention signals, soluble proteins are then secreted (dark arrows). Vesicular sorting to the inner hydrolytic compartments (lytic vacuoles in plants, lysosomes in animals; green arrows) requires signals that differ in plants and animals. In addition to lytic vacuoles, plant cells may also have protein storage vacuoles (PSV), which develop according to specific signals and protein sorting mechanisms (arrow in Fig. 5.9) and may be useful in the biotechnological production of recombinant pharmaceutical proteins. Vast amounts of proteins of cereals (e.g., the prolamin storage proteins) form protein bodies (PB) within the ER in aggregation processes that may also be exploited for biotechnology; the aggregation mechanisms are an area of active research and may involve prolonged interactions with chaperones, the formation of inter-chain disulfide bonds, and selective insolubility. Oil bodies and lipid droplets originate from the ER by the deposition of lipids in the space between the two monolayers of the ER membrane (lighter arrows).

Affinity tagged recombinant proteins are expressed and purified (~ >90% pure) using a one-step purification procedure from crude lysates or cell culture supernatants. Briefly, cells are lysed under native or denaturing conditions, (membrane proteins, inclusion bodies, etc) applied to 96 well plates, and incubated with metal-chelate affinity resin. The resin is washed to remove nonspecifically bound contaminants, and the protein of interest is eluted using either increased imidazole concentrations or low pH (Fig. 5.10).

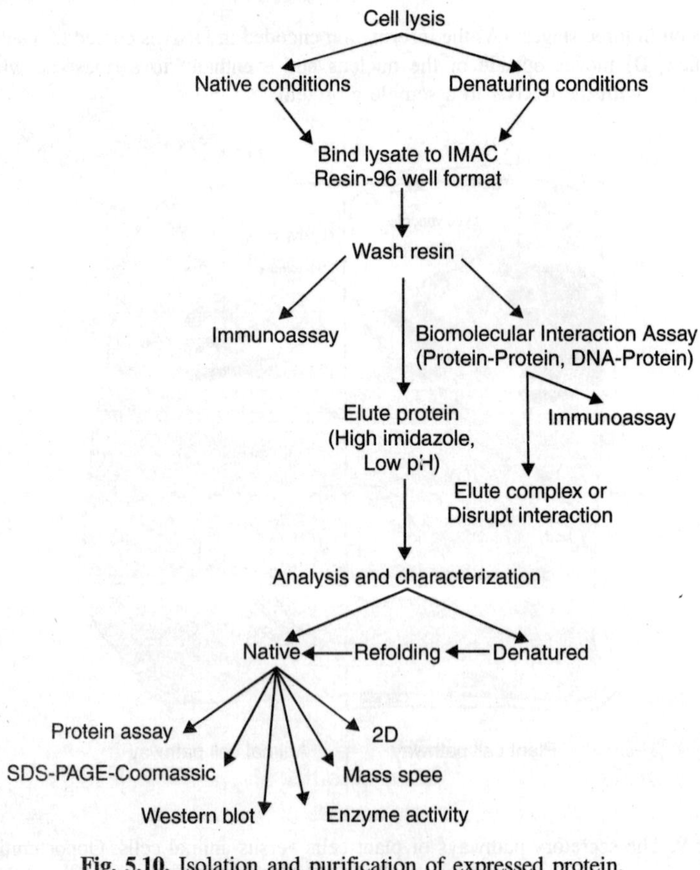

Fig. 5.10. Isolation and purification of expressed protein.

The process typically yields 1-50 μg purified protein. Alternatively, increased protein yields can be obtained using affinity spin columns or larger plates. The capacity of Ni affinity resin is reported to be in the range of 5-10mg/ml. The process is as follows:

Step 1. Gene selection and vector design

Isolating the gene of interest required for production of proteins in cell culture.

Step 2. Construction of expression vectors

The target sequence will be cloned into vector system. Based on the specific target protein, the sequence can be inserted into an expression cassette for a secreted or non-secreted protein. It is necessary to will conduct DNA sequencing and analysis to confirm correct coding and flanking sequences.

Step 3. Cell transfection, selection and cloning of target-expressing cells

The protein can be produced by transient burst or stable cell line expression depending on the client's specific needs. The expression of the target protein will be confirmed by SDS page or Western blot using an anti-tag antibody or client-provided antibody against the target protein.

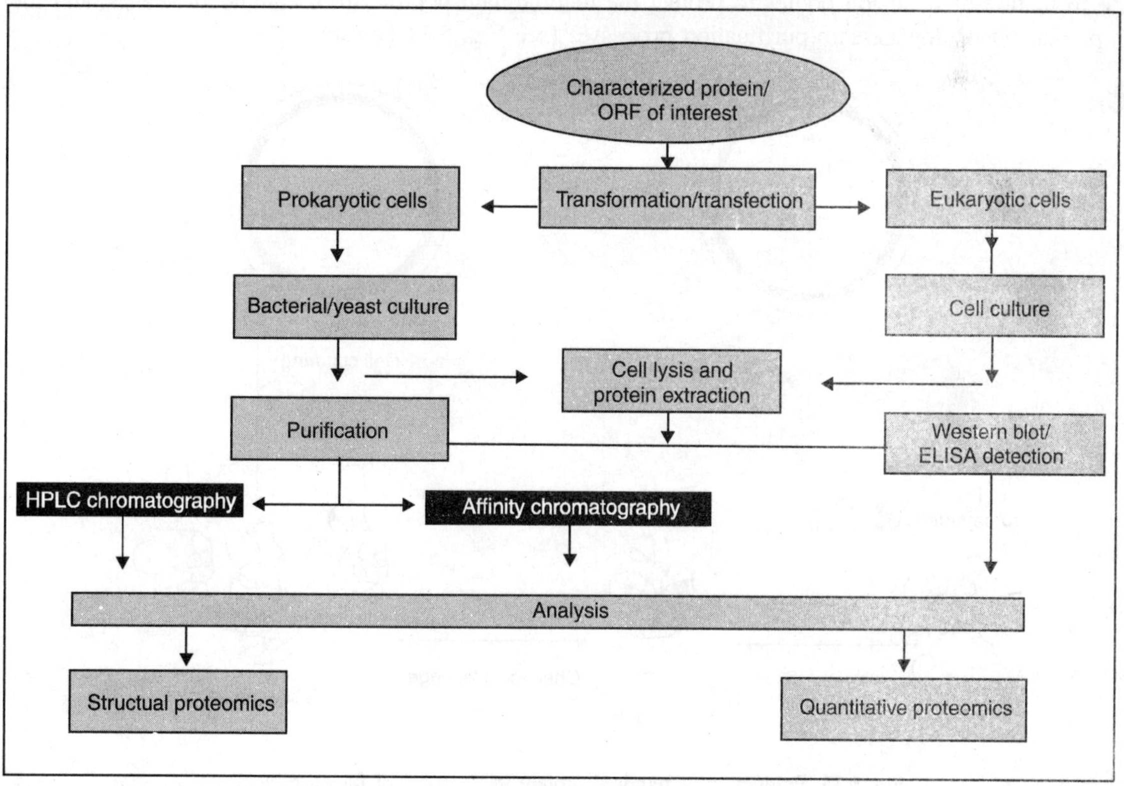

Step 4. Protein production and purification

Cell culture bioreactor will express the protein in a scale of up to 10,000 litre capacity. The protein will be purified using a tag-assisted affinity column or other chromatographic methods. Protein samples may be provided to clients if the protein function needs to be confirmed. The process includes following stages:

- Large scale growth of over expressing construct and cells
- Harvesting cells and cell lysis by sonication
- Protein purification on affinity tag resins
- Secondary purification on anion exchange resin or by gel filtration
- Dialysis and western blot analysis
- Delivery of proteins

5.12 OVER PRODUCTION

Fusion protein constructs are a commonly used strategy to increase the production efficiency of recombinant proteins however this approach has met with limited success for high-yield production of peptides. In this technology, novel carrier proteins were engineered as a series of protein fragments derived from *Staphylococcus* nuclease. These Small Fusion Carriers (SFCs), which confer peptide stability and solubility, allow highly efficient and low cost production of small or large recombinant peptides (2-100 amino acids residues) using any suitable bacterial expression systems. This high efficiency is achieved through the use of a small carrier protein that allows C- or N-terminal fusions, the formation of inclusion bodies to protect the recombinant peptide from cellular proteases, and the improvement of downstream purification processes (see Fig. 5.11 below).

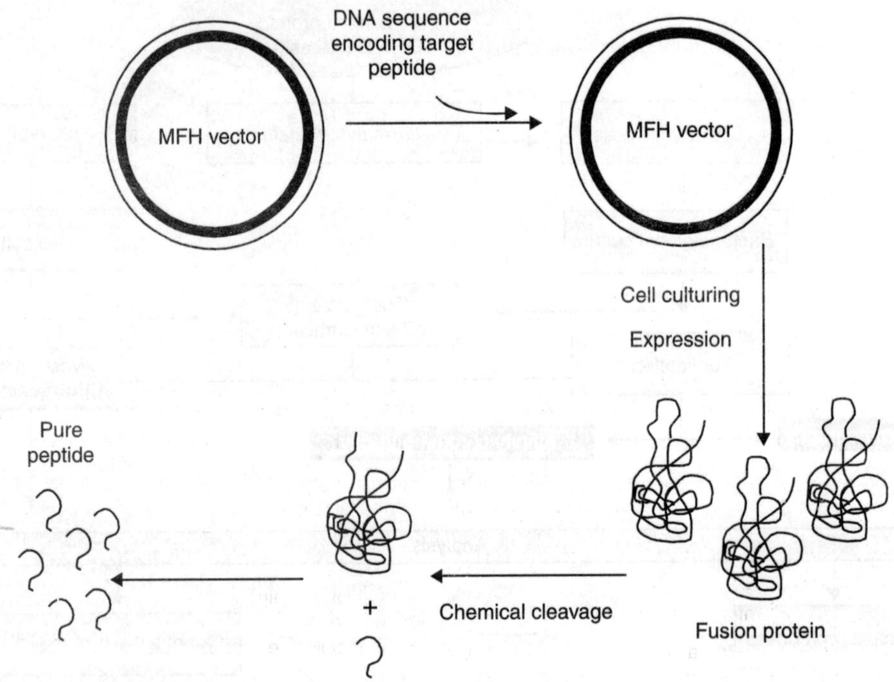

Fig. 5.11. Production of required protein in the form of fusion protein.

Highly Efficient Peptide Production

The SFC fusion approach to peptide production allows maximizing the yield and purity while minimizing the cost. This is achieved by reducing the size of the carrier, enhancing the stability and solubility of the carrier-peptide fusion, and improving carrier removal and peptide purification (yields of >10 mg/L were obtained in LB or M9 medium).

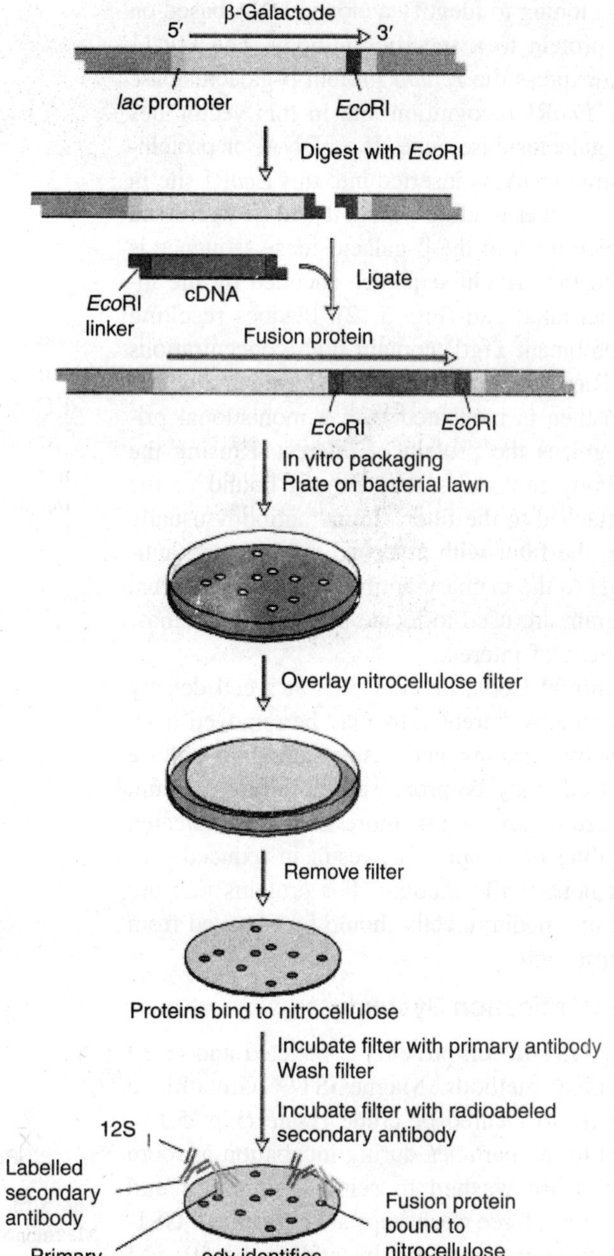

Fig. 5.12. Production and detection of required protein in the form of fusion protein.

Use of £f expression cloning to identify a cloned DNA based on binding of the encoded protein to a specific antibody. The £fgt11 vector was engineered to express the *E. coli* protein β-galactosidase at high levels. The only *Eco*RI recognition site in this vector lies near the 3′ end of the β-galactosidase gene. If a cDNA, or protein-coding fragment of genomic DNA, is inserted into this *Eco*RI site in the correct orientation and proper reading frame, it will be expressed as a fusion protein in which most of the β-galactosidase sequence is at the N-terminal end and the protein sequence encoded by the inserted DNA is at the C-terminal end (Fig. 5.12). Plaques resulting from infection with recombinant £fgtll contain high concentrations of such fusion proteins. These proteins can be transferred and bound to a replica filter, which then is incubated with a monoclonal primary antibody that recognizes the protein of interest. Rinsing the filter washes away antibody molecules that are not bound to the specific fusion protein attached to the filter. Bound antibody usually is detected by incubating the filter with a second radiolabeled antibody (dark red) that binds to the primary antibody. Any signals that appear on the autoradiogram are used to locate plaques on the master plate containing the gene of interest.

Purification from animal Cells. Process cells at a cell density of 2 × 10^6 cells/ml of culture. Adherent cells may be removed from the tissue culture vessel by scraping and resuspending in culture medium to this density. Cells may be processed in culture medium containing up to 10% serum. Processing more than the indicated number of cells per milliliter of sample may result in reduced protein yield and increased nonspecific binding. For proteins that are secreted into the cell culture medium, cells should be removed from the medium prior to purification.

MagneGST™ Protein Purification System

A cell culture expressing GST-fusion proteins is pelleted and lysed by enzymatic or mechanical methods. MagneGST™ Glutathione Particles are added directly to cleared or crude lysate (Fig. 5.13). GST-fusion proteins bind to the particles during incubation at room temperature or 4°C, then are washed to remove unbound and nonspecifically bound proteins; three wash steps are performed. GST-fusion protein is eluted from the particles by adding 10–50 mM reduced glutathione at pH 8.

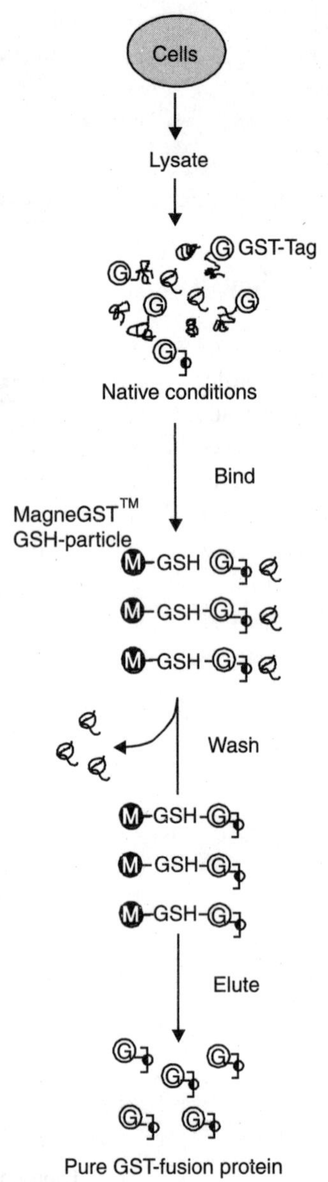

Fig. 5.13. Schematic diagram of the MagneGST™ Protein Purification System.

Characterization

In addition to verifying the sequences of gene clones, we also need to characterize the proteins produced and the processes used to produce them) to ensure their purity and biological behavior (*Quality Control and Quality Assurance*).

All proteins produced will be run through a battery of tests and screening procedures (*Biophysical Characterization*) to assess their quality and to provide initial insights into their structures. For each protein, molecular weight, stability, and proper folding must be determined. No single test will be sufficient to characterize every protein adequately and accurately. Instead, a combination of various spectroscopic, separation, and imaging techniques will be used. Some proteins, such as those involved in hydrogen production or cleanup of environmental contaminants, will be characterized further for biological function by assaying for specific enzymatic activity or binding properties.

Automated systems will simultaneously characterize hundreds of proteins for purity and, in some cases, function.

5.13 DOWNSTREAM PROCESSING OF EXPRESSED PROTEINS

Downstream processing or downstreaming is the extraction and purification of the desired end products of biofermentation processes or bioreactor productions. Such products might include cells, solvents or solutes. Various processes are available for the separation of cells from the fermentation broth in which they are grown, including flocculation, filtration, centrifugation, sedimentation or flotation. The procedure adopted depends on whether it is the cells, or the solution surrounding them, that contains the desired end products. In a large scale production unit, after completion of reaction, cell biomass must be separated from supernatant, which usually contains the desired product. Cell separation is usually achieved by centrifugation or by cross flow filtration. The product may then be purified by using suitable techniques. A range of sophisticated systems catering to clarification and purification of biological process fluids. These systems are available as completely automated/semi-automatic/manual-based with scalability to match individual requirement and a host of customer-friendly features like Sanitization and SIP as per cGMP guidelines.

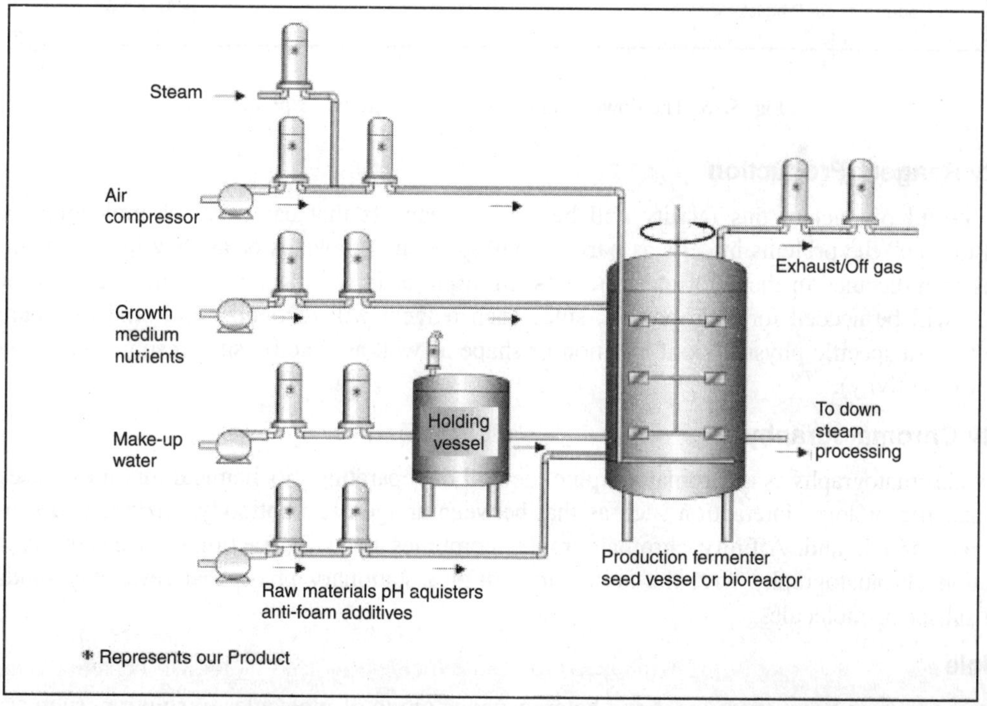

Fig. 5.14. Production of required protein in the form of fusion protein The production phase of protein is known as upstream processing and it consists of following steps and showed in brief in the figure.

After production phase or upstream processing (Fig. 5.14) the protein has to be separated from the cell culture medium or from the cells by lyses and is followed in the downstream processing and is shown in the following figure 5.15.

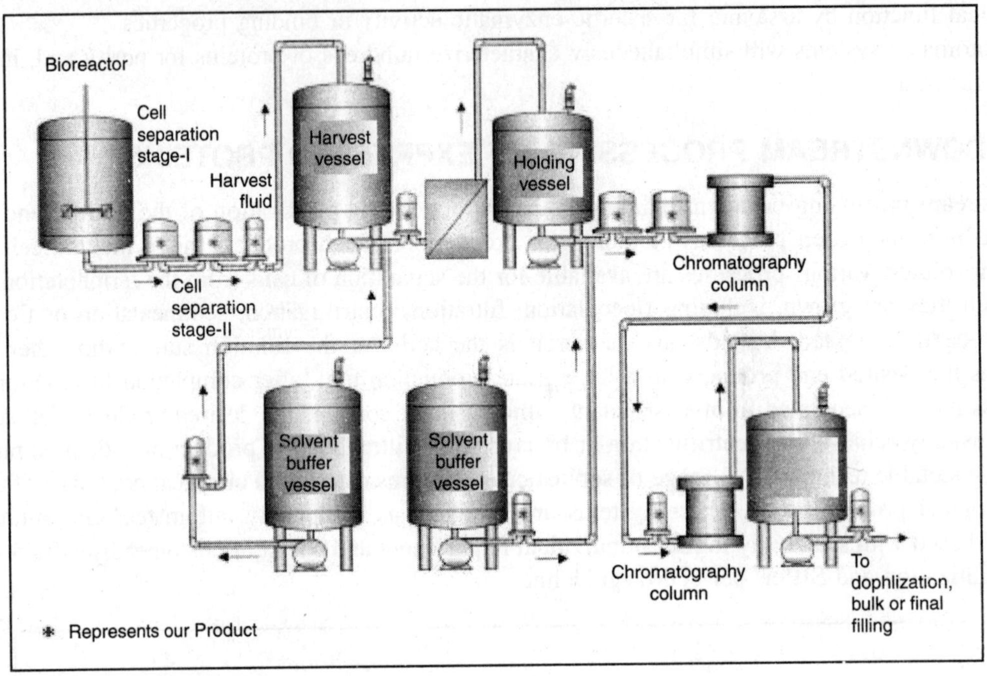

Fig. 5.15. The downstream processing of produced protein.

Affinity Reagent Production

A very useful product of this facility will be affinity reagents that can serve as molecular markers needed to "see" the proteins in cells as parts of multiprotein complexes or as they interact with other proteins or molecules in their normal functions. Multiple affinity reagents, produced by a variety of methods, will be needed for each protein, since each reagent will recognize and bind to a particular feature (e.g., a specific physical conformation or shape as well as specific sites responsible for protein function or activity).

Affinity Chromatography

Affinity chromatography is a chromatographic method of separating biochemical mixtures, based on a highly specific biologic interaction such as that between antigen and antibody, enzyme and substrate, or receptor and ligand. Affinity chromatography combines the size fractionation capability of gel permeation chromatography with the ability to design a stationary phase that reversibly binds to a known subset of molecules.

Principle

Usually the starting point is an undefined heterogeneous group of molecules in solution, such as a cell lysate, growth medium or blood serum. The molecule of interest will have a well known and defined property which can be exploited during the affinity purification process. The process itself can be

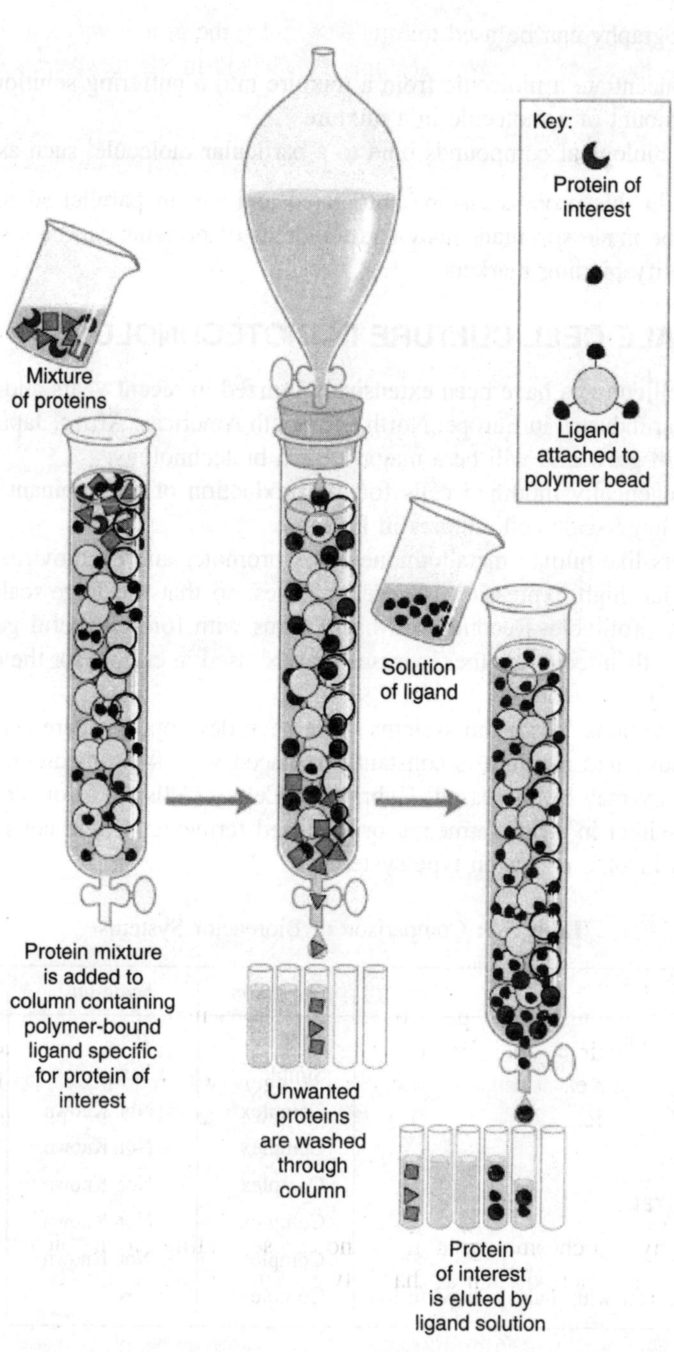

Fig. 5.16. Affinity chromatography for separation of protein from cell lysate.

thought of as an entrapment, with the target molecule becoming trapped on a solid or stationary phase or medium. The other molecules in solution will not become trapped as they do not possess this property. The solid medium can then be removed from the mixture, washed and the target molecule released from the entrapment in a process known as elution (Fig. 5.16).

Affinity chromatography can be used to:

1. Purify and concentrate a molecule from a mixture into a buffering solution
2. Reduce the amount of a molecule in a mixture
3. Discern what biological compounds bind to a particular molecule, such as drugs

Almost all steps in this process can be automated and run in parallel so millions of potential affinity reagents can be made simultaneously and hundreds of proteins can be screened against these large libraries to identify binding markers.

5.14 LARGE SCALE CELL CULTURE IN BIOTECHNOLOGY

Large scale animal cell cultures have been extensively utilized in recent years and they form the basis of a multimillion dollar industry in Europe, North and South Americas, Africa, Japan and in India. This industry, in future, will grow and will be a major area in biotechnology.

Availability of genetically modified cells for the production of recombinant compounds has increased the utility of large-scale cell cultures in industry.

Specific promoters like human metallothionein IIA promoter and baculovirus IIA promoter have been utilized to provide high expression of foreign genes, so that the large-scale cell cultures may become commercially profitable. Recombinant baculovirus with foreign useful genes have also been designed, so that the cells infected by these viruses may be used in culture for the commercial production of certain chemicals.

For this purpose, various 'perfusion systems' have been developed, where cells are "retained in a bioreactor, while conditioned medium is constantly replaced with fresh medium, so that cell density and reactor productivity may be increased. Hybridoma Cells—Cells used for the production of antibodies can be grown either in airlift fermenter or in stirred fermenter. These cells may also be immobilized and may then involve perfusion type systems.

Table 5.2: Comparison of Bioreactor Systems

Bioreactor	*Operation*	*Scale-up*	*Productivity*
1. Stirred-tank batch reactor	Simple	Yes	Low
2. Airlift batch reactor	Simple	Yes	Low
3. Hollow fibre reactor	Complex	Not Known	High
4. Ceramic matrix	Complex	Not Known	Intermediate
5. Glass bead reactor	Complex	Not Known	Intermediate
6. Fluidized bed reactor	Complex	Not Known	High
7. Membroferm	Complex	Not Known	High
8. Stirred tank/airlift reactor with dialysis or perfusion	Complex	Yes	High

5.15 BACULOVIRUS AS VERSATILE EXPRESSION VECTORS FOR PROTEIN EXPRESSION IN ANIMAL CELLS

Today, many thousands of recombinant proteins, ranging from cytosolic enzymes to membrane-bound proteins, have been successfully produced in baculovirus-infected insect cells. Yet, in addition to its value in producing recombinant proteins in insect cells and larvae, this viral vector system continues to evolve in new and unexpected ways. This is exemplified by the development of engineered insect cell lines to mimic mammalian cell glycosylation of expressed proteins (Fig. 5.17), baculovirus display strategies and the application of the virus as a mammalian-cell gene delivery vector. Novel vector design and cell engineering approaches will serve to further enhance the value of baculovirus technology.

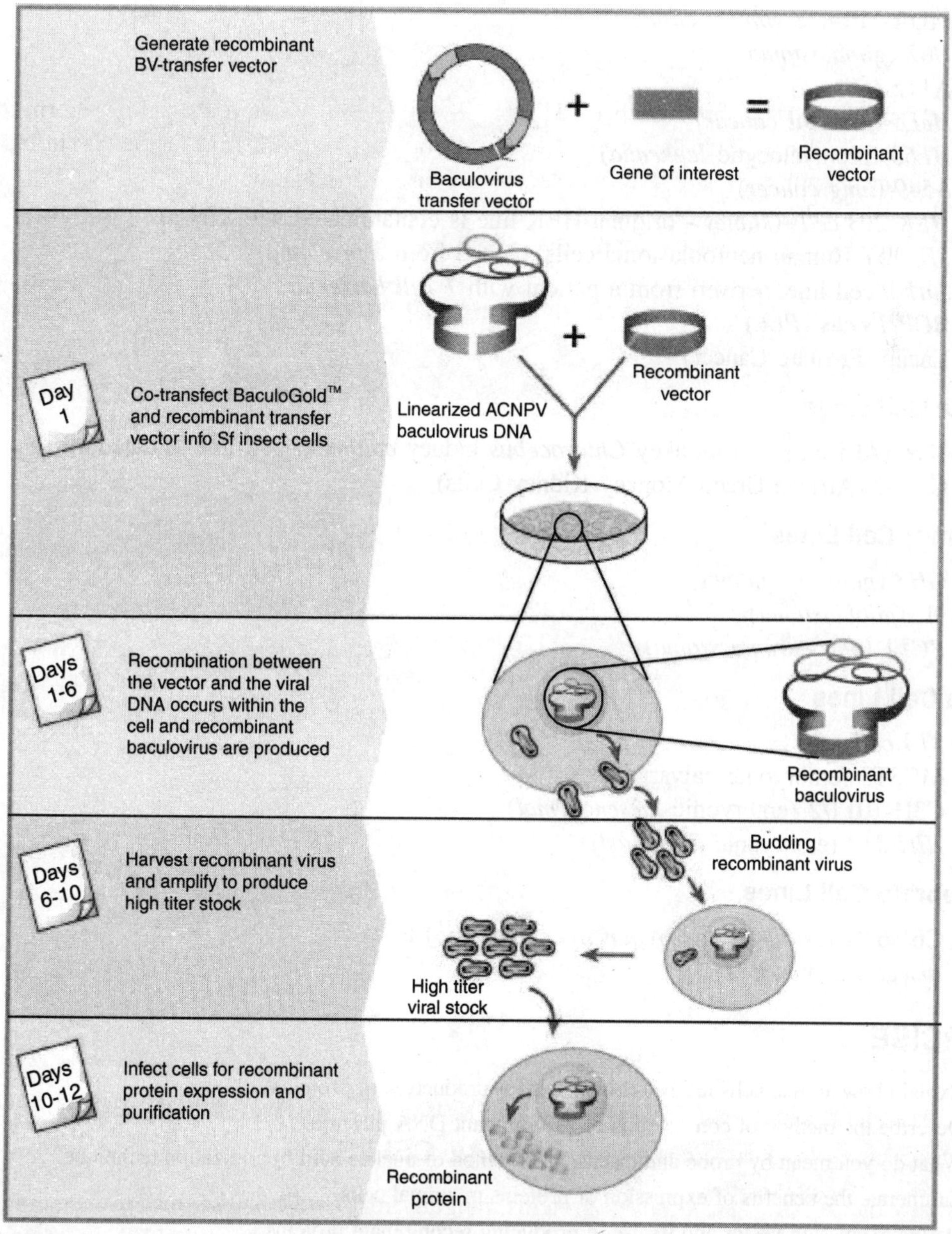

Fig. 5.17. Production of recombinant proteins by using baculovirus expression system.

Many animal cell lines are used for better and safe expression of recombinant proteins in the cell culture system. The following is the list of some important cell lines that are used in animal biotechnology.

Human Cell Lines

- *National Cancer Institute*'s 60 *cancer* cell lines
- MCF-7 (*breast cancer*)
- MDA-MB-438 (*breast cancer*)
- *U87* (*glioblastoma*)
- A172 (*glioma*)
- *HeLa* (*cervical cancer*)
- *HL60* (promyelocytic *leukemia*)
- *A549* (*lung cancer*)
- *HEK 293 cells* (*kidney* - original HEK line is contaminated with HeLa)
- *SHSY5Y* Human neuroblastoma cells, cloned from a *myeloma*
- *Jurkat* cell line, derived from a patient with *T cell leukemia*
- *BCP-1 cells* (*PEL*)
- Lncap (Prostate Cancer)

Primate Cell Lines

- *Vero* (African green monkey *Chlorocebus* kidney *epithelial* cell line initiated 1962)
- COS-7 (African Green Monkey Kidney Cells)

Rat Tumor Cell Lines

- *GH3* (*pituitary tumor*)
- 9L (*glioblastoma*)
- *PC12* (*pheochromocytoma*)

Mouse Cell Lines

- *3T3 cells*
- MC3T3 (embryonic calvarial)
- C3H-10T1/2 (embryonic *mesenchymal*)
- *NIH-3T3* (embryonic *fibroblast*)

Invertebrate Cell Lines

- C6/36 *Aedes albopictus* (*Asian tiger mosquito*) larva
- *Insect* cell line *Sf21*

EXERCISE

1. Explain how animal cells are better host cell for production of proteins.
2. Describe the method of construction of recombinant DNA plasmid.
3. What do you mean by probe and discuss the method of nucleic acid hybridization technique?
4. Enumerate the benefits of expression of proteins in animal cells.
5. Define expression vector and its use in producing recombinant proteins.
6. Mention the importance of promoter and regulatory elements in protein expression.

Chapter

6

Vaccine Production from Animal Cell Lines

6.1 INTRODUCTION

A preparation using a non-infectious element or relative of a particular virus or bacteria, and administered with the intention of halting the progress of an infection, or completely preventing it. Vaccine can also defined as an agent prepared to produce active immunity that usually kills microbes, attenuated live microbes, or variant strains of microbes and can induce antibody production without producing disease. One of the main aims of the World Health Organization (WHO) and the International Conference on Harmonization (ICH) is to reach international consensus on the basic acceptance criteria for biological products. These include general and international requirements for the production of different classes of biological substances such as vaccines, interferon, hormones, blood and its derivatives.

6.2 VACCINE

Vaccine is a suspension of attenuated or killed microorganisms (viruses, bacteria, rickettsiae or any pathogenic organisms), or of antigenic proteins derived from them, administered for prevention, amelioration, or treatment of infectious diseases. It is also defined as a preparation of a weakened or killed pathogen, such as a bacterium or virus, or of a portion of the pathogen's structure that upon administration stimulates antibody production (immune responses) against the pathogen but is incapable of causing severe infection.

Acellular vaccine a cell-free vaccine prepared from purified antigenic components of cell-free microorganisms, carrying less risk than whole-cell preparations.

Anthrax vaccine a cell-free protein extract of cultures of *Bacillus anthracis,* used for immunization against *anthrax*.

Attenuated vaccine a vaccine prepared from live microorganisms or viruses cultured under adverse conditions leading to loss of their virulence but retention of their ability to induce protective immunity.

Autogenous vaccine a vaccine prepared from microorganisms which have been freshly isolated from the lesion of the patient who is to be treated with it.

BCG vaccine a preparation used as an active immunizing agent against tuberculosis and in treatment of bladder cancer, consisting of a dried, living, avirulent culture of the Calmette-Guérin strain of *Mycobacterium bovis.*

Cholera vaccine a preparation of killed *Vibrio cholerae,* used in immunization against *cholera.*

Diphtheria and Tetanus toxoids and Pertussis vaccine (DTP) a combination of *diphtheria* and *tetanus toxoids* and *pertussis vaccine*; used for simultaneous immunization against diphtheria, tetanus, and whooping cough. When the pertussis vaccine is an acellular form, the combination may be abbreviated *DTaP*.

Diphtheria and Tetanus toxoids and Pertussis vaccine adsorbed and *Haemophilus* b conjugate vaccine a combination of *diphtheria toxoid, tetanus toxoid, pertussis vaccine*, and *b conjugate vaccine*; used for simultaneous immunization against *diphtheria, tetanus, pertussis*, and infection by *Haemophilus influenzae* type b.

***Haemophilus* b conjugate vaccine** (HbCV) a preparation of *Haemophilus influenzae* type b capsular polysaccharide covalently bound to *diphtheria toxoid* or to a specific diphtheria, meningococcal, or tetanus protein; it stimulates both B and T lymphocyte responses and is used as an immunizing agent in infants and young children.

***Haemophilus* b polysaccharide vaccine** (HbPV) a preparation of highly purified capsular polysaccharide derived from *Haemophilus influenzae* type b, which stimulates an immune response in B lymphocytes only; used as an immunizing agent in children.

Hepatitis A vaccine inactivated an inactivated whole virus vaccine derived from an attenuated strain of *hepatitis A virus* grown in cell culture.

Hepatitis B vaccine a preparation of *hepatitis B surface antigen*, derived either from human plasma of carriers of hepatitis B (*hepatitis B v. inactivated*) or from cloning in yeast cells (*hepatitis B v. [recombinant]*).

Heterologous vaccine a vaccine that confers protective immunity against a pathogen that shares cross-reacting antigens with the microorganisms in the vaccine.

Human diploid cell vaccine (HDCV) see *rabies v.*

Influenza virus vaccine a killed virus vaccine used in immunization against *influenza*; it is trivalent, usually containing two *influenza A virus* strains and *one influenza B virus* strain.

Live vaccine one prepared from live microorganisms that have been attenuated but that retain their immunogenic properties.

Lyme disease vaccine (recombinant OspA) a preparation of outer surface protein A (OspA), a cell surface lipoprotein of *Borrelia burgdorferi,* produced by recombinant technology; used for active immunization against *Lyme disease.*

Measles, mumps, and rubella virus vaccine live (MMR) a combination of live attenuated *measles, mumps*, and *rubella viruses*, used for simultaneous immunization against *measles, mumps*, and *rubella.*

Measles and rubella virus vaccine live a combination of live attenuated *measles* and *rubella viruses*, used for simultaneous immunization against *measles* and *rubella.*

Measles virus vaccine live a live attenuated virus vaccine used for immunization against *measles,* although it is generally administered as the combination *measles, mumps, and rubella virus vaccine.*

Meningococcal polysaccharide vaccine a preparation of capsular polysaccharide antigen of *Neisseria meningitidis,* used to provide immunity to *meningitis.*

Mixed vaccine *polyvalent v.*

Mumps virus vaccine live a live attenuated virus vaccine used in immunization against *mumps*; usually administered as the combination *measles, mumps, and rubella virus vaccine.*

Pertussis vaccine a preparation of killed *Bordetella pertussis* bacilli (whole-cell vaccine) or of purified antigenic components thereof (acellular vaccine), used to immunize against *pertussis*; generally used in combination with diphtheria and tetanus toxoids (*DTP* or *DTaP*).

Plague vaccine a preparation of killed *Yersinia pestis* bacilli, used as an active immunizing agent.

Pneumococcal heptavalent conjugate vaccine a preparation of capsular polysaccharides from the seven serotypes of *Streptococcus pneumoniae* most commonly isolated from young children, coupled to a nontoxic variant of *diphtheria toxin*; used as an active immunizing agent.

Pneumococcal vaccine polyvalent a preparation of purified capsular polysaccharides from the 23 serotypes of *Streptococcus pneumoniae* causing the majority of pneumococcal disease; used as an active immunizing agent.

Poliovirus vaccine inactivated (IPV) *Salk v.*; a suspension of formalin-inactivated polioviruses used for immunization against *poliomyelitis*.

Poliovirus vaccine live oral (OPV) *Sabin v.*; a preparation of a combination of the three types of live, attenuated polioviruses used as an active immunizing agent against *poliomyelitis*.

Polyvalent vaccine one prepared from cultures or antigens of more than one strain or species.

Purified chick embryo cell vaccine a preparation of inactivated rabies virus grown in cultures of chicken fibroblasts; used for pre- and post-exposure rabies immunization.

Rabies vaccine an inactivated virus vaccine used for pre- and postexposure immunization against rabies; it may be prepared from virus grown in human diploid cell culture (*human diploid cell v.*), that grown in cultures of chicken fibroblasts (*purified chick embryo cell v.*), or that grown in cultures of fetal rhesus lung and concentrated by adsorption to aluminum phosphate (*rabies v. adsorbed*).

Replicative vaccine any vaccine containing organisms that are able to reproduce, including live and attenuated viruses and bacteria.

Rotavirus vaccine live oral a live virus vaccine produced from a mixture of four *rotavirus* types grown in fetal rhesus diploid cells; used to immunize infants against rotaviral *gastroenteritis*.

Rubella and mumps virus vaccine live a combination of live attenuated *rubella* and *mumps viruses*, used for simultaneous immunization against *rubella* and *mumps*.

Rubella virus vaccine live a live attenuated virus vaccine used for immunization against *rubella*, usually administered as the combination *measles, mumps, and rubella virus vaccine*.

Sabin vaccine *poliovirus v. live oral.*

Salk vaccine *poliovirus v. inactivated.*

Subunit vaccine a vaccine produced from specific protein subunits of a virus and thus having less risk of adverse reactions than whole virus vaccines.

Typhoid vaccine any of several preparations of *Salmonella typhi* used for immunization against *typhoid fever*, including a parenteral heat- and phenol-inactivated bacteria vaccine, an oral live vaccine prepared from the attenuated strain Ty21a, and a parenteral vaccine prepared from typhoid Vi capsular polysaccharide.

Varicella virus vaccine live a preparation of live, attenuated *human herpesvirus 3* (*varicella-zoster virus*) used for production of immunity to *varicella* and *herpes zoster*.

Yellow fever vaccine a preparation of attenuated *yellow fever virus*, used to immunize against *yellow fever*.

Anti-idiotype vaccine

Antibody made to antigenic determinants located in the variable domains of immunoglobulin molecules. Proposed as a means of regulating antibody responses and also as a substitute antigen for vaccination.

Vaccine-associated sarcoma

See *sarcoma.*

Attenuated vaccine

A vaccine prepared from live microorganisms that have lost their virulence but retained their ability to induce protective immunity. Attenuated microorganisms including particularly bacteria and viruses may be found naturally or they may be produced in the laboratory, for example by adaptation to a new medium or cell culture or they may be produced by recombinant DNA technology.

Autogenous vaccine

A vaccine prepared from cultures of material derived from a lesion of the animal to be vaccinated, e.g. wart vaccine.

Bacterial vaccine

A preparation of attenuated or killed bacteria, used to immunize against organisms injected, or sometimes for pyrogenetic effects in treatment of certain noninfectious diseases.

Biosynthetic vaccine

A formulation containing a protective, noninfectious, immunogenic subunit produced in or by a biological system.

Caprinized vaccine

A vaccine, usually a virus, attenuated by serial passage through goats, e.g. caprinized rinderpest vaccine. In highly susceptible cattle this vaccine may cause significant reactions and lapinized vaccines are preferred.

Core vaccine

One that should always be included in the basic immunization program for the species.

Dead vaccine

Inactivated vaccine; one with organisms that have been killed.

DNA vaccine

DNA sequences that code for immunogenic proteins located in appropriately constructed plasmids which include strong promoters, which when injected into an animal are taken up by cells and the immunogenic proteins are expressed and elicit an immune response. No vaccines of this type are licensed and concerns about safety have not been resolved.

Heterotypic vaccine

One developed from a virus that is antigenically distinct but related to that causing the disease for which the animal is being immunized, e.g. measles vaccine used to protect dogs from canine distemper.

Homotypic vaccine

One developed from the same virus as that causing the disease the animal is being immunized against.

Human diploid cell vaccine

An inactivated rabies vaccine made from rabies virus grown on human embryo lung fibroblast cells.

Inactivated vaccine

See dead vaccine (above).

Killed Virus (KV) vaccine

See dead vaccine (above).

Live vaccine

A vaccine prepared from live, usually attenuated, microorganisms.

Vaccine lymph

Material containing vaccinia virus collected from vaccinial vesicles of inoculated calves; used for active immunization against smallpox.

Modified Live Virus (MLV) vaccine

See attenuated vaccine (above).

Polyvalent vaccine

One prepared from more than one strain or species of microorganisms.

Recombinant vaccine

One created by recombinant DNA technology.

Subunit vaccine

One containing only specific antigenic proteins of the infectious agent.

Synthetic peptide vaccine

Using synthetic short peptides which correspond with major epitopes of viral proteins to elicit a protective antibody response.

Virus-vectored vaccine

Use of viruses as vectors to carry selected genes from another virus for immunization.

6.3 DISEASE PREVENTION VACCINES

Prevention of diseases is the most desirable, most convenient and highly effective approach to health; this is achieved by vaccination or immunization using biological preparations called vaccines. Vaccines represent an invaluable contribution of biotechnology as they provide protection against even such diseases for which effective cures are not yet available. The effectiveness of vaccines may be appreciated from the fact that small pox, once a dreaded disease the world over, has been completely eradicated from the world; the last case of small pox was reported in 1977.

The various vaccines can be grouped under the following types:

(*i*) Conventional vaccines (live vaccines, inactivated pathogens),
(*ii*) Purified antigen vaccines, and
(*iii*) Recombinant vaccines (recombinant proteins/polypeptides, DNA vaccines).

Conventional Vaccines

Conventional vaccines consist of whole pathogenic organisms, which may either be killed (most bacterial vaccines and some viral vaccines), or live [hence called live vaccines; the virulence of pathogens is greatly reduced (attenuation); most viral vaccines]. Conventional vaccines, although highly effective and relatively easy to produce at low cost, suffer from the following limitations.

1. In many cases, live vaccines have to be used since killed pathogen vaccines are ineffective.
2. Live vaccines are generally based on cultured animal cells, hence expensive tissue culture set up is essential.
3. Live vaccines are heat labile.
4. Conventional vaccines carry a variable risk of disease development due to the occasional presence of active virus particles or reversion to virulence after replication in the vaccinated individuals.
5. In many cases, they are difficult to produce, e.g., hepatitis B virus does not grow in high titre in cultured cells.These limitations have prompted the successful development of vaccines based on purified antigens and of recombinant vaccines, which are rather costly, at least for the present.

Purified Antigen Vaccines

These vaccines are based on purified antigens isolated from the concerned pathogens. Since they do not contain the organism, the risk of pathogenicity is avoided. However, their cost is higher due to the steps involved in purification and vaccine preparation, and many of the isolated antigens are poorly immunogenic.

Successful examples of such vaccines are mostly from bacteria, e.g., vaccines based on polysaccharide antigens from the bacterial cell wall capsules of *Neisseria meningitis* (causing meningitis) and *Streptococcus pneumoniae* (causing pneumonia).

Many bacteria produce exotoxins, which are highly immunogenic. But these toxins produce toxic effects, the intensity of which decreases with storage and due to heat, formaldehyde and other chemicals.

Fortunately, most exotoxins that have lost their toxicity retain their immunogenicity; they are called toxoids and are used as effective vaccines, e.g., toxoids of the pathogens causing tetanus, diptheria, gangrene, etc. Precipitation of toxoids with alum enhances their immunogenicity. The toxoid vaccines are quite effective and cheap.

Recombinant Vaccines

A recombinant vaccine contains either a protein or a gene encoding a protein of a pathogen origin that is immunogenic and critical to the pathogen function; the vaccine is produced using recombinant DNA technology. The vaccines based on recombinant proteins are also called **subunit vaccines**.

The logic of such vaccines, in simple terms, is as follows. Proteins are generally immunogenic, and many of them are critical for the pathogenic organism. The genes encoding such proteins can be identified and isolated from a pathogen and expressed in *E. coli* or some other suitable host for a mass production of the proteins.

The concerned proteins are then purified and mixed with suitable stabilizers and adjuvants, if required, and used for immunization. The different steps involved in the development of a recombinant protein based vaccine may be simply summarised as follows.

(*i*) The first step is to identify a protein that is both immunogenic and critical for the pathogen.
(*ii*) The gene encoding this protein is then identified and isolated.
(*iii*) The gene is integrated into a suitable expression vector and introduced into a suitable host where it expresses the protein in large quantities.

(*iv*) The protein is then isolated and purified from the host cells.

(*v*) It is used for the preparation of vaccine. The host organisms used for expression of immunogenic proteins to be used as vaccines may be anyone of the following:

(1) A genetically engineered microorganism, e.g., yeast for the expression of hepatitis B surface antigen (HBsAg) used as vaccine against hepatitis B virus (approved for marketing in India)

(2) Cultured animal cells, e.g., HBsAg expressed in CHO (Chinese hamster ovary) cell line and C-127 cell line.

(3) Transgenic plants, e.g., HBsAg, HIV-l (human immunodeficiency virus-I) epitope (in experimental stages).

(4) Insect larvae; the gene is integrated into a bacculovirus genome, which is used to infect insect larvae. Often a very high quantity of the recombinant protein is produced. So far a large number of recombinant immunogenic proteins of pathogens have been produced and evaluated. In general, a majority of such proteins are ineffective or only poorly effective in immunization.

6.4 CONDITIONS FOR IDEAL VACCINES

1. It should not be tumorigenic, toxic or pathogenic, *i.e.*, it should be safe.
2. It should have very low levels of side effects in normal individuals.
3. It should not cause problems in individuals with impaired immune system.
4. It should not spread either within the vaccinated individual or to other individuals (live vaccines).
5. It should not contaminate the environment.
6. It should be effective in producing long lasting humoral and cellular immunities.
7. The technique of vaccination should be simple.
8. The vaccine should be cheap so that it is generally affordable. So far, such an ideal vaccine has not been developed.

6.5 PRODUCTION OF VACCINES IN ANIMAL CELL CULTURE SYSTEM

Production of FMD vaccines is the most important example of the use of large-scale cell cultures. There are several other vaccines including polio vaccine, bovine leukaemia virus (BLV) vaccines, rabies vaccines, etc., which are produced at commercial scale using cell cultures.

During vigorous purification procedures, the integrity and immunogenicity of viruses produced in cultured cells are largely destroyed. Consequently, most whole virus vaccines are relatively impure in comparison with other products, similarly derived from cultured cells.

Despite this, whole virus vaccines are derived from large scale animal cell cultures. However, there are examples, where it is not possible either to culture the whole viruses or to make a safe product from them.

Therefore and otherwise, using recombinant DNA technology it became possible to obtain viral subunits from animal cell lines. Vaccines derived from these proteins are purer, safer and more efficacious and are already being used for clinical trials.

Similarly, Vaccinia virus (originally used for small pox vaccines) may be used as a vector for foreign viral genes, which may express on infection of host cultured cells by Vaccinia.

This system has been tried and needs to be tried in future for live vaccine production. Using this system, recombinant 'Vaccinia' virus vaccines have actually been developed. One such rabies virus vaccine has been field tested in Europe and U.S.A (using cattle).

Since several years the cultivation of animal cells to produce complex proteins and vaccines is used in biotechnology. Several mammalian cell culture processes using hybridoma, CHO or BHK suspension cells have been subject of modeling approaches. Until now, no model has been proposed for growth and product formation of adherent cell lines, like MDCK cells, or for description of all relevant steps in production of cell culture-derived biologicals.

Vaccine production processes and are developing structured and non-structured models for design, monitoring and optimization of these complex systems. Compared to other mammalian cell culture processes, e.g. manufacturing of recombinant proteins or monoclonal antibodies, vaccine manufacturing usually involves a combination of two consecutive cultivation steps: 1. cell growth and 2. virus replication. As a result, additional influences on product yield have to be considered: washing steps, medium exchange, virus infectivity, multiplicity of infection, incubation time, virus replication rate, cell death rate, etc.

The current research is the development and optimization of integrated concepts to design and control vaccine production processes for the amelioration of viral yield and reproducibility in conjunction with efficacy, purity and safety of vaccines. With a process of influenza virus production as an example it consists of the following steps:

(a) Cultivation and scale-up of animal cells (Madin-Darby Canine Kidney epithelial cells; MDCK and Vero (african green monkey kidney cells)) using microcarrier-systems.

(b) Replication of different influenza viruses (equine, human, swine).

(c) Downstream processing to subunit or split virus vaccines (ISCOM-technology). An overview of the process is given in Fig. 6.1.

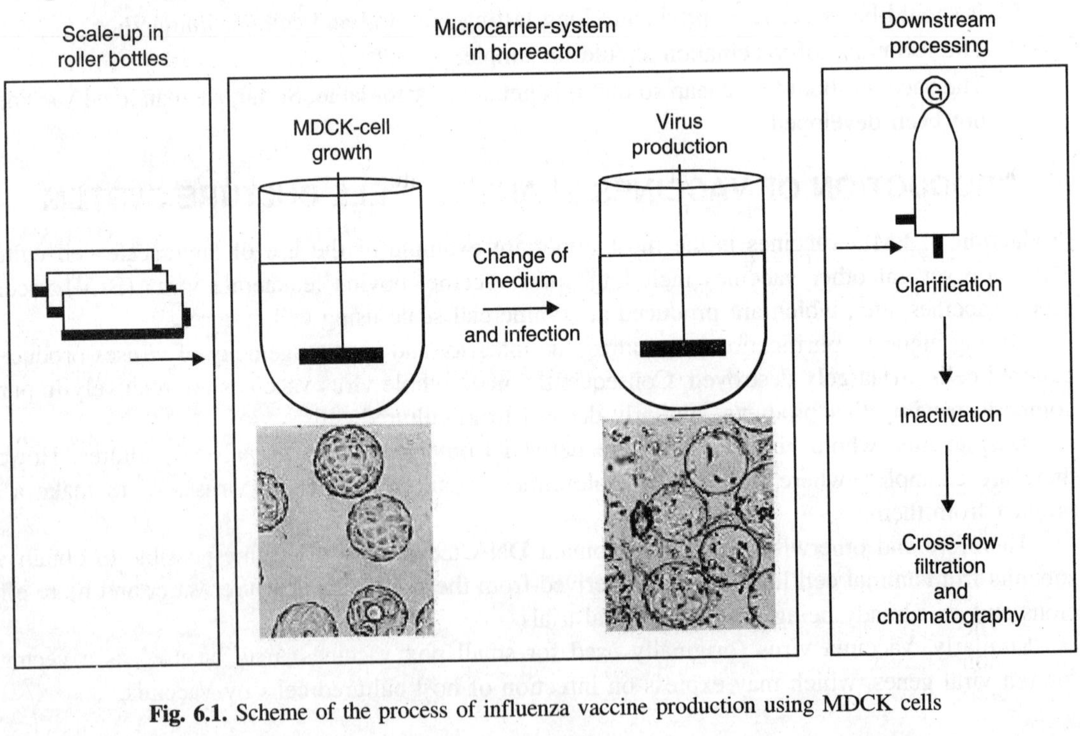

Fig. 6.1. Scheme of the process of influenza vaccine production using MDCK cells

A typical influenza virus production process starts with the cultivation of MDCK cells in flasks. After a scale-up into roller bottles the cells (inocculum: 2×10^5 cells/mL) are transferred into 5 to 20L bioreactors containing growth medium and microcarriers (e.g. Cytodex 1). The subsequent bioprocess is divided into two steps: (*a*) MDCK cell growth on microcarriers up to 1×10^6 cells/mL, (*b*) washing of cells/carriers for medium exchange and virus addition (equine influenza Newmarket 1/93 H3N8; obtained HA-value 2.1-2.4). For the cultivation of MDCK cells GMEM medium with additional FCS and peptone is used, whereas during virus replication FCS-free medium is needed. This results in different starting concentrations of amino acids in the culture media during these two steps.

Variations of cell culture medium (serum-free), bioreactors (disposable Wave bioreactor), virus subtypes (human influenza A and B) and host cells (Vero, human A549) are carried out giving experimental data for different process conditions.

Vaccines are clearly the most effective means of preventing infectious diseases and have been particularly successful in controlling viral infection. For example, global small-pox eradication has been the greatest achievement in this regard. However, many existing vaccines are not efficient and there are many diseases against which vaccines are not available at all.

6.6 INTERFERONS

Due to heavy demands for a variety of interferons for cancer therapy, recently interferon-α and interferon-β have been made in large quantities in Europe and Japan. For production of interferon-α 8000 litre cell culture vessels have been used by Wellcome Foundation. Similarly, interferon-β was produced by cells grown on microcarriers at scales of greater than 1000 litres.

Genetic engineering has also been used for interferon production. For instance, interferon genes were inserted in cultured mouse fibroblasts. In one case, these genes inserted in fibroblasts expressed constitutively for the production of glycosylated human interferon-β.

Special bioreactors have been used for the culture of these fibroblasts on suitable microcarriers, and yields as high as 5000 units of interferon-β /ml/day can be obtained.

6.7 MONOCLONAL ANTIBODIES

Production of monoclonal antibodies with the help of cultured cells has also achieved significance in industry, due to the increasing utility of these antibodies in

(*i*) in vitro diagnostics,
(*ii*) in vivo imaging,
(*iii*) therapy in humans and animals and
(*iv*) industrial applications like immunopurification.

Ten to twenty years ago, antibodies were available in milligram amounts, today due to increasing demand, antibodies are needed in multi kilograms per year.

Two approaches have been followed for the production of monoclonal antibodies:

(*i*) *in vivo* production using ascites tumours in mice or rats,
(*ii*) *in vitro* production through cell culture technique.

The second method is economically viable in view of the increasing demand. It has following additional advantages:

(*i*) it can be scaled up; for instance, one kilogram of antibody (which can be produced through cultured cells in a single fermenter), would require 20,000 mice, if made by ascites tumours;

(*ii*) the risk of contamination is greatly reduced.

Hybridoma Cells

Cells used for the production of antibodies can be grown either in airlift fermenter or in stirred fermenter. These cells may also be immobilized and may then involve perfusion type systems.

To increase reliability and decrease labour costs, high level of automation is also used in the process, and computer devices are used to monitor factors like pH, temperature, dissolved O_2, etc. After fermentation, cell and debris are removed by centrifugation and supernatant is used for isolation of antibodies.

In recent years the need for hybridoma cells has been obviated and recombinant DNA techniques are being utilized for the production of either the whole antibodies or only a fragment thereof, which is equally useful.

The genes or parts of genes are cloned, and inserted in cells which are cultured for the production of monoclonal antibodies.

EXERCISE

1. Define vaccine.
2. Explain various type of vaccines used for therapeutic purpose.
3. What do you mean by BCG vaccine and how it is prepared?
4. Differentiate between live and killed vaccines.
5. Explain various bio synthetic vaccines.
6. List out differences between sabin and salk vaccines.
7. Elucidate various attenuated vaccines used in medical field.
8. How recombinant vaccines are produced in animal biotechnology?
9. What are synthetic peptide vaccine and give suitable examples?
10. Discuss types and classification of disease prevention vaccines.
11. Enumerate limitations of conventional vaccines.
12. Mention conditions for usage of any substance as ideal vaccine.
13. Give an account of how vaccines can be produced in animal cell culture system.
14. Write a note on advantages of recombinant vaccine.
15. Define purified antigen vaccines and their limitations.

Chapter

7

Transgenic Animals

7.1 INTRODUCTION

A transgenic animal contains in its genome, a gene or genes introduced by one or the other technique of transfection. Transfection in turn is defined as the introduction of a DNA segment, either naked or integrated into a vector, into an animal cell. The difference between transfection and transformation is that transformation has long been used to describe the change of normal non-tumorous cells in culture to tumor like cells. Therefore, it became necessary to use the term transfection. Transgenesis is the process of introducing foreign or exogenous DNA into an animal's genome.

The process of introducing nucleic acids into eukaryotic cells by nonviral methods is defined as "transfection". Using various chemical, physical and biological methods, genes can be transferred into animal cells. This gene transfer technology is a powerful tool for studying gene function in the context of a cell. **Transgenics** is the science of intentionally introducing a foreign gene or genetic construct (series of genes and associated regulatory elements) into the genome of a target animal. Initial work involved a splicing technique to insert foreign genetic materials into mammalian cells maintained in culture. This *in vitro* work rapidly progressed into laboratory rodents, providing a more targeted and proactive approach for the establishment of new animal models for biomedical research. The results have been very successful and provide a unique and precise mechanism for the study of a variety of specific conditions or diseases with a genetic basis or influence.

Necessity for transgenesis

- Improve genetic features of domesticated animals
- Provide animal models for study of human diseases
- Animal pharming using farm animals for production of human pharmaceuticals in mammary glands
- Study the genes regulation, development of animals

7.2 DEVELOPMENT OF TRANSGENIC SCIENCE

First attempt of genetic manipulation of animals started during 1970s. Such experiments were conducted with *embryonal carcinoma* cells and *teratocarcinoma* cells to construct *chimeric* mice by Brinster, 1974; Mintz and Illmensee, 1975; Bradley *et al.*, 1984. In these chimeric animals, cultured cells derived from one strain of mouse were introduced into the embryos of another strain of mouse by direct embryo aggregation or by injection into the blastocyst stage embryo. The adult chimeric mouse could be produced by cellular contributions from several "parents" and would exhibit characteristics of each strain. McGrath and Solter, 1983 conducted another type of animal genome transfer involved the transfer of the entire nucleus from an embryo directly into the enucleated oocyte of a different recipient strain. These transgenic animals were produced without any recombinant DNA techniques and represent important milestones in the elucidation of genetic regulatory mechanisms in mammalian systems.

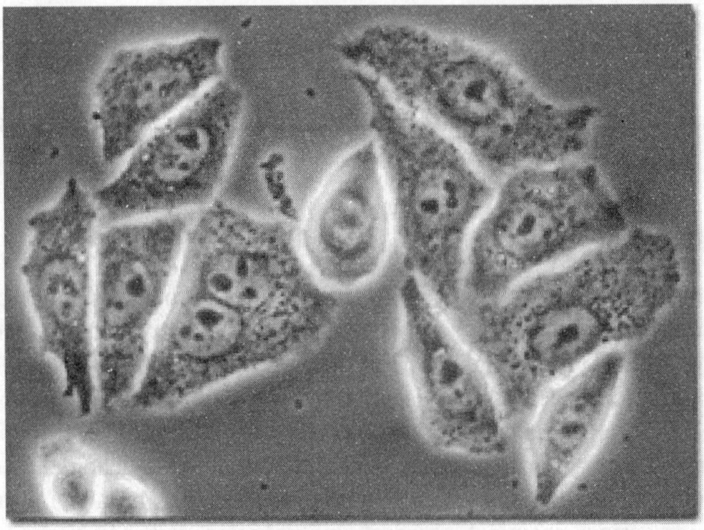

Fig. 7.1. Animal cells in culture

The next step in the evolution of transgenic technology was done by **Jaenisch and Mintz, 1974; Jaenisch, 1976.** They infected the preimplantation **mouse embryos with retroviruses**. The viral information was successfully transferred into the genome of the recipient animal. Later **Stuhlmann et al., 1984** the technique of utilizing retroviruses as vectors for specific foreign DNA sequences was soon developed. Retrovirus-mediated transgenesis produces a high degree of mosaicism; the size of the transgene sequence is limited, and the viral sequences may interfere with expression of the transgene.

Later many researchers developed several other techniques to produce transgenic animals. These include embryonic stem cell-mediated techniques (Gossler et al., 1986), transfer of entire chromosomal segments (*i.e.*, "transomic" mice; Richa and Lo, 1988) and gamete transfection in conjunction with *in vitro* fertilization (Lavitrano et al., 1989). Currently however, the technique of *pronuclear microinjection* is the most successful and most widely-used method of producing transgenic animals. Using this technique, trangene sequences up to about 50 kilobases (kb) in length from viral, prokaryotic, plant, invertebrate or vertebrate sources may be introduced into the mammalian genome, where they may be expressed in both somatic cells (Fig. 7.1) and germ cells.

7.3 METHODS OF TRANSGENESIS

The transfection of DNA can be either transient or stable. When performing a transient transfection the expression of transfected genes in cells is usually measured within 16 to 96 hours.

However a stable transfection integrates the DNA into the genome of the cell and therefore allows unlimited analysis of the gene function. One also can transfect several genes in parallel with the co-transfection method. The common method used to gene transfer in animals include

1. Calcium phosphate precipitation
2. DEAE-Dextran mediate
3. Liposomes
4. Electroporation
5. Gene-gun
6. Microinjection
7. Embryonic stem cells
8. Retroviral Vectors
9. Laser beam
10. Ultrasonication

1. Calcium Phosphate Precipitation Method

Calcium phosphate transfection, initially described in the early 1960s, was refined and systematically improved to result in a standard protocol which has changed little since the early 1970s.

Calcium phosphate co-precipitation became a popular transfection technique after Graham and van der Eb, (1973) work in animal cells. Calcium phosphate transfection is routinely used for both

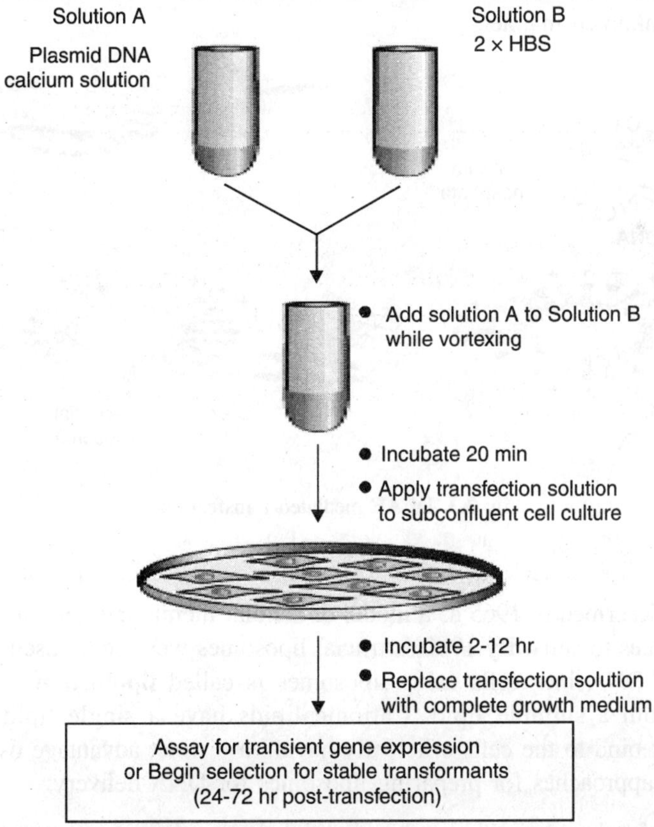

Fig. 7.2. Calcium phosphate mediated transfection

transient and stable transfection of a variety of cell types. In addition, calcium phosphate also appears to provide protection against intracellular and serum nucleases.

The DNA preparation to be used for transfection is first dissolved in a phosphate buffer and calcium chloride solution is added. This leads to the formation of insoluble calcium phosphate which co-precipitates with the DNA (Fig. 7.2). The calcium phosphate-DNA precipitate is added to the cells to be transfected.The controlled mixing generates a precipitate that is dispersed onto the cultured cells. The precipitate is taken up by the cells via endocytosis or phagocytosis.

2. Diethylaminoethyl-dextran (DEAE-Dextran) Method

This is (DEAE) one of the first chemical reagents used for transfer of nucleic acids into cultured mammalian cells and DEAE (Di Ethyl Amino Ethyl)-dextran was used by Vaheri and Pagano, (1965) and McCutchan and Pagano (1968). DEAE-dextran is another compound that interacts and delivers DNA into cells (Fig. 7.3). It is more reproducible than calcium phosphate transfection but only works in a select few cell lines. DEAE-Dextran (diethylaminoethyl-dextran) is water soluble and polycationic in nature. DEAE-dextran is a cationic polymer that tightly associates with negatively charged nucleic acids. An excess of positive charge, contributed by the polymer in the DNA: polymer complex, allows the complex to come into closer association with the negatively charged cell membrane. Results only in transient transfections. The exact mechanism of DNA uptake mediated by this technique is not understood. Uptake of the complex is presumably by endocytosis.

It is added to the DNA solution. DEAE-dextran brings about DNA uptake by the cells through endocytosis in a yet unknown manner.

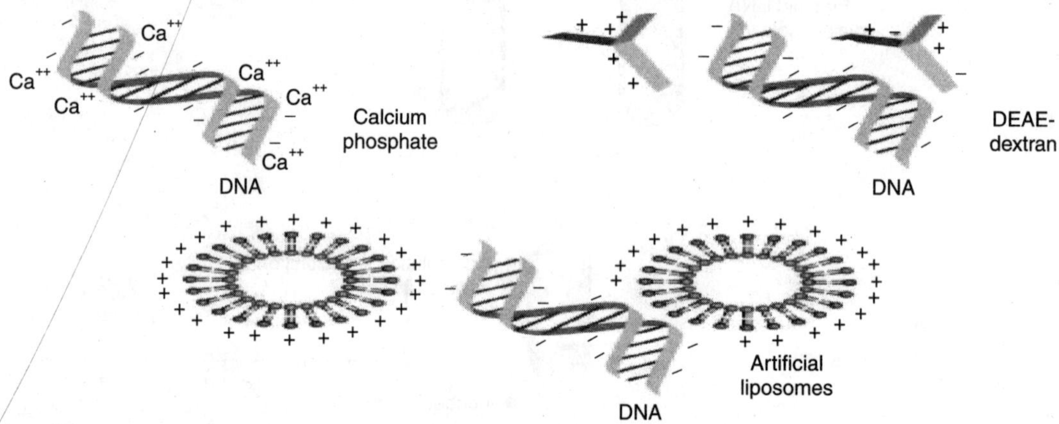

Fig. 7.3. DEAE mediated transfection

3. Lipofection

Liposomes were first described in 1965 as a model of cellular membranes and quickly were applied to the delivery of substances to cells. By 1980, artificial liposomes were being used to deliver DNA into cells. The delivery of DNA into cells using liposomes is called **lipofection**. Liposomes are small vescicles prepared from a suitable lipid. Cationic lipids have a single lipid bilayer membrane (unilamellar), and they bind to the cells efficiently having a distinct advantage over nonionic lipids.

There are various approaches for preparing liposomes for DNA delivery:

(*i*) By dispersion of a phospholipid like phosphotidyl choline (PC) in water by mechanical methods, like sonication. DNA of up to 1kb has been incorporated into small sonicated liposomes.

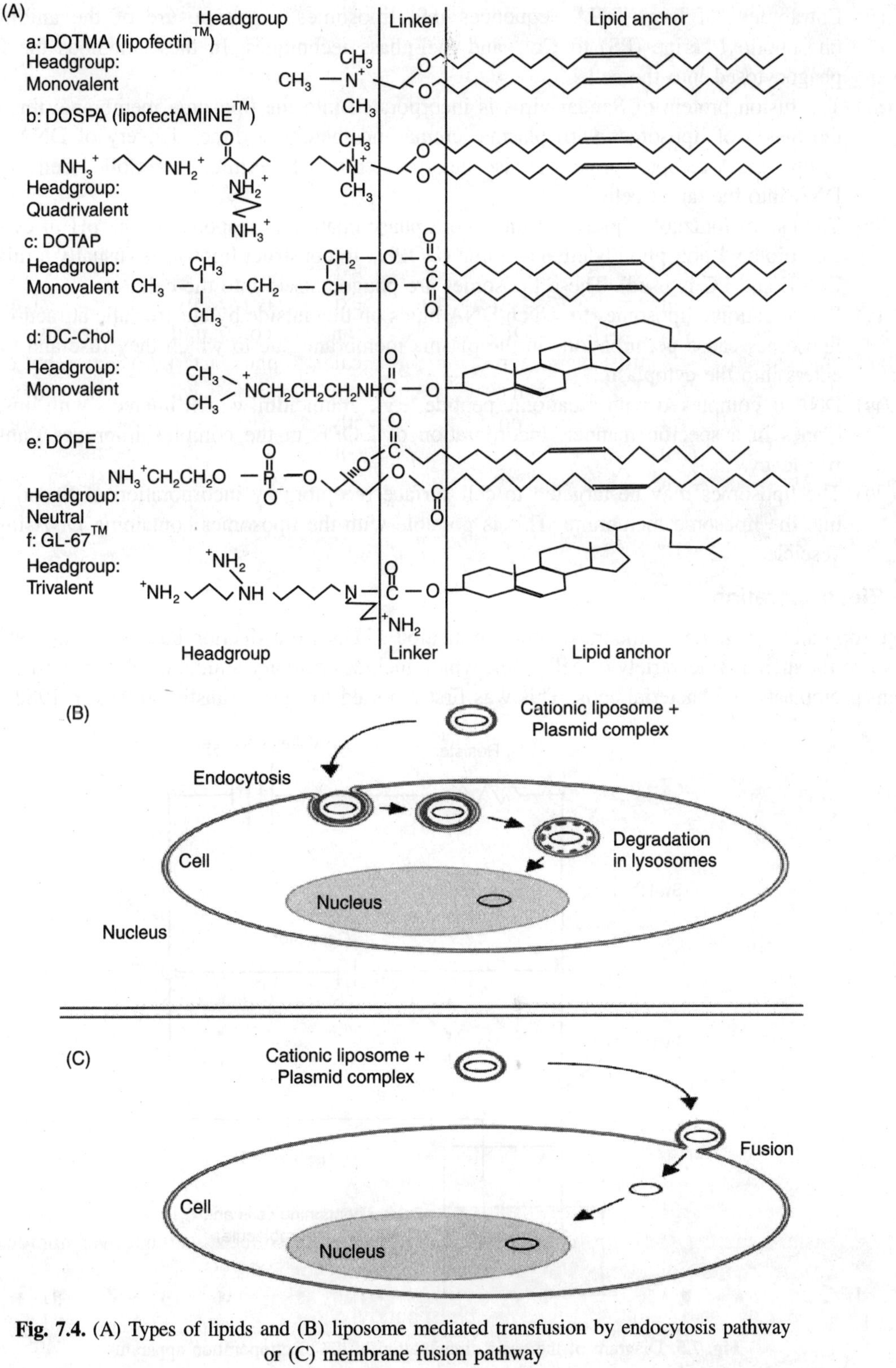

Fig. 7.4. (A) Types of lipids and (B) liposome mediated transfusion by endocytosis pathway or (C) membrane fusion pathway

(*ii*) Entrapment of large DNA sequences into liposomes, e.g.,exposure of the anionic lipid phosphotidyl serine (PS) to Ca^{++} and two-phase techniques. In these methods the DNA is phagocytosed into the cell.

(*iii*) The fusion protein of Sendai virus is incorporated into the liposome membrane; this enables the fusion of liposomes with plasma lemma and thereby a direct delivery of DNA into the cytoplasm. A receptor protein is also incorporated which permits a controlled delivery of the DNA into the target cells.

(*iv*) The use of ionizable lipids that undergo a phase change in response to the pH of cytoplasm, e.g., dioleoyl phosphotidylethanolamine (DOPE), to construct liposomes (usually a mixture of DOPE and PC is used). These liposomes are phagocytosed into the cell.

(*v*) Use of cationic liposomes to which DNA binds on the outside by electrostatic attraction. These liposomes cause perturbations in the plasma membrane due to which they fuse and the DNA enters into the cytoplasm.

(*vi*) DNA is complexed with a cationic peptide, e.g., gramicidin, which interacts with lipid membranes in a specific manner. Incorporation of DOPE in the complex improves transfection frequency.

(*vii*) The liposomes may be targeted to cell surface receptors by incorporation of ligand proteins into the liposome membrane. This is possible with the liposomes containing DNA inside the vescicle.

4. Electroporation

Electroporation (Fig. 7.5) is the most versatile method of DNA transfection because it has been shown to work for such a wide variety of cell types, which includes primary animal cells from tissue isolates, plant protoplasts and bacterial cells. This was first reported for gene transfer studies in 1982. In this

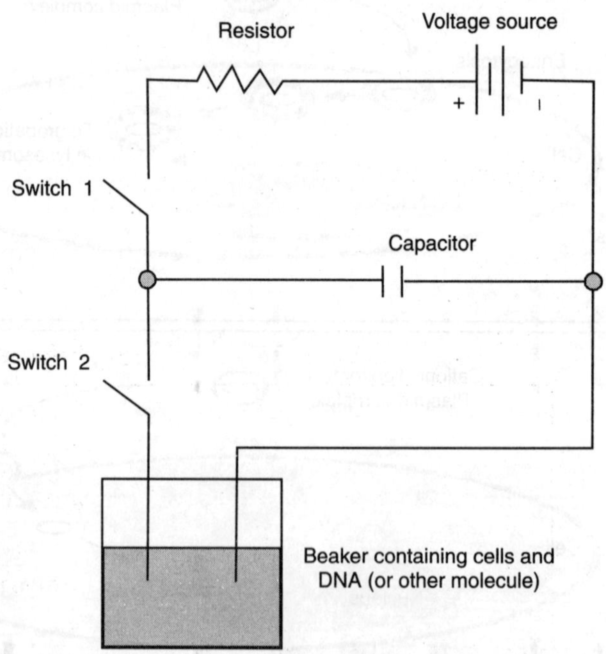

Fig. 7.5. Diagram of the basic circuit setup of the electroporation apparatus

method, the solution containing cells and DNA is exposed for a very short period of time (few milli-seconds) to a very high voltage gradient of 4000-8000 V/cm. This induces transient pores in the cell membranes (Fig. 7.6A) through which DNA seems to enter the cells. Treatment of cells with colcemid before they are electroporated increases the frequency of transfection. This is most likely due to the arrest of cells at metaphase and the associated absence of a nuclear envelope or to an unusual permeability of the plasma membrane.

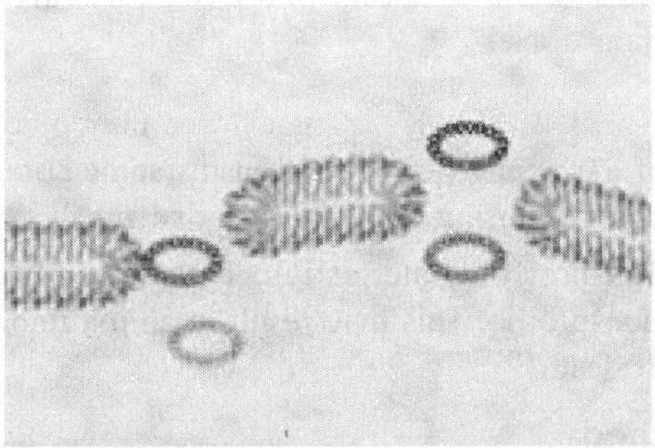

Fig. 7.6. (A) formation of transcient pores in the cell membrane during high voltage gradient

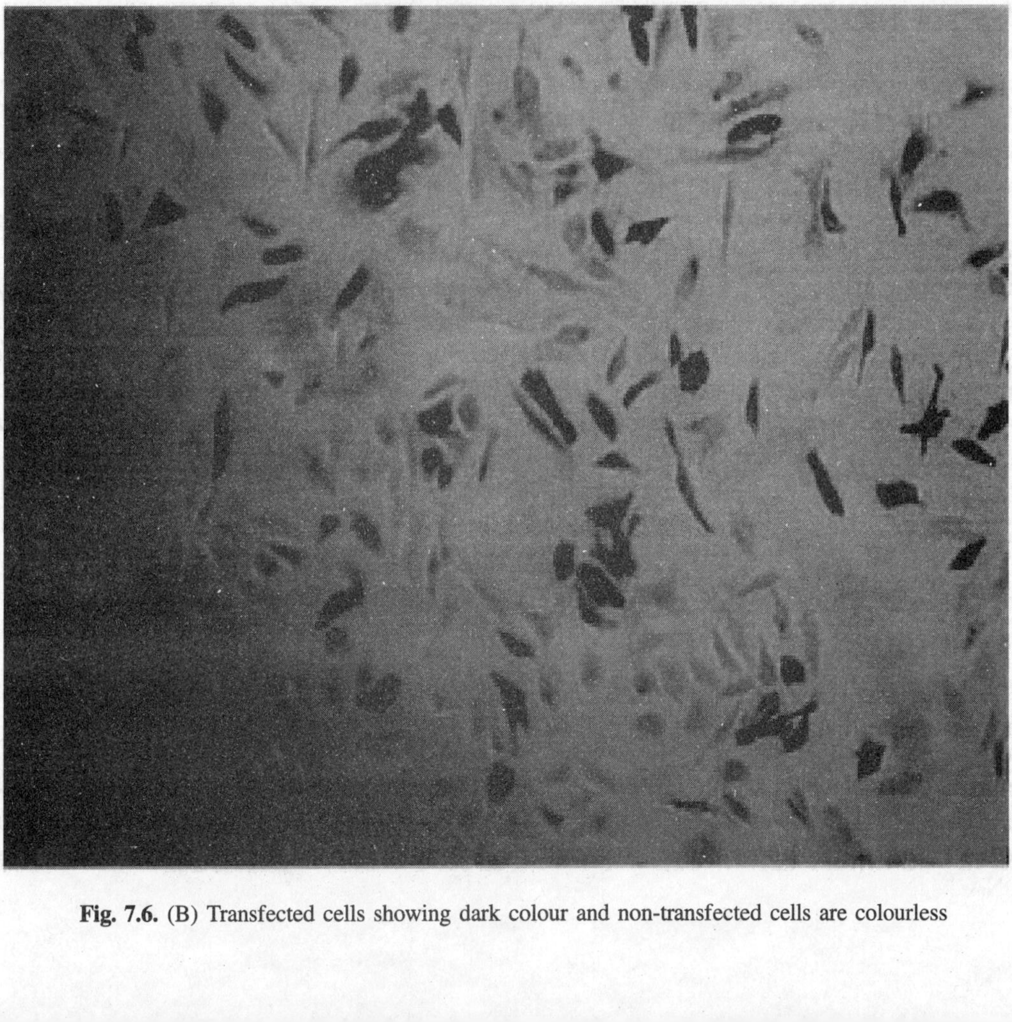

Fig. 7.6. (B) Transfected cells showing dark colour and non-transfected cells are colourless

Advantages

- **Versatility**: Electroporation is effective with nearly all cell and species types.
- **Efficiency:** A large majority of cells take in the target DNA or molecule. In a study on electrotransformation of *E. coli*, for example, 80% of the cells received the foreign DNA.
- **Small Scale:** The amount of DNA required is smaller than for other methods.
- ***In vivo***: The procedure may be performed with intact tissue. A paper published in *Developmental Biology* showed the successful transfer of a DNA construct with a fluorescent reporter gene into intact mouse brain tissue.

Disadvantages

Cell Damage: If the pulses are of the wrong length or intensity, some pores may become too large or fail to close after membrane discharge causing cell damage or rupture.

Nonspecific Transport: The transport of material into and out of the cell during the time of electropermeability is relatively nonspecific. This may result in an ion imbalance that could later lead to improper cell function and cell death.

5. Particle Gun Delivery Method

Since its initial introduction (in plants), it has been modified by Wolff in 1994, to transfer genes into mammalian cells both *in vitro* and *in vivo*. This method is also known as gene gun method, it is used in biolistic genetic transformation. Tiny metal particles e.g., gold or tungsten, coated with genes having desired characteristics, such as nutritional value are put into a particle gun and fired directly into cells. It uses high gas pressure to accelerate the DNA coated particles against a porous plate, which separates the gold particles and the DNA. The gold particles remain behind on the plate, while DNA are shot into the targeted cells (Fig. 7.7A).

This technique usually consists of basically two types of projectile systems, macro and microprojectile. The DNA sequence of interest is precipitated and coated on micrometals such as tungsten or gold particles of size 1-3 μm. This is called as microprojectile. These DNA coated micrometals are then placed before a plastic or bullet known as macroprojectile. Inside the gene gun barrel both the projectile systems are placed and with the help of a firing device in a vacuum condition macroprojectile are fired against retained plate. High pressurised nitrogen is applied to drive the macroprojectile which along with the microprojectile hits the retaining plate. The retaining plate allows microprojectile to move and retains the macroprojectile. Under high explosive discharge the high velocity microprojectile hits the tissues or cells placed next to it on the stage (Fig. 7.7B). Once microprojectile enters the cells, the DNA gets dissociated from the micrometals (Tungsten or gold particles) and integrate into host DNA. The cells present only in line with the microprojectile movement get damaged. However, rest of the cells survive and are transformed efficiently.

6. Ultrasonication

It is a simple method facilitating DNA uptake by directing ultrasound on the target cells. Small pieces of tissue or cells and known amount of exogenous DNA are taken in a vial, containing appropriate medium.

The microtip of sonicator is immersed in the suspension and pulses of ultrasound is delivered by sonicator. The vigorous vibration in the medium and violent collapse of bubbles generate high pressure and shock waves may result in sporadic localized rupture in the membrane and it can uptake exogenous DNA.

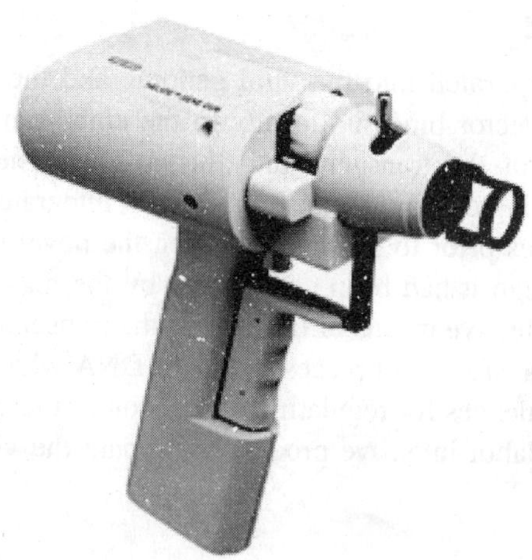

Fig. 7.7 (A) Biorad gene-gun used for animal cells transfection

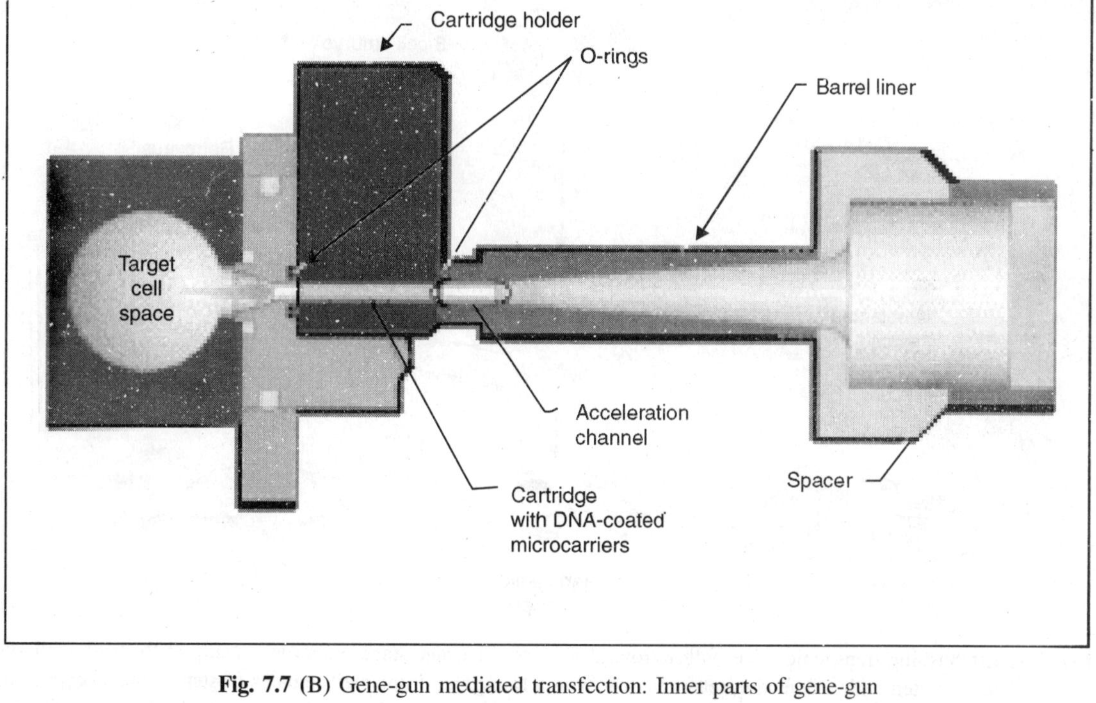

Fig. 7.7 (B) Gene-gun mediated transfection: Inner parts of gene-gun

7. Microlaser

The basic principle in this technique is localized rupture of cell membrane. A microlaser beam is focused into the light path of a microscope and it can be used to burn holes into the cell walls and membranes. The localized rupture or burn hole region permits the entry of exogenous DNA. The extent of damage on the membrane is reversible.

8. Retroviral Vector Method

The gene(s) of interest is incorporated into the viral genome and the virus is then used to infect an early stage embryo. The viral vector binds uniformly to the embryonic cells and acts as a vehicle to allow transfer and integration of the transgene into the host genome. An advantage to using viral vectors is that usually only a single copy of the transgene is integrated into the genome. If the viral transfection is applied to oocytes prior to fertilization, then the novel gene will be present in all cells of the resulting embryo as though it had been contributed by the maternal germline.

Retroviral vectors are an effective means of integrating the transgene into the genome of a recipient cell. However, they can only transfer small pieces (8 kb) of DNA which because of the size constraint may lack essential adjacent sequences for regulating expression of the transgene. Another disadvantage of this system is the time and labor-intensive process to prepare the viral vector.

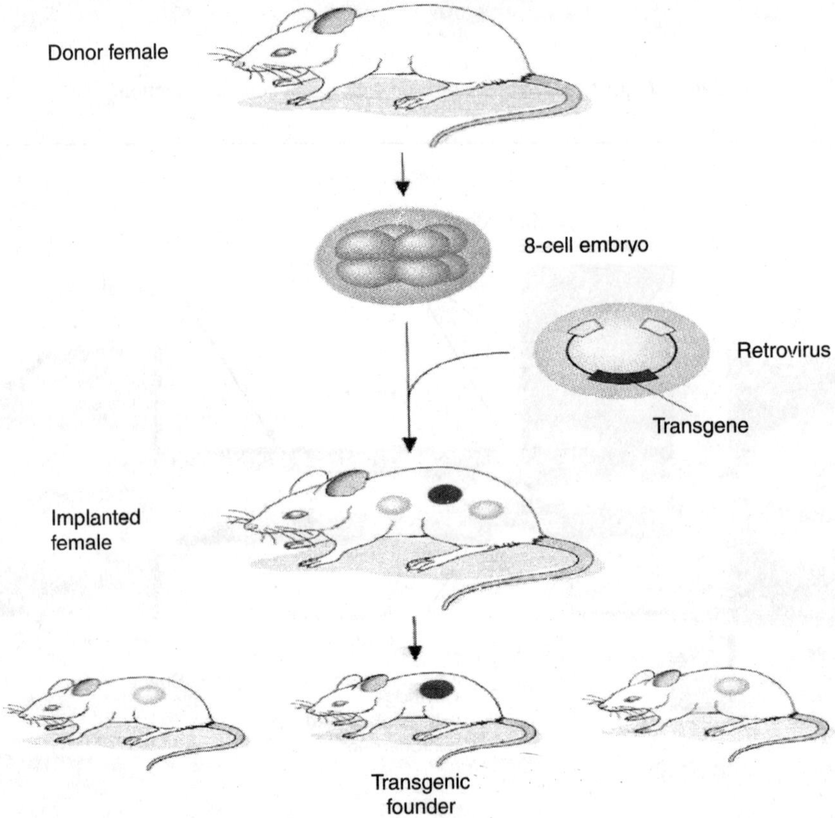

Fig. 7.8. Establishing transgenic mice with retroviral vectors. Cleavage-stage embryos, usually at the eight-cell stage, are infected with a defective retrovirus carrying a transgene. Implanted females (foster mothers) give birth to transgenic pups. Matings are carried out to determine which pups have the transgene in their germline cells. Transgenic lines can be established from these founder transgenic animals.

There is further major drawback to the use of retroviral vectors. Although these vectors are designed to be replication defective, the genome of the retroviral strain (helper virus) that is needed to create large quantities of the vector DNA can be integrated into the same nucleus as the transgene. Despite special precautions, helper strain retroviruses could be produced by the transgenic organism. Consequently, for applications in which either a commercial product is to be synthesized by the transgenic organism is to be used as food, it is absolutely necessary that there be no retroviral contamination. Therefore, because alternative methods are available, retroviral vectors are rarely used for creating transgenic animals that have a commercial use.

9. DNA Microinjection or Pronuclear-Mediated Transgene Microinjection

Because of the disadvantages of the retroviral vector method mentioned above, microinjection of DNA is the currently preferred method for producing transgenic mice. Most existing transgenic animal lines have been produced using DNA microinjection or pronuclear microinjection. In this approach DNA solution is injected directly in to the nucleus of a cell or into the male pronucleus of a fertilized ovum. This technique involves the injection of genetic material into an early-stage embryo to create what are called germ-line transgenic animals. Gordon et al., 1980 first successfully produced transgenic mice using pronuclear microinjection technique. Palmiter et al., 1982 first showed visible phenotypic change in transgenic mice for expressing rat growth hormone sequence.

1. The number of available fertilized eggs that are to be inoculated by microinjection is increased by stimulating donor females to super-ovulate. They are given an initial injection of pregnant mare serum and another injection about 48 hrs later of human chorionic gonadotropin. A superovulated mouse produces about 35 eggs instead of the normal number of 5-10.
2. These females are mated and then sacrificed. The fertilized eggs are flushed from their oviducts.
3. The eggs are microinjected – usually immediately.

The microinjected transgene construct is often in a linear form and free of prokaryotic vector sequences. In mammals, after, entry of the sperm into the egg, both the sperm nucleus (male pronucleus) and female exist separately. After the female nucleus completes its meiotic division to become a female pronucleus, nuclear fusion (karyogamy) occur. The male pronucleus, which tends to be larger than the female pronucleus, can be located by using a dissecting microscope. The egg can then be oriented and held in place while the DNA is microinjected. Sometimes, several hundred male pronuclei can be inoculated.

After inoculation, 25 to 40 eggs are implanted microsurgically into a foster mother, who has been made pseudopregnant by being mated to a vasectomised male.

For several hours following the entry of the sperm into the oocyte or egg, the male and the female pronuclei can still be seen individually under a normal light microscope and they have not fused yet into a so called zygote or fertilized egg. The foreign DNA may be injected into either pronuclei with no difference in results; however, the DNA is typically injected into the male pronucleus because it is **slightly larger** and **closer to the oocyte surface** (Fig. 7.9). Transgenic mice are constructed by injecting cloned DNA into fertilized mouse eggs; those eggs that survive are then implanted in foster females to develop to term. The gestation period of the mouse is 19-21 days. Pups are ready for weaning at 3-4 weeks of age, and reach sexual maturity at 6 weeks (females) to 8 weeks (males). The minimum elapsed time between injection of the construct and readiness for breeding of the transgenic founders is 9 to 11 weeks. For most experimental purposes researchers will want to use offspring of a transgenic animal, rather than the founder animal itself (Fig. 7.10).

A. The double-stranded DNA components of the transgene are combined enzymatically to yield a transgene expression cassette.
B. Transgene cassettes are inserted into plasmid vectors and cloned.
C. Transgene-bearing plasmids are transfected into cultured eukaryotic cells to evaluate expression of the transgene.

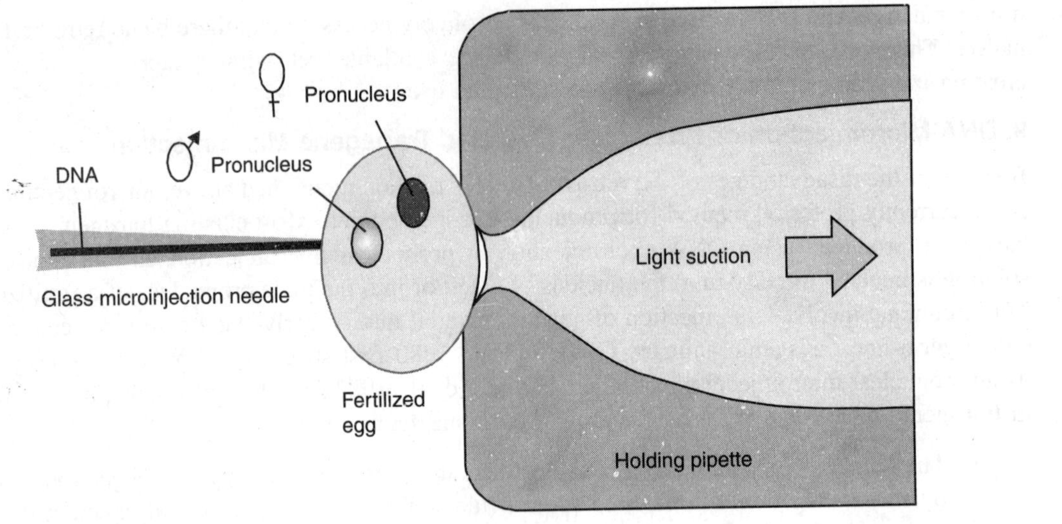

Fig. 7.9. The production of transgenic animals by microinjection of DNA into fertilized eggs

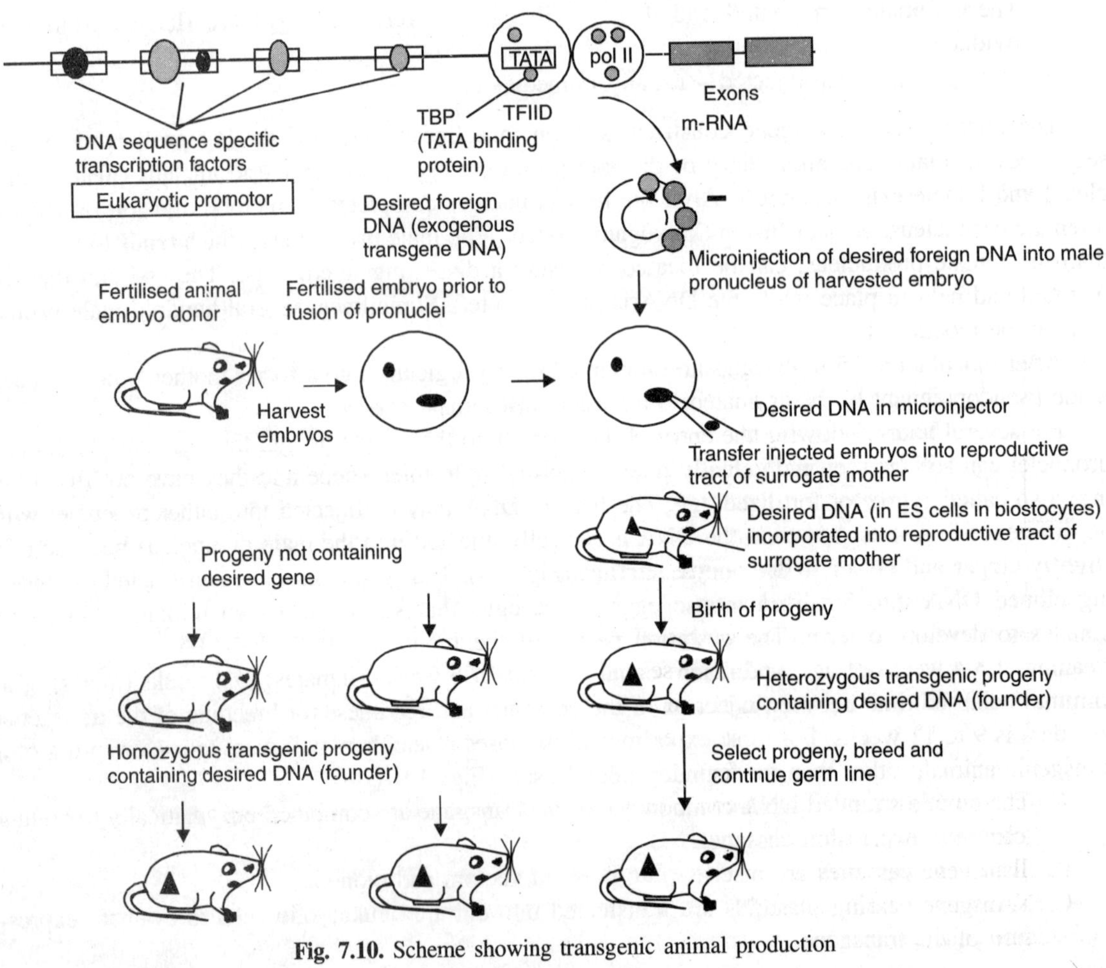

Fig. 7.10. Scheme showing transgenic animal production

D. Plasmid-free transgene fragments are introduced directly into embryonic pronuclei.
E. Manipulated embryos are placed in the reproductive tract of a pseudopregnant recipient.
F. The genomic DNA of live-born pups is analyzed for the presence of the transgene DNA sequence.
G. The offspring is screened to confirm a successful integration of the gene of interest for use in further studies.

10. Embryonic Stem Cell Method

Cells from the blastocyst stage of a developing mouse embryo can proliferate in cell culture and still retain the capability of differentiating into all other cell types – including germ line cells—after they are reintroduced into another blastocyst embryo. Such cells are called pluripotent embryonic stem (ES) cells (Fig. 7.11). When in culture, ES cells can be readily engineered genetically without altering their pluripotency. With this system, for example, a functional transgene can be integrated at a specific site within a dispensable gene in the genome of ES cells. Then, the genetically engineered cells can be selected, grown, and used to generate transgenic animals. In this way, the randomness of integration that is inherent with the DNA microinjection and retroviral vector systems is avoided.

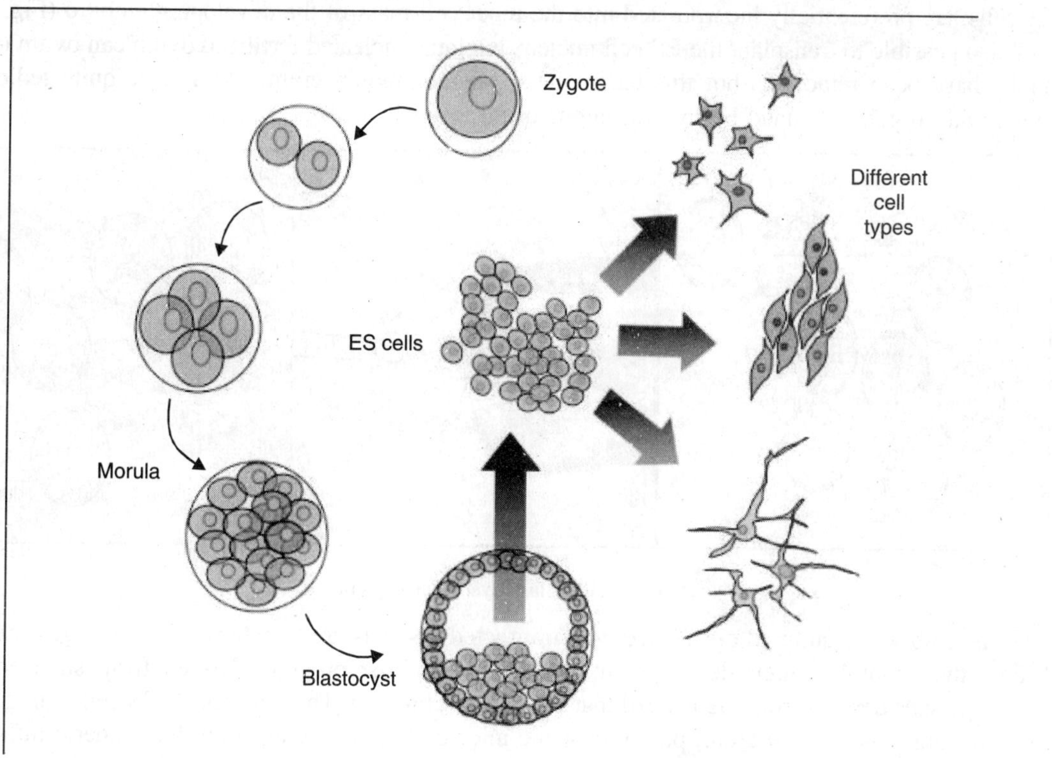

Fig. 7.11. Zygote development and embryonic cells formation

After transfection of ES cells in culture with a DNA vector that is designed to integrate within a specific chromosomal location, some cells will have DNA integrated at non target (spurious) sites; in other cells, integration will occur at the target (correct) site. Moreover, in most of the ES cells, the input DNA will not be integrated at all. To enrich for the cells with DNA integrated at the target site, a procedure called **positive-negative selection** is implemented. This strategy uses positive selection for selection for cells that have vector DNA integrated anywhere in their genomes and negative selection against the vector DNA sequence that is integrated at spurious sites.

Embryonic stem (ES) cells of mice are pluripotent cultured cells derived from early pre implantation embryos. These cells can be maintained and multiplied *in vitro* long enough to permit the various manipulation for gene transfer. The cultured pluripotent ES cells are transfected with the appropriate transgene construct by a suitable tranfection technique. Tranfected ES cells are identified and selected, generally by employing a selectable marker gene and are cloned.

The production of transgenic animals using transfectd ES cell clones may be achieved in one of the following two ways:

- Injection
- Co-culture

Injection of Genetically Manipulated ES Cells

The cloned ES cells are injected into the blastocoel of blastocyst stage embryos; the blastocysts are obtained from donor females in a similar manner as that for microinjection. Alternatively, the Zona-Pellucida of eighth cell to morula stage embryos is removed, and the morulae are co-cultured with the ES cells; are preferentially incorporated into the inner cell mass of the developing embryo (Fig. 7.12). It is also possible to transplant the ES cell nucleus into an enucleated fertilized ovum can ovum (whose nuclei have been removed), but this technique is not commonly employed as it is quite tedious although all progeny obtained by this technique are transgenic.

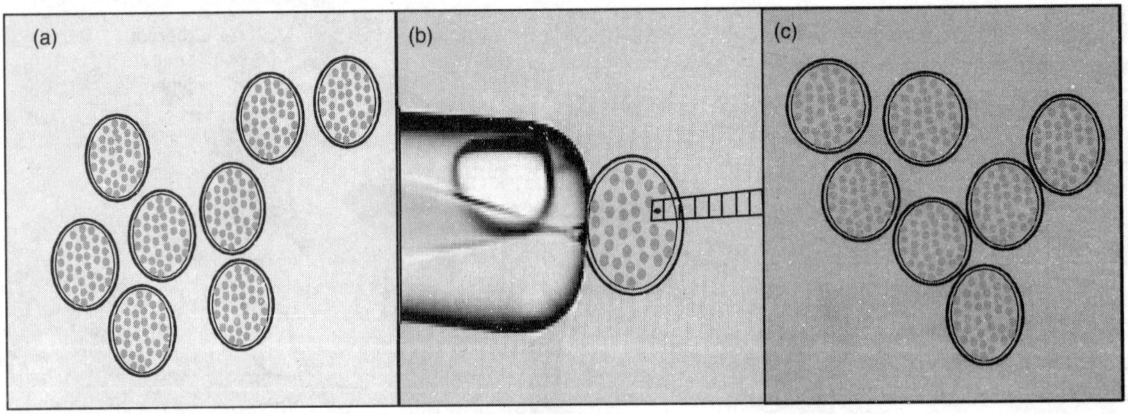

Fig. 7.12. Blastocyst microinjection

The embryo co cultured of injected with tranfected ES cells are transferred into surrogate mothers. Where they complete their development. About 30% of the progeny derived from such embryos contain tissues derived from the ES cell that is they are chimeric. The pluripotent ES cells can give rise to germ cells as well. Therefore, pure transgenic mice can be recovered from the chimeric mice using suitable breeding scheme.

The entire process can be summarized as follows:

1. **DNA Prepartion:** Using recombinant DNA methods, build DNA containing the structural gene (e.g., the insulin gene), vector DNA to enable the molecules to be inserted into host DNA, promoter and enhancer sequences to enable the gene to be expressed by host cells.
2. **Transform ES cells in culture** Expose the cultured cells to the DNA so that some will incorporate it.

3. **Selection of successfully transformed cells.**
4. **Injection of these cells into the inner cell mass (ICM) of mouse blastocysts.**
5. **Embryo transfer:** Prepare a **pseudopregnant** mouse (mating a female with a vasectamized male). Mating elicits the hormonal changes to make her uterus receptive. Transfer the embryos into uterus. They **implant** successfully and develop into healthy pups <1/3.
6. **Testing offspring:** Remove a tissue from the tail and examine its DNA for the desired gene. No more than 10–20% will have it, and they will be heterozygous for the gene.
7. **Establish a transgenic strain:** Mate two heterozygous mice and screen their offspring for the 1:4 that will be **homozygous** for the transgene. Mating these will found the transgenic strain (Fig. 7.13).

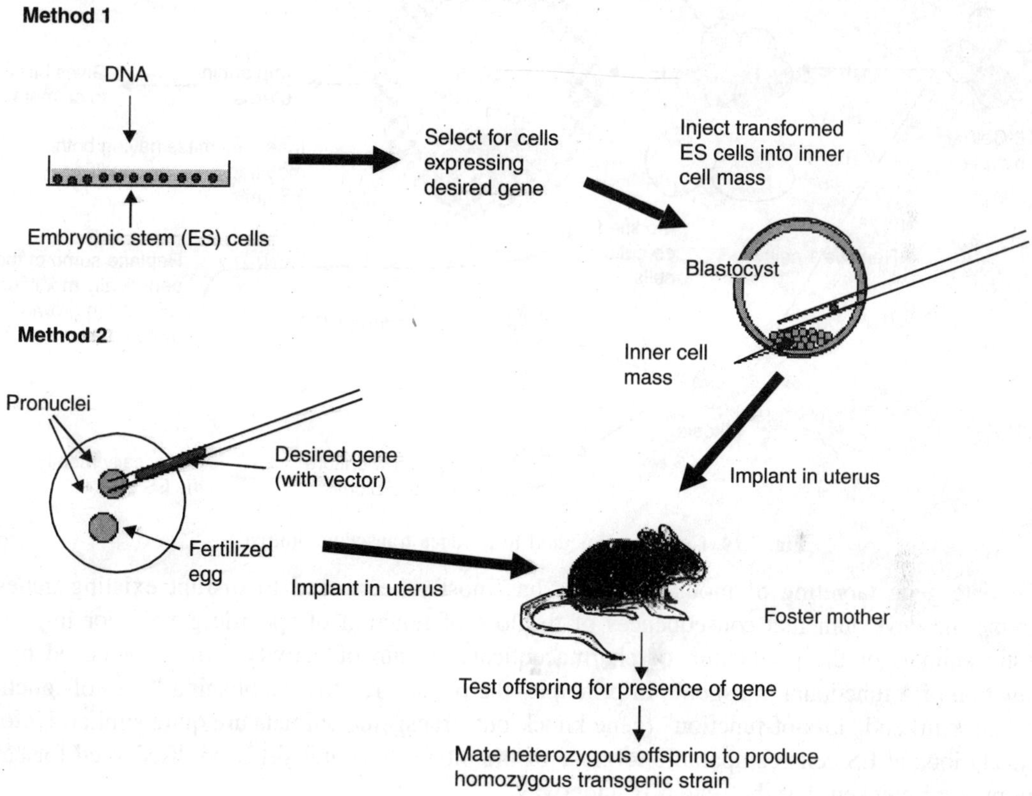

Fig. 7.13. Detailed process of transgenic animal production by injection of genetically modified ES cells

Co-culture of Genetically Manipulated ES Cells into Normal Blastocyst

In co-culture method first ES cells are introduced with the required transgene and thus they express the required gene character. Then these genetically manipulated ES cells are then cultured along with the normal morula (eight cell stage embryo) in the culture condition (Fig. 7.14). This genetically manipulated ES cells becomes the part of developing normal embryo after implanting in the surrogate mother. The offspring express the character for the transgene.

The target site should be located in a section of genomic DNA that encodes no essential products so that, after integration of the input DNA, there is no interference with any developmental or cellular functions. Moreover, it is essential that the transgene be integrated into a part of the genome that does not prevent it from being transcribed. Researchers are currently looking for these kinds of sites.

Embryonic stem cells carrying an integrated transgene can be cultured and inserted into blastocyst stage embryos, and these embryos can then be implanted into pseudopregnant foster mothers. From those progeny in which the genetically engineered ES cells have contributed to the formation of germ line cells, transgenic lines can be established by first mating founder transgenic mice to animals from the same strain and then crossing transgenic litter mates to create transgenic mice that are homozygous for the transgene.

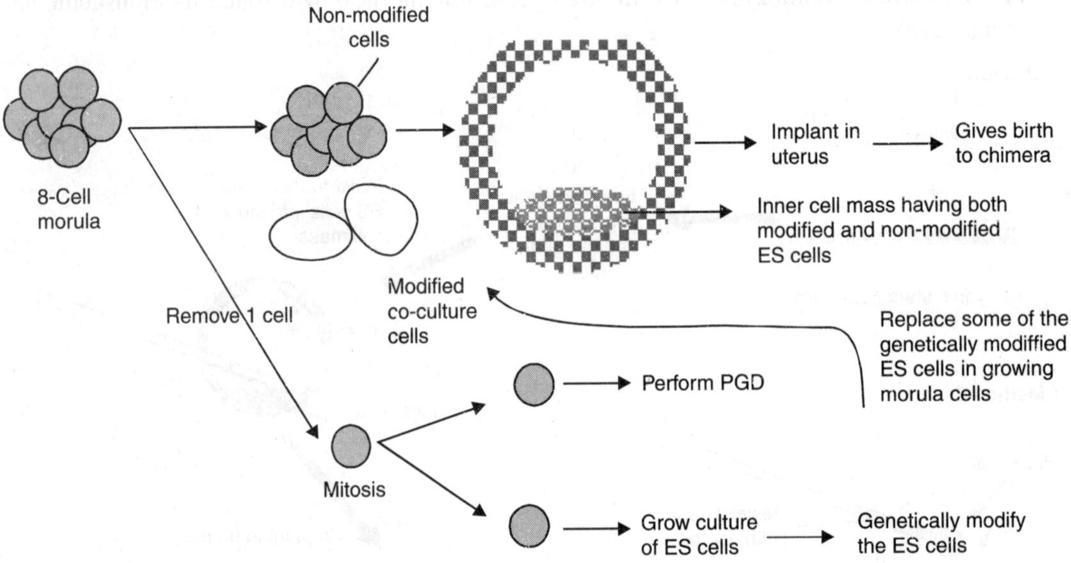

Fig. 7.14. Co-culture method to produce transgenic animal

Of date gene targeting of mouse ES cells has mostly been used to disrupt existing genes to determine the developmental consequences of the loss of function of specific genes. For improving domestic animals or the production of pharmaceuticals, a gain of activity will be achieved by the introduction of a functional transgene. In principle, the procedures for establishing "gain-of-function" (gene knock-in) and "loss-of-function" (gene knock-out) transgenic animals are quite similar. Unfortunately, pluripotent ES cells comparable to those of the mouse have not yet been discovered for cattle, sheep, pigs, or chicken; but the search is underway.

7.4 VECTORS USED FOR TRANSFECTION

Vectors used for transfection are SV40 vectors, ***Drosophila*** P element vectors, retrovirus vectors, bovine papilloma virus and also non replicating viruses such as the passive transforming SV40 vectors (Table 7.1). The latter are used to carry DNA sequences into the host cells for integration into the animal genome.

Drosophila P elements have been developed as valuable vectors for this invaluable genetic material. The 31 bp inverted repeat border and the neighbouring sequences of the transposable P elements are combined with a suitable ***E. coli*** plasmid, e.g. ***pUC*** 8, to produce a shuttle vector. DNA insert of up to 40 kb can be placed between the two border sequences. The recombinant P vector is injected into ***Drosophila*** larvae along with a helper P element which produces transposase. Transposase enables the transposition of recombinant P element from the recombinant vector into the ***Drosophila*** genome thereby bringing about the integration of DNA insert as well.

Bacculovirus vectors have been developed for transfection of insects. The nuclear polyhedrosis viruses e.g. ***Autographa california*** nuclear polyhedrosis virus (AcNPV) and ***Bombyx mori*** nuclear polyhedrosis virus (BmNPV) have been exploited for this purpose. The nuclear polyhedron gene has a very strong promoter, and the polyhedron gene is not needed for NPV replication. Hence this region is replaced by the required gene of interest.

Linearized vectors are far more readily integrated into the host genome than the circular vectors. The presence of additional vector –DNA along with the integrated gene construct interferes with the expression of introduced gene or transgene. Therefore, it is often desirable to introduce the transgene with a minimum of vector-DNA associated with it.

The inserted DNA is known as the **transgene** (Fig. 7.15). Conventional recombinant DNA techniques are used to construct the transgene so that the desired gene product will be expressed in the desired location. Typical transgenes contain transgene cassette nucleotide sequences that correspond to the gene of interest, with all the components necessary for efficient expression of the gene, including a transcription-initiation site, the 5' untranslated region, a translation-initiation codon, the coding region, a stop codon, the 3' untranslated region, a polyadenylation site and a promoter. A transgene must be present in proper orientation in relation to and in association with the various sequences required for its efficient transcription and translation in the host cells. The sequences are listed as follows:

- An efficient promoter/enhancer
- The translation initiation codon (AUG in mRNA and ATG in DNA)
- The chain termination codon(s)
- Transcription termination sequence
- Polyadenylation cleavage/addition site.

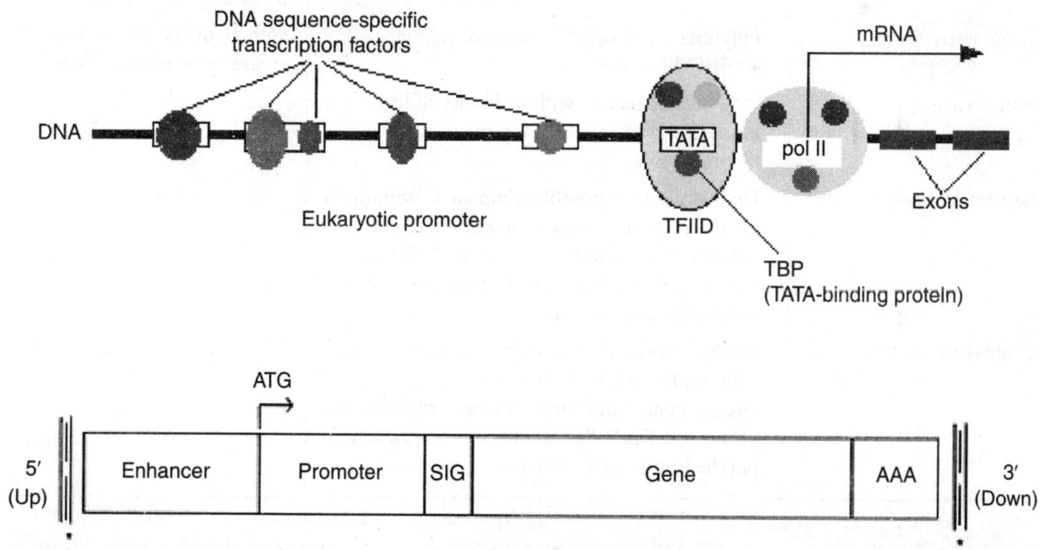

Fig. 7.15. Structure of transgene cassettes in eukaryotic cells

Table 7.1: Different Vectors used in Transgenic Animal Production

Vector	*Derived from*	*Features*
1. SV40 vectors		
(a) Early region replacement vectors	Replacement of large-T gene of SV40	Produce virions, which infect host cells Transient gene expression Mammalian cells are hosts of SV40
(b) Late region replacement vectors	Replacement of VP1, VP2, and VP3 genes of SV40, e.g., SVGT-5	
(c) Plasmid vectors	Origin of replication and large-T gene of SV40	—
(d) Shuttle plasmid vectors	Plasmid vector [item 1(c)] plus pBR322 origin and amp' gene, e.g., pSV2, pSV3, etc.	—
	Rous sarcoma virus promoter in place of SV40 early promoter, e.g., pRSV.	Strong expression of the marker gene
(e) Passive transfecting vectors	SV40 transcription regulatory and poly-adenylation sequences, plus pBR322 origin and amp gene	Shuttle vectors; used for gene integration in mammalian cell
2. BPV vectors	Bovine papilloma virus "Transforming region" + pBR322 sequences.	Shuttle vectors; often pBR322 sequence deleted prior to transfection; pl. Like v
3. Retrovirus vectors	pBR322 + retrovirus sequences	Shuttle vectors; integrates as provirus into mammalian genome; produces virions
4. Polyomavirus vectors	Polyoma virus origin and early region + pBR322 sequences	Similar to SV4O vectors; mouse cells used as host
5. Vaccinia virus	DNA insert placed within the thymidine kinase gene of virus by a process of recombination.	
6. P element vectors	Drosophila transposable element P; minimum of 31 bp inverted repeat borders and the neighbouring regions, plus an E. coli vector, e.g.. pUC8; DNA insert of up to 40 kb placed within the two borders	
7. Bacculovirus vectors	Nuclear polyhedroma virus (NPV) polyhedrin gene replaced by DNA insert; e.g., AcNPV (Autographa californica nuclear polyhedroma virus) and BmNPV (Bombyx mori nuclear polyhedroma virus) vectors	

Table 7.2: Various Types of Promoters used for Transgenic Animal Production

Promoter/Enhancer Sequences	*Source*	*Used in Animal Species*	*Gene Expression in the Tissue*
Metallathionein promoter	1. Fish 2. Mouse 3. Sheep 4. Human	Fish Fish, mouse, pig Sheep Rabbit, pig	
SV40 early promoter	SV40	Mammalian cells, fish	All tissues
RSV promoter	Rous sarcoma virus (RSV)	Mammalian cells, fish	All tissues
Thymidine kinase promoter	Herpes simplex virus (HSV)	Mammalian cells	All tissues
SV 40 enhancer	72 bp sequence having tandem repeats near SV40 origin	Enhances transcription of all genes transcribed by RNA polymerase II	All tissues
Anti-freeze protein promoter	Fish	Fish	
β-casein promoter	Mice	Goat	Mammary tissue
β-lactoglobulin promoter	Sheep	Sheep	Mammary tissue
Prolactin promoter	Cow	Pig	Mammary tissue

Different promoters can be used to cause gene expression in all tissues of the body (non-specific) or only in specific tissues (Table 7.2 and 7.3). The tissue specific expression promoters are listed in the following table.

Table 7.3: The different Promoters used to cause Gene Expression only in Specific Tissues of Transgenic Animals

Promoter	*Gene Expression in*
(beta)-Actin promoter	Many tissues of the transgenic animal
Simian virus 40 T antigen promoter	Many tissues of the transgenic animal
Adipocyte P2 promoter	Fat cells
Myosin light-chain promoter	Muscle
Amylase promoter	Acinar pancreas
insulin promoter	Islets of Langerhans beta cells
Beta-lactoglobin promoter	Mammary glands

7.5 SPECIFIC APPLICATION OF TRANSGENIC MICE

Transgenic Mice in Immunology

The ability of the immune system to distinguish and mount a response to myriad foreign antigens while recognizing self as non antigenic has always been one of the most fascinating areas of biology. Information about the molecular genetics of the genes for immunoglobulins (Ig) and the major histo-compatibility complex (MHC) has increased dramatically and a major contribution to such information has come from studies with transgenic mice. Introduction of rearranged Ig genes into transgenic mice has helped enormously to clarify mechanisms of rearrangement of Ig genes and the importance of this phenomenon to the regulation of their expression.

The expression of rearranged heavy chain genes in T-cells indicates that the inability of the T-cell to express heavy chain gene is related to the absence of gene rearrangement in these lineages. This result is all the more intriguing since T-cells are able to rearrange the T-cell receptor gene, which shares several features with the Ig gene. One of the classical concepts in immunology, namely, the rearrangement of K genes inhibits rearrangement of genes (isotopic exclusion) was challenged by the finding that the presence of transgenic rearranged K genes did not, in all areas, result in isotopic exclusion. This observation led to the hypothesis that two separate B cell lineages exist, one that is capable of rearranging both L and * genes and one that can only rearrange K genes.

Transgenic Mice in Oncology

The study of transferred oncogenes has always been hampered by the fact that cell lines in culture have already been transformed to an abnormal phenotype. Thus, the transforming potential of an oncogenic sequence of nucleotides in a cell can be enhanced or obscured by pre-existing prolirerative abnormality. Our ability to insert oncogenes and protooncogene into embryos and to study their effects in the normally differentiating cells of an instant organism has circumvented this problem, and results of such studies have made an enormous contribution to our understanding of neoplastic disease and its relationship to aberrant gene expression.

Transgenic Mice as Animal Models for Human Disease

Animal models for human illnesses are useful for studying the pathogenesis of diseases as well as for developing and testing new therapies. Many human diseases either don't exist in animals or are only developed by higher mammals, making models scarce and expensive to researchers. Human diseases can be induced in transgenic mice by expression of transferred genes, or by insertional disruption of endogenous sequences. Some examples of models created by transgene expression are listed below:

- Hepatitis B is a human disease that lacks a readily workable animal model, introduction of HBs Ag gene into mice results a transgenic mice that mimic the carrier state with production of HBs Ag in liver but with an absence of disease.
- Progressive mulitfocal lenkencephalopathy can be recreated mice by injection of the Jc viral genome. Another disease of the nervous system, neurofibromatosis, can be induced in mice by introduction of the gene for tyrosine amino transferase (TAT) of HTLV-1.
- Osteogenesis imperfect is a dominant disorder that results from a mutation of the gene for Al (I) collagen. This simple resultant amino-acid substitution results in change in the helical structure of mature collagen. This condition has been induced in mice by microinjection of a mutant gene for collagen Al (I) created in the laboratory by site directed mutagenesis.

Transgenic Mice as Models for Gene Therapy

Genes can be inserted into transgenic animals and function to alleviate disease states. Such model system can be of great importance in improving our understanding of the potential for gene transfer as an approach to treatment of disease.

Mice with growth hormone deficiency are markedly reduced in size and males suffer from infertility. Introduction of the growth hormone gene into these animal leads to growth which exceeds that of normal animals and restores male fertility. The pattern of release of growth hormone that results from transgene function is apparently inconsistent with female fertility.

Mice lacking genes at the MHC region are unable to respond to the synthetic antigen poly (glu-lys-ple). When the gene for MHC la is inserted into transgenic animals, transgene function restores immunoresponsiveness to antigen. Insertion of either the mouse or human beta-goblin gene can reduce the severity of beta-thalassemia in mice. In these experiments, the product of the human globins gene was able to associate effectively with the mouse A chains and it actually functioned, better than the transferred mouse gene beta-globins in palliating the thalassemic state.

Mice with a deficiency in gonadotropin releasing hormone (GnRH) are infertile and exhibit profound perturbations of their reproductive endocrine functions, cloning of the GnRH gene and its transfer into mice has resulted in restoration of normal endocrine function and in fertility.

Transgenic Mice as Models for Genetic Engineering

Genetic engineering of resistance to chemotherapy is one techniques that can assists in increasing the therapeutic margin of a drug. In cancer therapy the toxin doses of most antineoplatic drugs overlap with their therapeutic doses, so that the intensity and duration of feug regiments are of necessity, limited, production of transgenic animals that are systematically resistant to the effects of a drug provides a test system the intensive chemotherapy can improve the overall survival of tumor bearing hosts of whether treatment would just select for highly resistant tumors.

Introduction of the gene for the receptor for low density lipoprotein (LDL). Under control of the MT-1 promoter, resulted in inducible expression of the gene with a resultant decrease in levels of free LDL's within the circulation of the transgenic animals. This experiment suggests a potential strategy for a genetic approach to the control of high levels of LDL, which could eventually reduce the risk of antherosclerosis.

The successful production of transgenic rabbits, pigs and sheep by microinjection suggests that in the future. Genetically engineered farm animals may play an important role in agriculture and medicine. The isolation of tissue plasminogen activatior (TPA) from the milk of transgenic mice often linkage of the TPA gene to the promoter for the acidic protein strongly suggests that transgenic animals might be exploited as factories for production of valuable products. Which because of post-translational modification and other reasons cannot readily be produced in bacteria.

Antisense Genes in Transgenic Mice

Another method for neglecting gene function involves the use of antisense transcripts. When genes are cloned in reverse orientation with respect to the promoter, RNA may be produced from the non coding strand. This RNA presumably by forming a heteroduplex with the sense RNA can block translation of cytoplasmic mRNA. This antisense genes can be used to obliterate production of proteins from specific genes in transgenic animals. The feasibility of this approach has recently been demonstrated by the transfer of an antisense construct of the gene for myelin basic protein (MBP) into mice. Interference with the production of MBP resulted in dysmyelination. Although this is still infancy, it has great potential for future experiments.

Transgenic Sheep

The process of using transgenic animals to produce human drugs, is staking its claim in a lucrative world market (Tables 7.4 and 7.5). Transgenic animals are animals which have been genetically transformed by splicing and inserting foreign animal or human genes into their chromosomes. The inserted gene, when successful, enables an animal to make a certain pharmaceutical protein in its milk, urine, blood, sperm, or eggs, or to grow rejection-resistant organs for transplant.

Table 7.4: Transgenic Sheeps Produced for Drug Synthesis

Animal	*Drug/Protein*	*Use*
Sheep	Alpha1 anti trypsin	Deficiency leads to emphysema
Sheep	CFTR	Treatment of cystic fibrosis
Sheep	Tissue plasminogen activator	Treatment of thrombosis
Sheep	Factor VIII, IX	Treatment of hemophilia

Transgenic sheep have been produced to achieve better growth and meat production as well as to serve as bioreactors. Example human genes for blood clotting factor nine and for alpha antitrypsin have been transferred in sheep and expressed in mammary tissue, this was achieved by fusing the genes with the mammary tissue. Specific promoter of the bovine beta-lactoglobulin gene. Human growth hormone gene has also been introduced in sheep in order to promote growth and meat production. The transgenic animals showed improvements in body weight gains, feed efficiency lean meat/fat ration for composition etc., they also showed several undesirable joint pathology skeletal defects, gastric ulcer, infertility etc., these problems have been attributed to chronic over expression of growth hormone trans gene. This gains support from the fact that in many cases long term administration of elevated levels of growth hormones produces similar abnormalities.

Table 7.5: Important Gene Transfers in Animals of Economic Importance

Gene Transferred	*Objective*	*Achievements*	*Remarks*
Cattle, goat, sheep and swine Human genes: α1-antirypsin, tissue plasminogen activator, blood clotting factor IX, and protein C	Gene Farming	Genes expressed in mammary tissue; proteins secreted in milk in functional form	Experimental stages
Sheep Bacterial genes cysE and cysM (genes concerned with cysteine biosynthesis)	Improved wool production/quality	Genes expressed in transgenic animals	
Mice and Swine Human haemoglobin and specific circulating immunoglobulins (antibodies)	Proteins from blood serum for blood transfusion and disease diagnosis	Genes expressed; proteins released in blood serum	Experimental stages
Swine and Sheep Human growth hormone	Increased and desirable body growth	Improve body weight gain, feed efficiency, etc.	Joint pathology, skeletal defects, ulcers, etc.
Rabbits Human genes: Interleukin 2, growth hormone, tissue plasminogen activator, etc. Bovine gene: - α - Lactoglobulin	Molecular farming or gene farming	Recombinant proteins harvested from milk	Experimental stage
Fish Salmon or rainbow trout growth hormone Anti-freeze protein gene, α-globin gene, etc.	Increased body growth Variable	Up to 60% increase in size genes expressed in transgenics	Growth improved by selection. Transgenes are stably inherited

Increased wool production and improved quality are important objectives of transgenic sheep research efforts. Towards this end, genes involved in the biosynthesis of amino acids important in keratin biosynthesis are being identified and cloned. Keratin proteins are found in wool for example, two bacterial genes involved in cystein biosynthesis is E(sevine acetyl transferase) and cysM(o-acetylserine sulphydrolase) have cloned and introduced in sheep the transgenic sheep do express these genes, but much further work is needed to improve wool production and for quality of transgenic sheep.

7.6 APPLICATIONS OF TRANSGENIC ANIMALS

There are numerous potential applications of transgenic methodology for developing new or altered strains of agriculturally important livestock. In agriculture, the use of transgenic animals is being investigated as a potential way of developing breeding stock to enhance production traits such as growth rate and feed efficiency, mitigate environmental impacts or increase disease resistance. There have also been reports of the production of transgenic cows with altered casein, and others that have an increased resistance to mastitis, with the secondary benefit of reducing dependence on anti-microbial agents. Growth hormone genes have been inserted into sheep to produce faster-growing lambs, with increased feed efficiency for a net increase in productivity. Transgenic pigs have been produced by inserting in their genome a phytase gene that is expressed in salivary glands, enabling the swine to digest phosphorus more efficiently, which reduces the excretion of phosphorus into manure (Table 7.6).

Table 7.6 Transgenic Animals and the Stated Purposes for which they are being Developed

Type of Animal	*Modification*	*Intended End Use*
Cattle	Produce pharmaceutical proteins Increase disease resistance Increase milk and meat production Produce organs which are less likely to be rejected	Pharmaceutical production Human consumption Human consumption Xenotransplantation
Chickens	Produce pharmaceutical proteins in eggs	Pharmaceutical production
Goats	Produce pharmaceutical proteins Produce industrial proteins	Pharmaceutical production Production of new materials
Monkeys	Develop rhesus monkeys as research models for human diseases	Research
Pigs	Boost lean meat production Produce pharmaceutical proteins Create a research model for human diseases Decrease phosphorous content in pig manure to improve feed efficiency and environmental protection Produce organs which are less likely to be rejected	Human consumption Pharmaceutical production Research Human consumption Xenotransplantation
Sheep	Produce pharmaceutical proteins Increase milk and wool production	Pharmaceutical production Human consumption, production of fibre

(Contd.)

Type of Animal	*Modification*	*Intended End Use*
Fish	Increase growth rates and food conversion efficiency	Human consumption
	Increase disease resistance or cold tolerance	Human consumption, ornamental and companion animals
	Alter the colour of a fish under certain conditions	Ornamentals and companion animals, Environmental detectors of pollutants
	Engineer fish with sterile or deleterious genes	Biological control of non-indigenous species
	Produce pharmaceutical proteins	Pharmaceutical production

Production of biopharmaceuticals (animal 'pharming') has been achieved by generating transgenic animals that express novel proteins in secretions such as milk or seminal plasma. A range of polyclonal and monoclonal human antibodies, industrial biochemicals such as spider silk, and other proteins for treatment of human diseases have been generated through animal 'pharming'. The latest achievement has been the 'double-knockout pig', which may eventually lead to the production of organs and cells such as pancreatic islet cells to treat diabetes, or kidneys for transplantation into humans; and a knockout for prion protein to prevent infection and transmission of spongiform encephalopathies (e.g. Scrapie or BSE). Potential human health applications of this technology include producing biopharmaceuticals, and generating organs, tissues and cells for xenotransplantation. It is important to note, however, that the production capability of genetically selected and/or genetically modified animals will be realised only when their true genetic potential is attained through appropriate environmental and management measures.

Transgenic animals offer a plethora of benefits that will improve consumer health and nutrition, as well as animal welfare and productivity:

- Increased nutritional value.
- Quality assurance.
- Higher efficiency production.
- Stronger disease resistance.
- Improved animal welfare-less disease and longer lifespan.

7.7 TRANSGENIC ANIMALS FOR ADVANCING HUMAN HEALTH

For decades, animals have been used to produce human pharmaceuticals. Horses, pigs, rabbits and other species have been enlisted to produce such products as anti-venoms, biologics to prevent organ transplant rejection and the blood thinner heparin. Biotechnology now allows us to modify genes in these animals so that the drug proteins are more compatible with human biochemistry.

Animal production also offers the most efficient, practical way to produce certain drugs that are difficult to make in sufficient quantities using other methods. For example, animals can make human antibodies to deadly infectious diseases if they are modified with human immune genes.

Transgenic technology can also be used to make animal organs more compatible for transplant into humans, a process called xenotransplantation. Heart valves from pigs are already used to replace damaged valves in human hearts. If xenotransplantation could be perfected with the help of transgenics, hundreds of thousands of lives could be saved each year.

7.8 ANIMAL WELFARE AND THE ENVIRONMENT

Biotechnology animals and their products are extremely valuable, often fragile, and animal welfare is a priority for everyone working with these animals. Animal-made pharmaceuticals, for example, cannot be produced in sick animals and so every effort must be made to ensure animal welfare.

By making animal welfare a priority, transgenic technology (as well as other animal biotechnology) offers tremendous potential benefits for animals. Specifically, it can help cut animal mortality and disease, as well as reduce or eliminate painful standard agricultural practices like castration and dehorning.

Transgenic technology can also be used to mitigate environmental impacts of livestock production. The EnviroPig™, for example, dramatically lowers levels of phosphorus pollution. Such applications underscore the industry's commitment to environmental protection.

Most of these technologies are being developed in domesticated animal species. Since these animals live on farms and do not mate with wildlife, the risk to the environment is miniscule. Moreover, the transgenic animals used to produce pharmaceuticals are expressly excluded from use in the food supply. Scientists and the biotechnology companies who sponsor them use genetic engineering (GE) on farm animals in an attempt to achieve higher productivity (e.g.. meat, milk, wool), a modification of characteristic traits (e.g.. larger muscles, faster growth rates), an improved resistance to diseases, an improved uptake of nutrients or an improved adaptation to specific environmental conditions. The technology is also applied to laboratory animals so that these may be more 'useful' for animal testing and experimentation purposes, again by exhibiting the desired traits. Sheep, cattle, pigs, chickens and 35 species of fish have already been genetically modified in attempts to increase their production of milk and meat beyond the limits their bodies can bear without damage. They have also been engineered to secrete therapeutic protein products for human medicine; a process called 'gene pharming'.

Transgenic animals have been used for years to study gene function and to create models for the study of human diseases. This approach has become still more justified after the complete sequencing of several genomes. Transgenic animals are ready to become industrial bioreactors for the preparation of pharmaceuticals in milk and probably in the future in egg white. Improvement of animal production by transgenesis is still in infancy.

EXERCISE

1. What do you mean by transgenic animals?
2. Define transgene.
3. Explain how transgenic animal science developed.
4. Mention the necessity of transgenesis in animals.
5. Enumerate different methods of transgenesis.
6. Describe various methods used for transgenic animal production.
7. Give an account of lipofection and mention different lipids used in lipofection.
8. Briefly write about electroporation and add a note on its advantages and disadvantages.
9. Discuss the structure of gene-gun with its advantages.
10. Differentiate between ultrasonication and microlaser transfections.
11. Write about retroviral mediated method of transgenic animals production.
12. Elucidate how DNA microinjection method is used for animal transgenesis.
13. What are embryonic cells and how these cells are used for production of transgenic animals?

14. List out the differences of injection and co-culture methods employed for transgenic animals production.
15. Write a note on different vectors used in animal transgenics.
16. What are promoters and list out the common promoters used in animal transgenic techniques?
17. How transgenic mice helps in animal biotechnology?
18. Name different transgenic sheeps that are produced in animal biotechnology.
19. List the important gene transfers made in animals with their economic importance.
20. Enumerate transgenic animals applications in biotechnology.
21. Discuss how transgenic animals helps in advancing human health.
22. What is meant by animal pharming? Mention its advantages.
23. Explain how animal welfare and the environment can be improved by transgenic animals.
24. Describe the different transgenic fish that are produced for human welfare.
25. Elucidate various transgenic pigs produced in biotechnology.

Part B

Immunology

Chapter

8

Immune System and Immunity

8.1 INTRODUCTION

Immunology is the study of specific immunity and how the immune system responds to specific infectious agents. In simple way immunology can be defined as study of immunity to infection.

The vertebrate organisms have remarkable ability to fight against infections caused by varieties of organisms and cells. This resistance shown by the host towards infectious disease is known as immunity (*immunis* meaning "exempt", literally means free from burden).

8.2 THE IMMUNE SYSTEM AND IMMUNITY ALONG WITH HISTORICAL PERSPECTIVES

The immune system is similar to other body systems (respiratory or reproductive systems) in having number of different cell types, tissues and organs. Many of these cells organized into lymphoid organs or glands. The infection from microbes can occur at any site of the body, the immune system has moving force of cells in the blood stream. These cells are prepared to attack the invading microbe whenever it enters the body. Even though cells of immune system are separated from each other they maintain communication through cell contact and molecules secreted by them.

First reference of immunity was seen in 430 B.C. from **Thucydides**, when describing plague in Athens he wrote only those who have recovered from plague could nurse sick because they would not contract the disease second time. In fifteenth century the first recorded crude attempt to induce immunity done by Chinese and Turks. In 1718 **Lady Mary Wortley Montagu** by observing variolation (the process of inhaling the dried crusts from smallpox pustules or inserting into small cuts in the skin), she followed same method for her own children. English physician **Edward Jenner**, in 1798 made first scientific attempt of immunization. He extracted the contents of a pustule from the arm of cowpox-infected milkmaid and injected it to an eight-year-old boy. When he subsequently inoculated with smallpox, the boy showed no symptoms of the disease. This technique of inoculating with cowpox to protect against smallpox spread rapidly throughout Europe.

Louis Pasteur carried out further work on immunization. He showed that if culture of chicken cholera bacteria were allowed to age for two to three months, the bacteria cause mild cholera and allowed the chickens' recover. He then found that fresh cultures of bacteria failed to produce cholera that has been inoculated with old attenuated (weaken) cultures. Louis Pasteur called this attenuated culture as **vaccine** (vacca meaning "cow").

In 1881 **Louis Pasteur** extended the same method to anthrax bacillus (*Bacillus anthracis*). One group of sheep he vaccinated with heat attenuated bacteria and kept another group without vaccinated. When both group inoculated with the fresh bacteria, only vaccinated sheep lived whereas all unvaccinated sheeps died. This forms basis for beginning of immunology. Later in 1885 he vaccinated first human, Joseph Meister a young boy who had rabies. Later the boy recovered from the disease and became a custodian at the Pasteur Institute, Paris.

The research work of **Emil von Behring** and **Shibasaburo Kitasato** gave first insight into the mechanism of immunity earning Nobel Prize in 1901. They demonstrated that serum (liquid left over after coagulation) from animals previously immunized with diphtheria could transfer the immunity to unimmunized animals. **Elvin Kabat** in 1930s showed a fraction of serum known as gamma globulin was responsible for all these activities. This gamma globulin is now called **immunoglobulin** and active molecules in immunoglobulin is known as **antibodies**. As antibodies are present in the body fluid or humors this type of immunity is called humoral immunity.

Elie Metchinkoff in 1883 showed that certain white blood cells known as phagocytes were responsible for immunity by ingesting microbes and other foreign materials. Merrill Chase further confirmed this in 1940s by transferring immunity against the tuberculosis bacteria in guinea pigs.

Types of Immunity: There are two types immunity based on the immunity shown before exposure to foreign substances (antigen) or later. They include innate (non-specific) immunity and adaptive (specific) immunity.

8.3 INNATE (NON-SPECIFIC) IMMUNITY

It is the resistance that an individual possesses by virtue of his genetic make-up and inherently present in the host. It doesn't require prior contact with the microorganisms. This acts as first line of defense. The various non-specific defense mechanisms include physical barriers (skin, mucous membrane), chemical barriers (gastric juices, salivary juices), biological barriers (microflora) and general barriers.

8.4 ADAPTIVE (SPECIFIC) IMMUNITY

It is the resistance obtained by recognizing and selectively eliminating specific foreign microorganisms and molecules. This acts as second line of defense. It requires prior contact with the microorganisms and has long time memory. This leads to a more rapid removal of the microbe on its second and third time of entry.

Acquired immunity refers to the type of specific immunity a host develops after exposure to a suitable antigen or after transfer of antibodies or lymphocytes from an immune donor. Acquired immunity can be obtained actively or passively by natural or artificial means.

Naturally acquired active immunity occurs when an individual's immune system contacts an antigenic stimulus such as an infection during normal activities. The immune system responds by producing antibodies and sensitized lymphocytes that inactivate or destroy the antigen. The immunity produced can be either lifelong (e.g., chicken pox, measles) or last for only a few years (e.g., tetanus).

Naturally acquired passive immunity involves the transfer of antibodies from one host to another. For example, some of pregnant women's antibodies pass across the placenta to her foetus. If the female is immune to diseases such as polio or diphtheria, the placental transfer also gives the foetus and newborn immunity to these diseases. Certain other antibodies can pass from the female to her offspring through colostrum. This kind of immunity generally lasts only a short time.

Artificially acquired active immunity results when an animal is given an antigen preparation to induce the formation of antibodies and activated lymphocytes. A vaccine is a preparation of killed microorganisms or inactivated bacterial toxins that are administered to an animal to induce immunity artificially.

Artificially acquired passive immunity results when antibodies that have been produced either in an animal or by specific methods *in vitro* are introduced into a host. Although this type of immunity is immediate, it is short lived and lasts only a few weeks to a few months. For example, Botulinum anti toxin is produced in a horse and is given to a human suffering from botulism food poisoning.

Based on the type of materials or cells involved immunity can be classified into two types, namely, humoral (antibody–mediated) immunity and cellular (cell-mediated) immunity. Both these type are branches of specific immunity.

Humoral (antibody–mediated) immunity humor means body fluid. This type of immunity is obtained by the action of soluble proteins called antibodies that occur in the body fluids and on the plasma membrane of B lymphocytes. Circulating antibodies bind to bacteria, toxin or any foreign proteinaceous material and make them neutralized or marking them for degradation by various mechanisms.

Cellular (cell-mediated) immunity this type of immunity is obtained from the action of specific kinds of T lymphocytes that directly attack cells infected with viruses or parasites, transplanted cells or organs and cancer cells. T lymphocytes can lyse these cells or release chemicals (cytokines) that enhance specific and nonspecific defenses (phagocytosis, inflammation).

8.5 ORGANS AND CELLS OF THE IMMUNE SYSTEM

The immune system has many different types of organs and tissues that are present throughout the body. The blood vessels and lymphatic vessels connect these organs and forms functional system as a whole.

Organs of the Immune System

The organs involved in immune system can be divided into two types based on their function as primary and secondary lymphoid organs (Fig. 8.1).

Primary lymphoid organ: The primary (or central) lymphoid organs are thymus and bone marrow in case of mammals. They provide appropriate microenvironment for the development and maturation of lymphocytes (a type of white blood cells/leukocytes). The **thymus** is a flat, two lobed organ present above the heart. Each lobe is surrounded by a capsule and consists of number of lobules and separated by strands of connective tissue called **trabeculae** (Fig. 8.2). Each lobule is having outer cortex and central medulla. The cortex consists of more number of immature thymocytes or T cells and medulla contain only a few T cells. Both cortex and medulla are having number of epithelial cells, interdigitating dendritic cells and macrophages. Some of the epithelial cells present in the outer cortex known as nurse cells surround as many as 50 thymocytes. Thymus gland provides a site for T cells maturation and selection. T cells functions in cellular immunity of the body. Improper functioning of thymus results in decreased circulating T cells and cellular immunity with an increase in the infectious disease.

In mammals the **bone marrow** acts as primary lymphoid organ, whereas in birds a specialized organ called **bursa of Fabricius** (a gut associated lymphoid tissue) helps in maturation, division and diversification of B cells. B cells produce various types of antibodies and functions in the humoral immunity of the body.

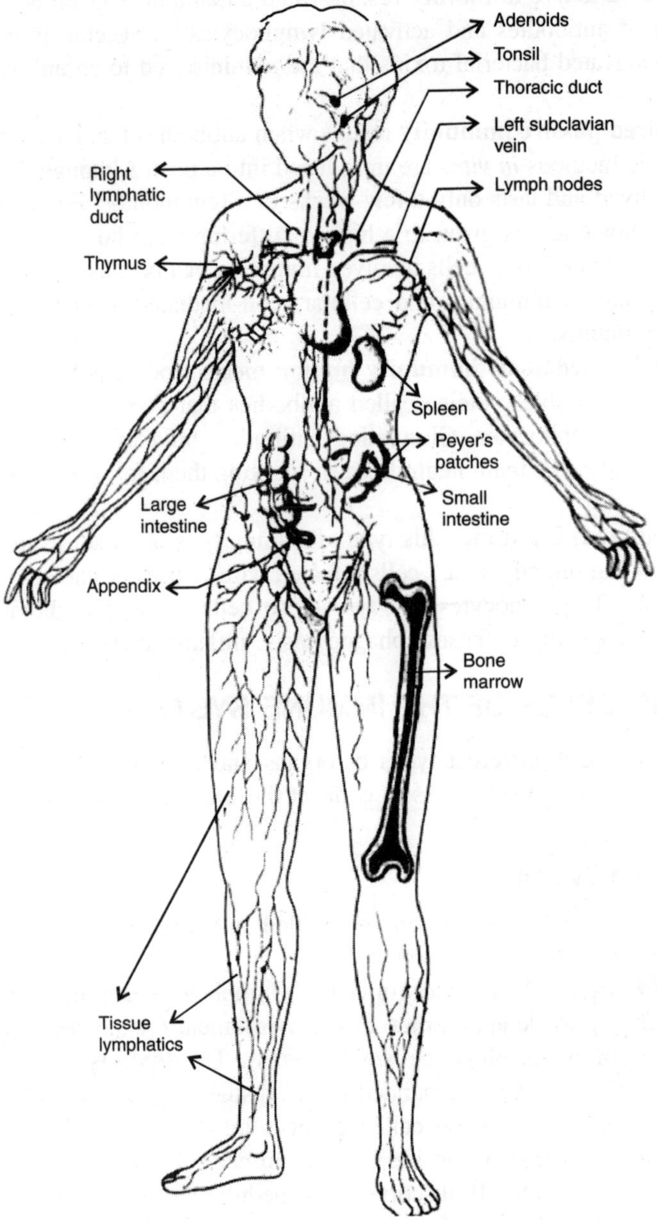

Fig. 8.1. The location of human lymphoid system in the body. The primary organs (bone marrow and thymus); secondary organs and tissues involved in immune response. These structurally and functionally diverse lymphoid organs and tissues are interconnected by the blood vessels (in figure it is not shown) and lymphatic vessels through which lymphocytes circulate.

Secondary lymphoid organ: The secondary lymphoid organs are lymph nodes, spleen and mucosal-associated lymphoid tissue (MALT). **The MALT is present in Peyer's patch (small intestine),**

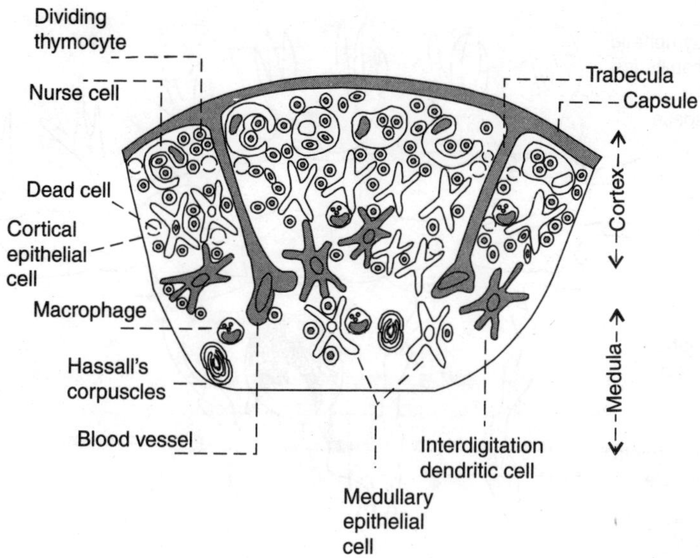

Fig. 8.2. Diagram showing cross-section of a portion of the thymus (primary lymphoid organ), indicating several lobules separated by connective tissue strands (trabeculae). The densely populated outer cortex is thought to contain many immature thymocytes (small), which undergo rapid proliferation coupled with an enormous rate of cell death. Also present in the outer cortex are thymic nurse cells (big), which are specialised epithelial cells with long membrane extensions that surround as many as 50 thymocytes. The medulla is sparsely populated and is thought to contain thymocytes that are more mature. During their stay within the thymus, thymocytes interact with various stromal cells, including cortical epithelial cells, medullary epithelial cells (tan), interdigitating dendritic cells, and macrophages). These cells produce thymic hormones and express high levels of class I and class II MHC molecules. Hassalls corpuscles, found in the medulla, contain concentric layers of degenerating epithelial cells.

tonsils, adenoids, appendix lining of upper airways, bronchi and genital tract. The **lymph nodes** are encapsulated bean shaped tissue consists of lymphocytes, macrophages and dendritic cells. Blood liquid part called **plasma** has the capacity to seeps out through the capillaries but some liquid portion remains in the surrounding tissue spaces, this fluid is called **lymph**. This lymph is collected by lymphatic capillaries and then by larger lymphatic vessels (Fig. 8.3). Lymphatic vessels finally attaches to the lymphnode by afferent lymphatic vessel. Lymphnodes anatomically having three regions (Fig. 8.4). The outermost layer cortex contains B cells, dendritic cells and macrophages. These cells aggregates together and forms primary follicle. After antigen exposure it converts into secondary follicle with a centre structure called germinal centre. Beneath the cortex is the paracortex having T cells and dendritic cells. In the center medulla, having plasma cells, which are engaged in antibodies production. The lymphnodes help in trapping the antigen from local tissues and presents to the T and B cells. The efferent lymphatic vessel carries out the antibodies from lymphnode.

The **spleen** is a big, oval organ present in the upper left abdominal cavity. It is having outer capsule and inner number of compartments separated by trabeculae (Fig. 8.5). Compartments are composed of outer red pulp containing many red blood cells (site of old and defective RBCs are removed) with macrophages and white pulp containing spleenic artery and T cells. The white pulp surrounds the spleenic artery and forms periarteriolar lymphoid sheath having large number of T cells. Both red and white pulp are separated by diffuse marginal zone rich in B cells. Spleen filter blood and traps blood-borne antigens.

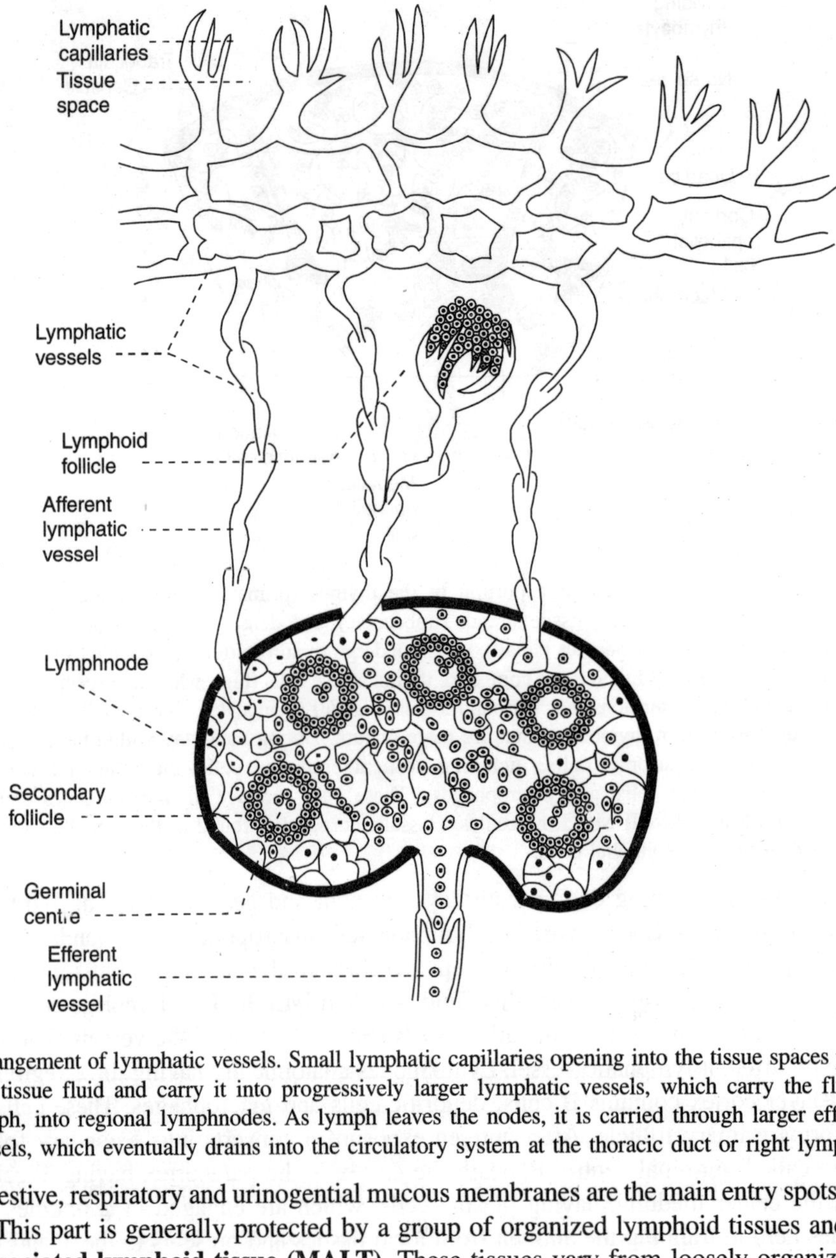

Fig. 8.3. Arrangement of lymphatic vessels. Small lymphatic capillaries opening into the tissue spaces pick up interstitial tissue fluid and carry it into progressively larger lymphatic vessels, which carry the fluid, now called lymph, into regional lymphnodes. As lymph leaves the nodes, it is carried through larger efferent lymphatic vessels, which eventually drains into the circulatory system at the thoracic duct or right lymph duct.

The digestive, respiratory and urinogential mucous membranes are the main entry spots for infectious pathogens. This part is generally protected by a group of organized lymphoid tissues and is named as **mucosal-associated lymphoid tissue (MALT)**. These tissues vary from loosely organized clusters of lymphoid cells in Lamina propria of intestinal villi to well developed tonsils, appendix and Peyer's patches (present in submucosa). MALT will be having large number of antibody producing plasma cells.

8.6 CELLS OF THE IMMUNE SYSTEM

Leukocytes or WBCs are the major cells involved in immune response. All blood and lymph cells are derived from stem cells, which are pluripotent (capacity to develop into many cell types) in nature (Fig. 8.6). These cells are known as hematopoietic stem cells. Later these cells differentiate into two

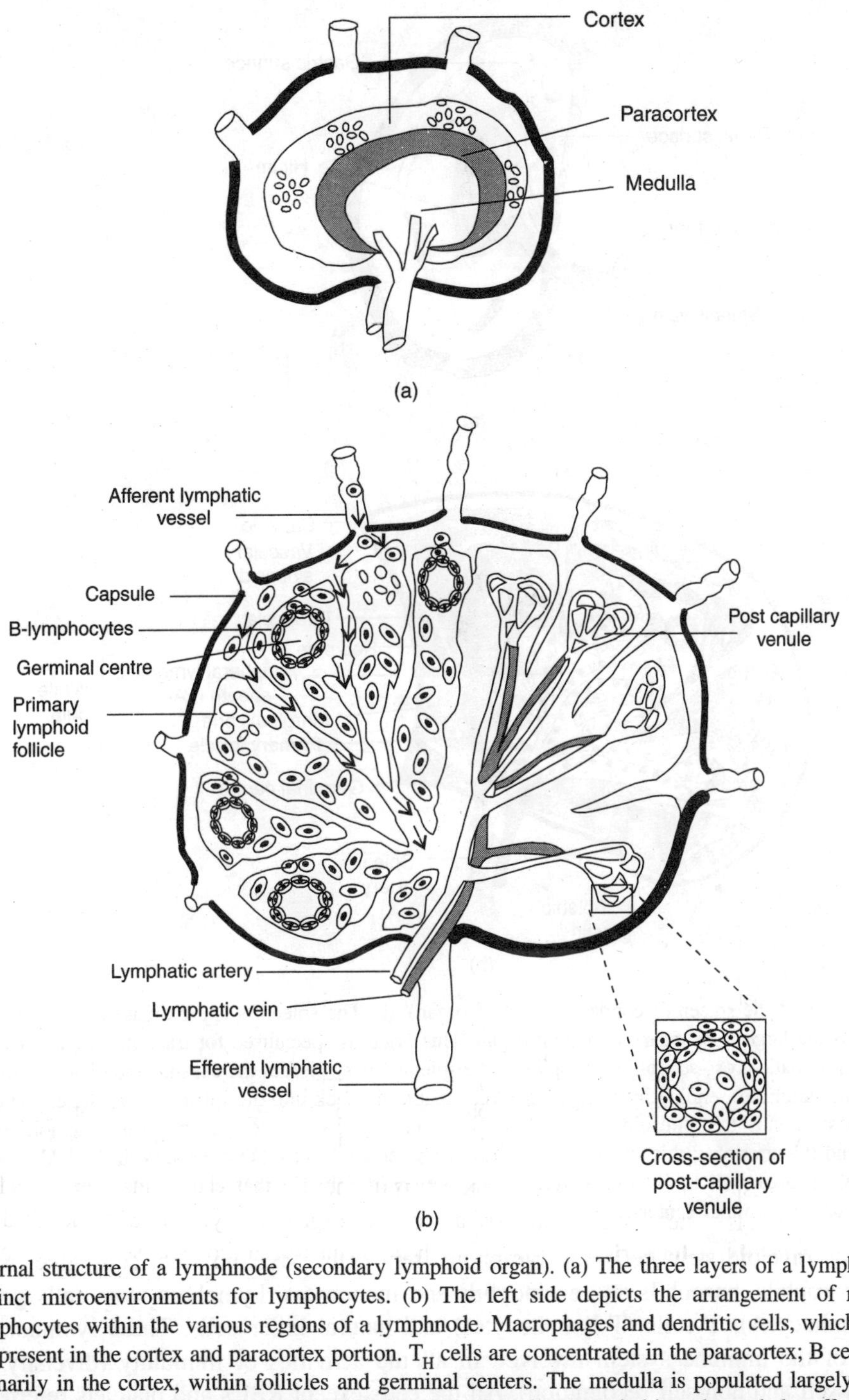

Fig. 8.4. Internal structure of a lymphnode (secondary lymphoid organ). (a) The three layers of a lymphnode support distinct microenvironments for lymphocytes. (b) The left side depicts the arrangement of reticulum and lymphocytes within the various regions of a lymphnode. Macrophages and dendritic cells, which trap antigen, are present in the cortex and paracortex portion. T_H cells are concentrated in the paracortex; B cells are located primarily in the cortex, within follicles and germinal centers. The medulla is populated largely by antibody-producing plasma cells. Lymphocytes circulating in the lymph are carried into the node by afferent lymphatic vessels; they either enter the reticular matrix of the node or pass through it and leave by the efferent lymphatic vessel. The right part of (b) shows the lymphatic artery and vein and the post-capillary venules. Lymphocytes in the circulation can pass into the lymphnode from the post-capillary venules by a process called extravastation (inset).

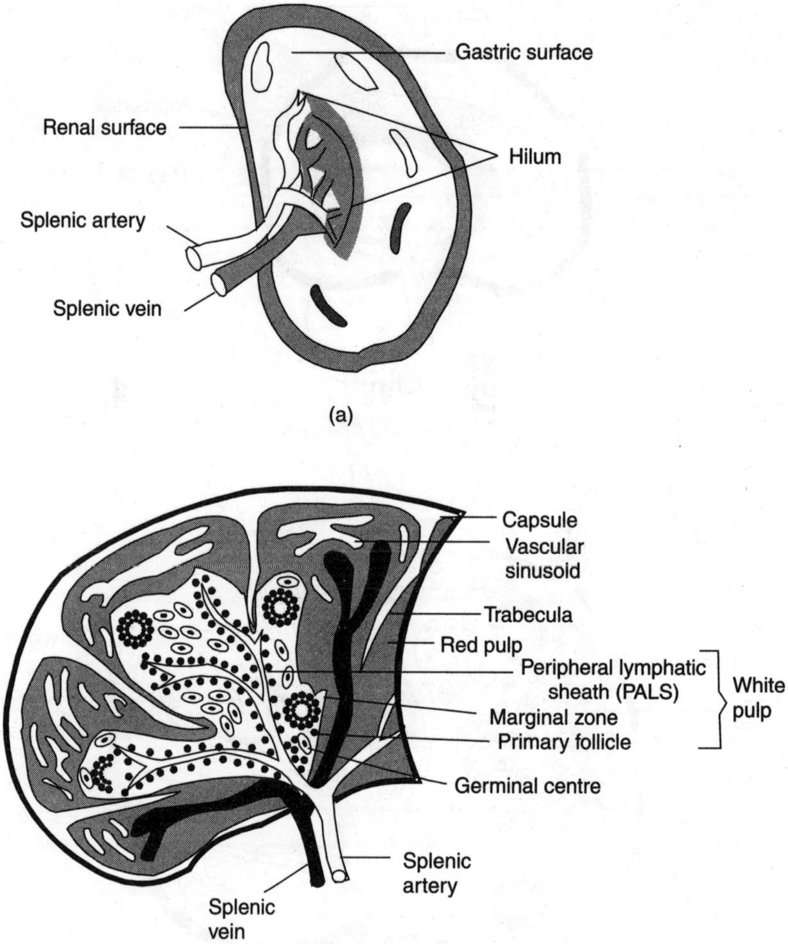

(b)

Fig. 8.5. Structure of the spleen (secondary lymphoid organ). (a) The spleen is approximately 5 inches long in adults, and is the largest secondary lymphoid organ. This organ is specialised for trapping blood-borne antigens. (b) Diagrammatic cross-section of the spleen. The splenic artery pierces the capsule and divides into progressively smaller arterioles, ending in vascular sinusoids that drain back into the splenic vein. The erythrocyte-filled red pulp surrounds the sinusoids. The white pulp forms a sleeve, the periarteriolar lymphoid sheath (PALS) around the arterioles; this sheath is made up of numerous T cells. Associated with the PALS is the marginal zone, an area rich in B cells that contains lymphoid follicles that can develop into secondary follicles containing germinal centers.

types, namely, myeloid stem cells and lymphoid stem cells. Myeloid stem cells give rise to RBCs, platelets, neutrophils, basophils and eosinophils and monocytes. Lymphoid stem cells give rise to B cells (B lymphocytes), T cells (T lymphocytes) and natural killer cells. These lymphocytes are the major cells of the immune system involved in all the activities of immunity (diversity, specificity, memory and self and non-self recognition). All the cells except RBCs and platelets referred as WBCs (leukocytes). The neutrophils, basophils and eosinophils are rich in granular materials in cytoplasm and hence they are known as granulocytes. The monocytes, B cells (B lymphocytes), T cells (T lymphocytes) and natural killer cells are not rich in granular materials and are known as agranulocytes.

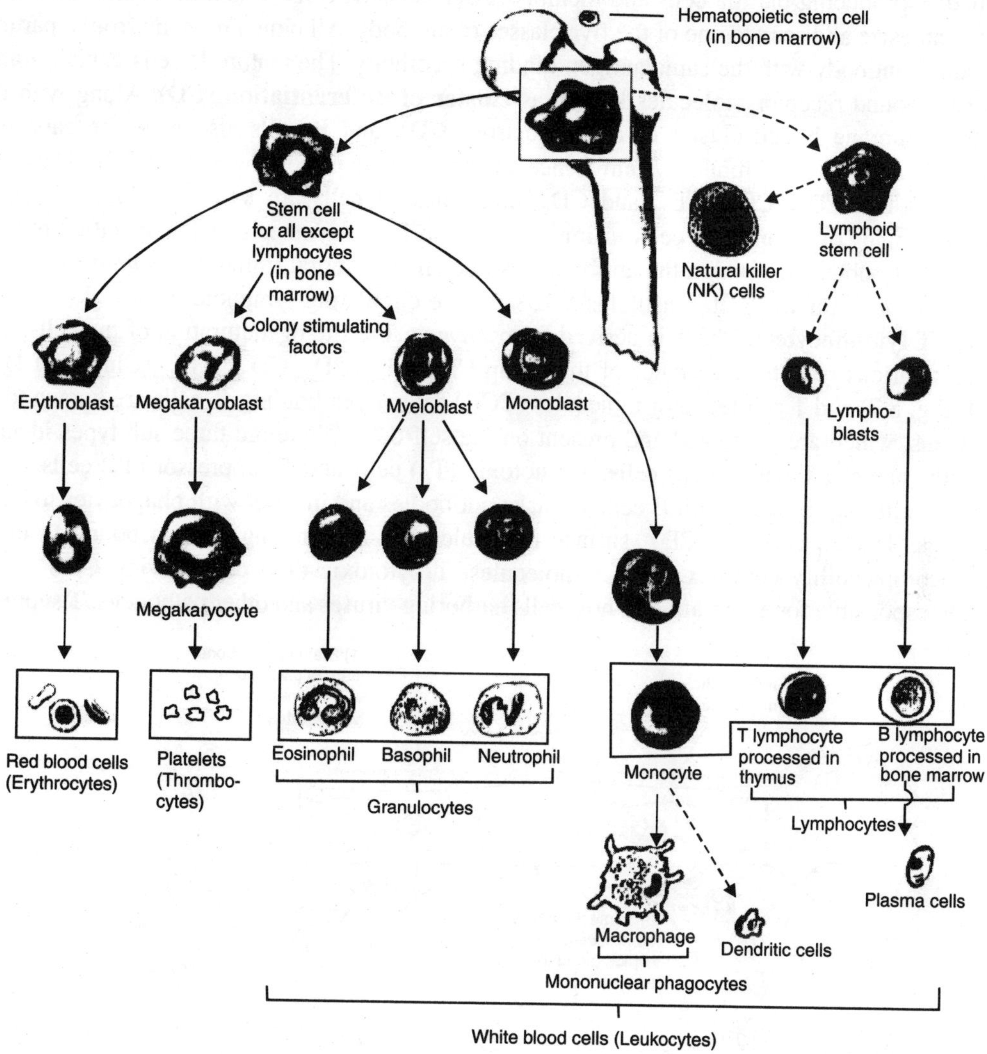

Fig. 8.6. The different types of blood cells in human. Pluripotent stem cells in the bone marrow divide to form two cells lineages: (1) the lymphoid stem cell give rise to B cells that become antibody-secreting plasma cells. T cells that become activated T cells, and natural killer cells. (2) The common myeloid progenitor cell gives rise to the granulocytes (neutrophils, eosinophils, basophils), monocytes that give rise to macrophages and dendritic cells, are as unknown precursor that gives rise to mast cells, megakaryocytes that produce platelets, and the erythroblast that forms erythrocytes (red blood cells). The dashed lines indicate steps that have not been very well studied.

8.7 AGRANULOCYTES

B cells (B lymphocytes): The B is derived from *bursa* of Fabricius in birds or bone marrow the site maturation of the cells (Fig. 8.7). This cell types is always having surfaced, membrane bound antibodies (~ 1.5×10^5 antibodies per B cell). These cells after coming in contact with the antigen differentiate into antibodies producing plasma cells and memory B cell. Plasma cells, which lack membrane-bound antibody, synthesize and secrete one of the five classes of antibody. All clonal progeny from a particular B cell produce antibody with the same antigen binding specificity. The mature B cells contain number of membrane bound receptor molecules known as **cluster of differentiation (CD).** Along with these number of membrane bound cluster of differentiation (**CD**), and B cells also have the capacity to presenting the antigen to immune system hence called antigen presenting cell (APC). These cells contain CD_{40}, CD_{45}, CD_{35}, CD_{21}, CD_{80} and CD_{86} molecules. B cells are known to bind the soluble antigens. The major function of B cells in immune system is the synthesis of the specific antibodies and the immunity obtained through the antibodies is known as humoral immunity (*humor* means *body fluid* contained by antibodies). It constitutes 5-15% of the circulating lymphoid cells.

T cells (T lymphocytes): The T is derived from *Thymus*, the site of maturation of the cells. These cells express various receptor molecules of the group CD_3, CD_4, CD_8, CD_{28}, CD_{45} (Cluster of Differentiation) (Fig. 8.7) and T cell receptor molecules (TCR) for antigen binding. T cells are known to bind to the antigens, which are processed and present on the self-cells. There are three sub types identified in the T cells, namely, T helper (T_H) cells, T cytotoxic (T_C) cells and T suppressor (T_S) cells.

T helper cells (T_H) interact with B cells to make antibodies and interact with phagocytes to destroy the pathogens. The T_H cells has CD_4^+ surface molecules and recognizing antigen bound to class II major histocompatability complex (MHC) molecules. T cytotoxic (T_C) cells having $CD8^+$ surface markers are responsible for elimination of host cells harboring viruses and other pathogens. T suppressor

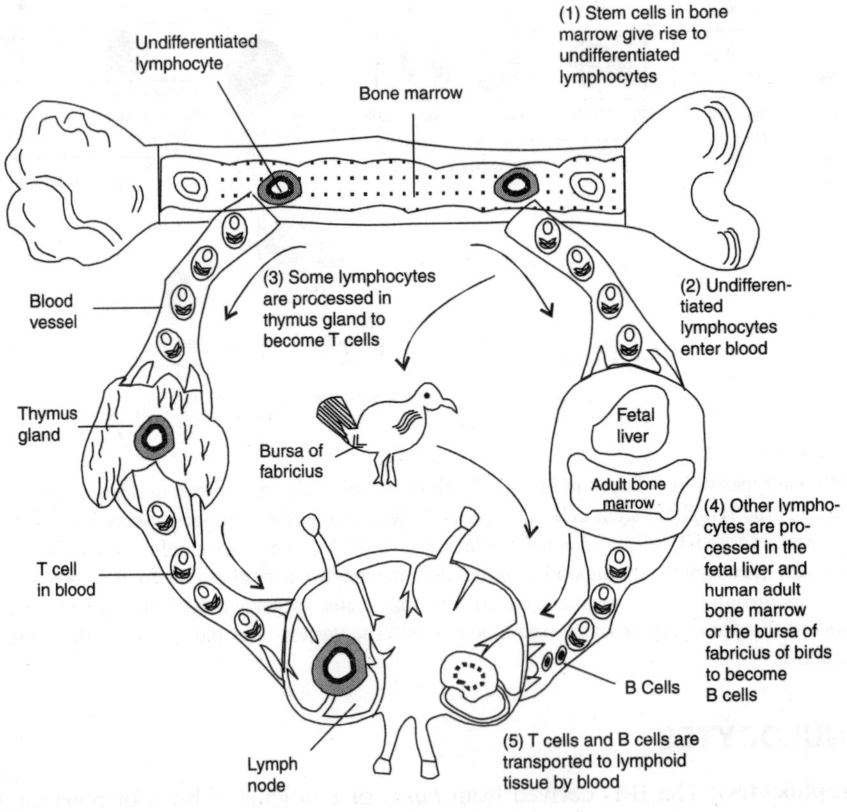

Fig. 8.7. Diagrammatic representation of lymphocyte development in lymphoid organs: Bone marrow produces undifferentiated lymphocytes, which, after processing develop into T and B cells.

(T_S) cells are known to suppress the functions of B and T cells, but still dispute is going on between immunologists regarding its exact function. The immunity obtained through the various T cells is known as cellular immunity.

Natural killer cells (NK cells): The lymphocytes that are not having any type of surface receptors and fail to synthesize antibodies are called null cells. Many of the null cells that are large, granular lymphocytes are called natural killer cells. It accounts for up to 15% of blood lymphocytes. These cells constitute 5%-10% of lymphocytes in human peripheral blood. The NK cells destroy tumor cells and virus infected cells. It also produces gamma interferon (proteins involved in protection against viral infections) and bears CD_2, $CD_{16,}$ CD_{56}, killer activatory receptor (KAR) and killer inhibitory receptor (KIR) molecules on the surface.

Monocytes: Monocytes are mononuclear cells and function as phagocytic cells in blood. During phagocytic activity they engulf particles including infectious agents and finally destroy them. These cells contain characteristic horse shoe-shaped nucleus and are larger than most circulating lymphocytes. Monocytes come out of the blood vessels and differentiate into **macrophages** in the specific tissues. Such as alveolar macrophages in the lung, histiocytes in connective tissues, kupffer cells in the liver, mesangial cells in the kidney, microglial cells in the brain and osteoclasts in bone (Fig. 8.8).

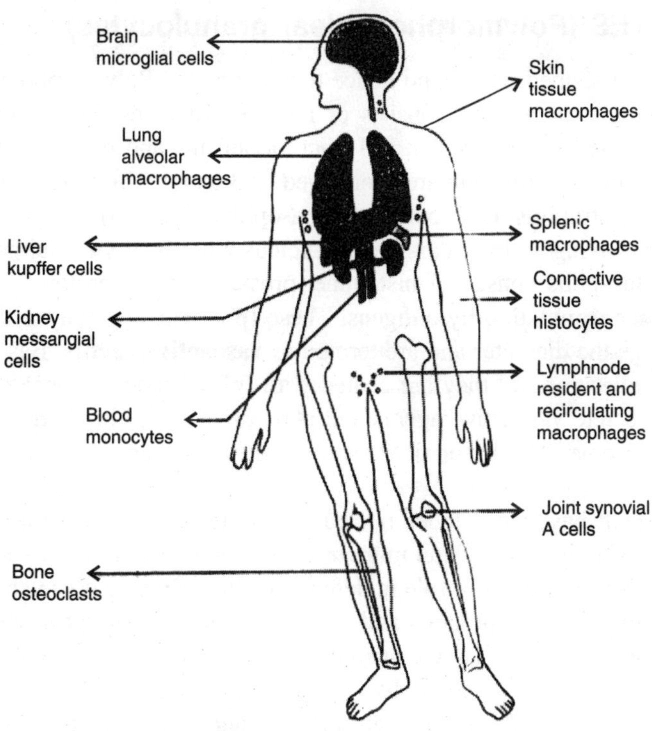

Fig. 8.8. The various Monocyte-Macrophage System in human body. This system consists of tissue (within the liver, spleen, and lymphnodes) having "fixed" or immobile phagocytes that have separate names depending on their location in the body.

Antigen presenting cells (APCs): These are heterogeneous group of leucocytes, which stimulate the immune response. Some of them induce the activity of T_H cells and few communicate with other leucocytes. Antigen presenting cells are primarily found in the skin, lymphnodes, spleen and thymus.

These cells after recognition of antigen they bind to the respective antigens and present to the immune system for further action. The Langerhans' cells, interdigitating cells, dendritic cells, B cells and macrophages can act as APCs, which are known to have class II MHC molecules. Follicular dendritic cells can differentiate between self cells and foreign cells. These APCs don't bear class II MHC molecules on the surface, but binds to antigen via CD21, CD35 (complement receptors that attaches complement associated immune complex) and Fc receptors. Recently another type of APC has been identified and they are called germinal centre dendritic cell (GCDC). They are involved in phagocytosis and activated by α-interferon, heat shock proteins and tumor necrosis factors that are released in response to microbial infection.

Null Cell

These are the population of lymphoid cells that do not have characteristics of either T or B cells. As they lack the specific surface markers of T or B cells, these are called null cells. Because these cells do not contain antigen binding receptor on surface, they lack exact immunologic specificity and memory. Null cells can be distinguished from them by the presence of cytoplasmic granules and large side. Most of the natural killer cells belong to the category of null cells.

8.8 GRANULOCYTES (Polymorphonuclear granulocytes)

Granulocytes have irregular shaped nuclei and hence they are called Polymorphonuclear granulocytes. The nucleus are of different shapes and consists of two to five lobes. The cytoplasmic matrix has granules that contain reactive substances that kill microorganisms and enhance inflammation. These cells are produced in the bone marrow and are short-lived (2-3 days). Three types of granulocytes have been identified, namely, neutrophils, eosinophils and basophils (Fig. 8.9).

Basophils: These are having an irregular-shaped nucleus with two lobes and granules stain bluish-black with basic dyes. Basophils consist of histamine, prostaglandins, serotonin leukotrienes in their granules and releases after stimulation by antigens. These (pharmacologically active) compounds act on blood vessels to change the diameter and are termed as **vasoactive agents**. Basophils have receptor for immunoglobulin E molecules and they are coated with IgE antibodies. Once coated, antigens can trigger the basophils to secrete vasoactive agents and play a major role in allergenic response such as asthma, hay fever and eczema. The basophils constitute the less than 1% of total circulating white blood cells.

Eosinophils: These cells have a two lobed nucleus connected by a slender thread of chromatin and granules stain red with acidic dyes. They can migrate from blood stream into tissue spaces and acts as phagocytic cells. Eosinophils play a major role in defense against parasitic organisms such as protozoan and helminthes by secreting cationic proteins and reactive oxygen metabolites which damage plasma membrane of the parasitic organisms. They constitute 1% to 3% of the total leukocytes.

Neutrophils: These cells have three to five lobed (multi lobed) nucleuses, hence also called poly morphonuclear leukocyte (PMN) and contain primary and secondary granules in the cytoplasm. Primary granules are made up of peroxidase, lysozyme and various lytic enzymes, whereas secondary granules have collagenase, lactoferrin and lysozyme. Thus they act as a major phagocytic and micro-bicidal cells. Neutrophils constitute 50% to 70% of the circulating leukocytes. These cells commonly first cells that arrive at the site of inflammation. Movement of circulating neutrophils into tissue through blood capillary is known as **extravasation**.

Mast cells: These are formed in bone marrow and found in connective tissue after differentiation. They contain granules with histamines and other active substances that contribute to the inflammatory response. Mast cells along with basophils play an important role in the development of allergies and hypersensitivities.

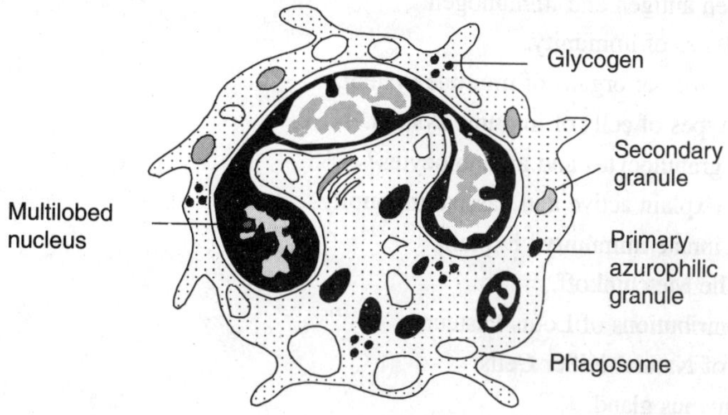

(a) Neutrophil

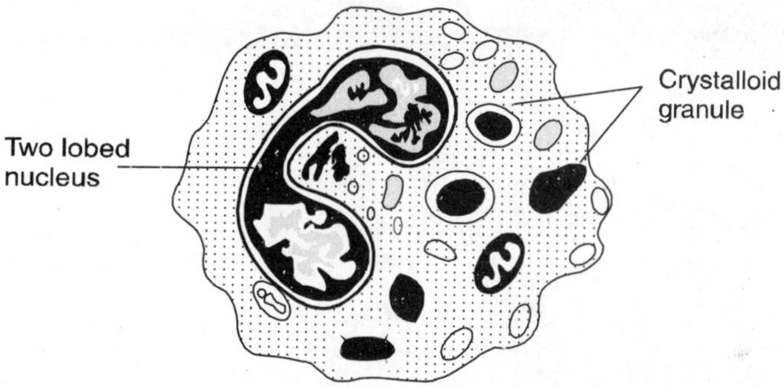

(b) Eosinophil

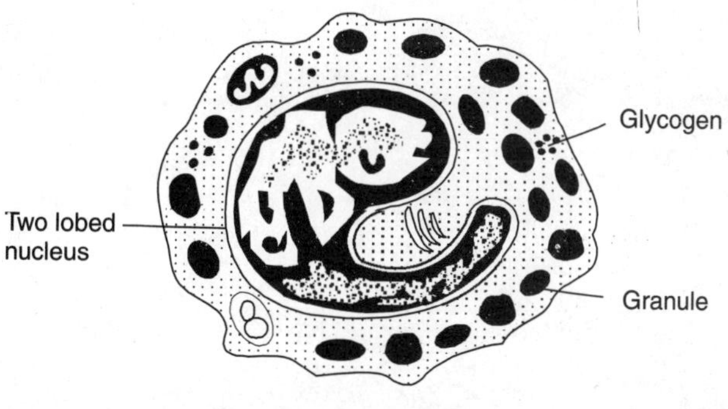

(c) Basophil

Fig. 8.9. Typical morphology of granulocytes. Notes differences in the shape of the nucleus and in the number and shape of cytoplasmic granules.

EXERCISE

1. Define immunity.
2. Distinguish between antigen and immunogen.
3. Explain different types of immunity.
4. Describe different types of organs of immune system. (BU April 2001, Oct. 2002)
5. Elucidate various types of cells of immune system.
6. What is meant by granulocytes and agranulocytes? (BU Oct. 2002)
7. Define immunity. Explain active and passive immunity. (BU April, 2002)
8. What is meant by innate immunity? Explain. (BU April, 2002)
9. Write a note on Elie Metchinkoff. (BU Oct. 2002)
10. Give a note on contributions of Louis Pasteur. (BU Oct. 2001)
11. Write an account of Natural Killer Cells. (BU April, 2001)
12. Briefly explain Thymus gland. (BU April, 2001)

Chapter

9

Antigen, Antibody and Their Structure

9.1 INTRODUCTION

An **antigen** (**anti**body **gen**erator) is defined as any substance, which induces immune system to produce an antibody. An antigen reacts specifically with antibody in an observable manner. Most antigens are large, complex molecules with a molecular weight generally greater than about 10,000 Daltons. Chemically antigens may be proteins, nucleoproteins, polysaccharides or glycolipids. The ability of a molecule to function as an antigen depends on its size and structural complexity.

Each antigen can have several **antigenic determinant sites** or **epitopes**. Epitopes are the regions or sites in the antigen that bind to the antigen binding site of a specific antibody or with a T cell receptor. Antibodies are formed most readily in response to determinant that project from the foreign molecule or to terminal residues of a specific polymer chain. The number of antigenic determinant sites (epitopes) on the surface of an antigen is its valence and it determines the number of antibodies binding with the antigen. If one epitope is present then the antigen is said to be monovalent, if two it is referred as bivalent and so on.

Many small organic molecules are not antigenic by themselves but can become antigenic, if they bond to a larger carrier molecules (such as protein). These molecules are referred as **haptens** (*haptein* means to grasp). Hapten cannot stimulate antibody formation, but can react with antibodies once formed for this reason they are also known as **partial antigen**. Penicillin is best example of hapten. Penicillin by itself is not antigenic, however when it combines with certain serum proteins of sensitive individuals it acts as antigen.

9.2 IMMUNOGEN AND ANTIGEN

Thus we can separate immunogen and antigen as different molecules. **Immunogen** is a molecule that provoke immune system to produce antibodies or other cellular response. Whereas **antigen** is a molecule which has the capacity to react with the antibody but cannot stimulate the antibody formation. Haptens are antigens, but not immunogens. Generally both immunogen and antigen are often used synonymously. **Immunogenicity** is the ability of the substance to induce a immune response to produce antibodies or cellular response and **antigenicity** is the ability of the substance to react with antibodies or cell surface receptors (cells of cellular mediated immunity). All molecules that have the property of immunogenicity also have the property of antigenicity. But all antigens are not immunogens.

Some compounds are known to act as immunostimulants i.e., enhance the immunogenicity of the antigens and such compounds are called **adjuvants**. Some of the example for adjuvants include aluminium hydroxide or phosphate, Freund's incomplete adjuvant (protein antigens in water oil emulsion) and Freund's complete adjuvant (Freund's incomplete adjuvant + Muramyl peptide). Adjuvants along with humoral responses they induces high degree of cellular immunity and prolongs the antigenic stimulus. Certain antigens provoke a drastic immune response by stimulating T cells non-specifically and such antigens are known as **superantigens**. The bacterial proteins are example for superantigens (staphylococcal enterotoxins).

Any substance to be an effective antigen is based on the following properties:

(*a*) Foreigness
(*b*) Molecular weight
(*c*) Chemical composition and heterogenicity
(*d*) Susceptibility to the degradation by tissue enzymes
(*e*) Genotype of the immune system
(*f*) Dosage and route of exposure
(*g*) Capacity to bind to carrier molecules

9.3 ANTIBODIES

Antibodies or **immunoglobulin (Ig)**, on the other hand, are produced rapidly in response to an antigen and can recognize and bind to the antigen that caused its production. Antibodies are chemically glycoproteins and present in the blood serum, tissue fluids and mucosal surfaces of vertebrate animals. Serum glycoproteins are separated by electrophoresis and gammaglobulin fraction consists of heterogenous mixture of immunoglobulins. These immunoglobulins differ from each other in molecular structure, charge, amino acid composition and carbohydrate content.

Antibody Structure

Immunoglobulins are large glycoproteins with molecular weights ranging from 150,000 to 900,000. Rodney R. Porter and Gerald M. Edelman discovered the immunoglobulin structure, for their work they got Nobel Prize in 1972. All immunoglobulins are having basic "Y" shaped structure and consists of two similar light chains (220 amino acids each, 25 kDa) and two similar heavy chains (about 440-450 amino acids each, 50-70 kDa) which are covalently interconnected by disulfide bond near the hinge region (Fig. 9.1). The stem region has polysaccharide side chain. This forms the monomeric structure of an antibody. Each arm of the Y shaped is composed of one light chain and one heavy chain, joined by the one disulfide bond. But stem consists of only two heavy chains joined by the two disulfide bonds. Both heavy (H) chain and light (L) chain contain two different regions. Constant regions ($\mathbf{C_L}$ and $\mathbf{C_H}$) have amino acid sequences that do not vary significantly between the antibodies of the same class. The variable regions ($\mathbf{V_L}$ and $\mathbf{V_H}$) from different antibodies do have different sequences. It is the variable regions ($\mathbf{V_L}$ and $\mathbf{V_H}$) that when folded together form the antigen binding sites.

The papain enzyme cleave each monomeric unit (Y-shaped structure) near hinge region into two fragments which bind to the antigen (Fab: antigen binding fragment) and one fragment which does not bind to antigen but undergoes crystallization (Fc: fragment crystallizable) (Fig. 9.1c). The Fc fragments are composed of only of constant regions ($\mathbf{C_H}$**2-3**), whereas the Fab fragments have both constant ($\mathbf{C_H}$**1**) and variable regions ($\mathbf{V_L}$ and $\mathbf{V_H}$). Even within variable region some part has more variation and these regions are referred as hyper-variable regions (Fig. 9.1c) or complementarity-determining regions.

Within each chain a loop is formed called domain by intrachain disulfide bonds consists of 60 amino acids.

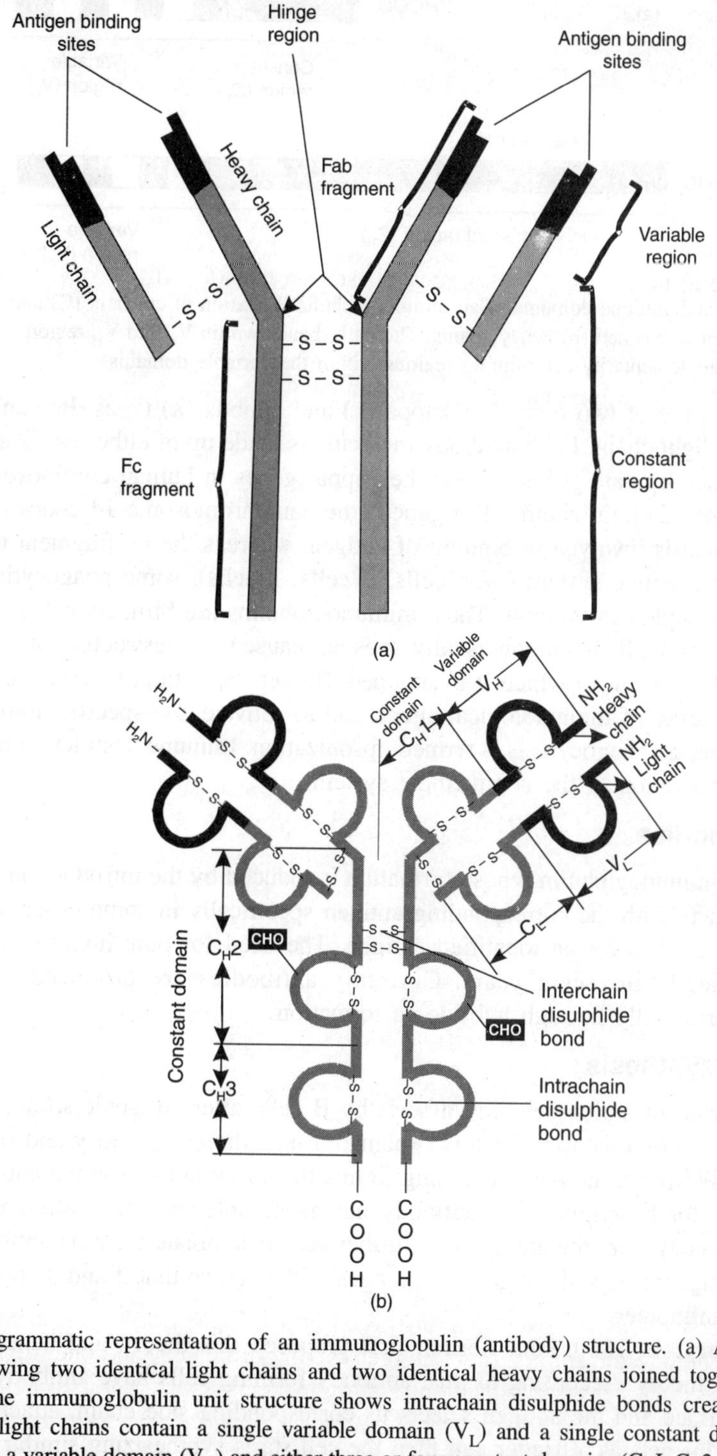

Fig. 9.1 (a & b). Diagrammatic representation of an immunoglobulin (antibody) structure. (a) An immunoglobulin molecule showing two identical light chains and two identical heavy chains joined together by disulphide bonds. (b) The immunoglobulin unit structure shows intrachain disulphide bonds create loops that form domains. All light chains contain a single variable domain (V_L) and a single constant domain (C_L). Heavy chains contain a variable domain (V_H) and either three or four constant domains (C_H1, C_H2, C_H3 and C_H4). The variable regions (V_H, V_L), when folded together in three-dimensions, form the antigen-binding sites

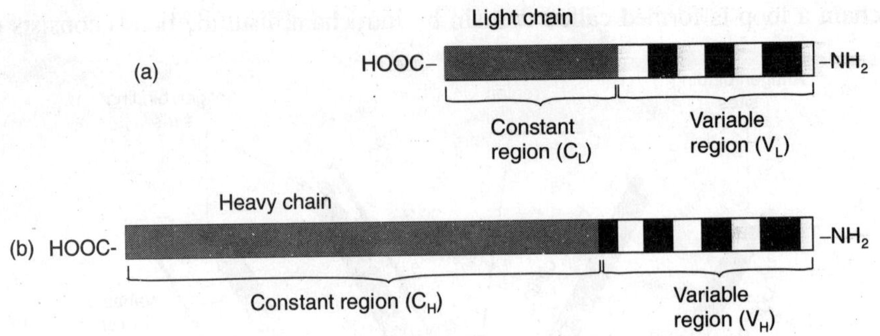

Fig. 9.1 (c). Constant and variable domains of an immunoglobulin. Location of constant (C) and variable (V) domains within (a) light chains and (b) heavy chains. The dark bands within V_L and V_H region represent hypervariable regions or complementarity-determining regions within the variable domains.

The light chains are of two types i.e., kappa (κ) and lambda (λ) types. Its constant region determines the type of a light chain. Each antibody molecule is made up of either two kappa or two lambda light chains. The two separate genes encode the kappa (genes in human chromosome 2) and lambda (human chromosome 22) light chains. The gene of human chromosome 14 codes the heavy chain.

The Fab fragment is involved in binding of antigen, whereas the Fc fragment helps in binding to various cells of the immune system (mast cells, B cells, T cells), some phagocytic cells, or the first component of the complement system. Thus, immunoglobulins are bifunctional in nature. The simple binding of an antibody with an antigen usually does not cause total destruction of the antigen or of the microorganism, cell or agent to which it is attached. Rather the antibody serves to mark and identify the target for wide array of immunological attack and to activate non-specific immune responses that can destroy the target by phagocytosis is termed opsonization. Immune destruction also is promoted by antibody-induced activation of the complement system.

Source of Antibodies

An antibody is an immunoglobulin whose formation is induced by the introduction of an antigen in an animal body. It reacts with the corresponding antigen specifically in some observable way, that each antibody has binding sites for an identified antigen. The need for pure homogeneous antibodies has increased dramatically in recent years. Currently antibodies are produced either naturally by immunization or artificially through hybridoma formation.

Antibodies Biosynthesis

The exact mechanism of antibodies biosynthesis by B cells after antigenic stimulation is completely understood. There are two theories for this explanation i.e., directive theory and selective theory.

Directive or Template theory: According to this theory an antigen in the antibody forming cells acts as a template for biosynthesis of antibody having complementary configuration. The antibody formed then dissociates from the antigen that further acts as template for next antibody synthesis. It is considered that antigen brings about genetic changes in the cell so that it and its progenies continue to produce specific antibodies.

Selective theory: In 1880, Paul Ehrlich first proposed this theory and hence is also known as Ehrlich side chain theory. According to this, antibody forming cells have antibody molecules as side chains on their surface and the antigen selects its corresponding side chain, attaches to it and finally knocks it down. This acts as a trigger and then the cell starts synthesizing similar side chains repeatedly, which are dislodged to free circulation and unite with the antigen.

9.4 CLASSES OF IMMUNOGLOBULINS (ANTIBODIES) AND THEIR SUBCLASSES

Immunoglobulins are classified into five types based on the antigenic character of heavy chains. These are Immunoglobulin G (IgG), Immunoglobulin M (IgM), Immunoglobulin D (IgD), Immunoglobulin A (IgA) and Immunoglobulin E (IgE). The heavy chains of an antibody are designated as γ heavy chain in IgG, μ in IgM, α in IgA, δ in IgD and ε in IgE. An individual person has all the five types of immunoglobulins (Table 9.1).

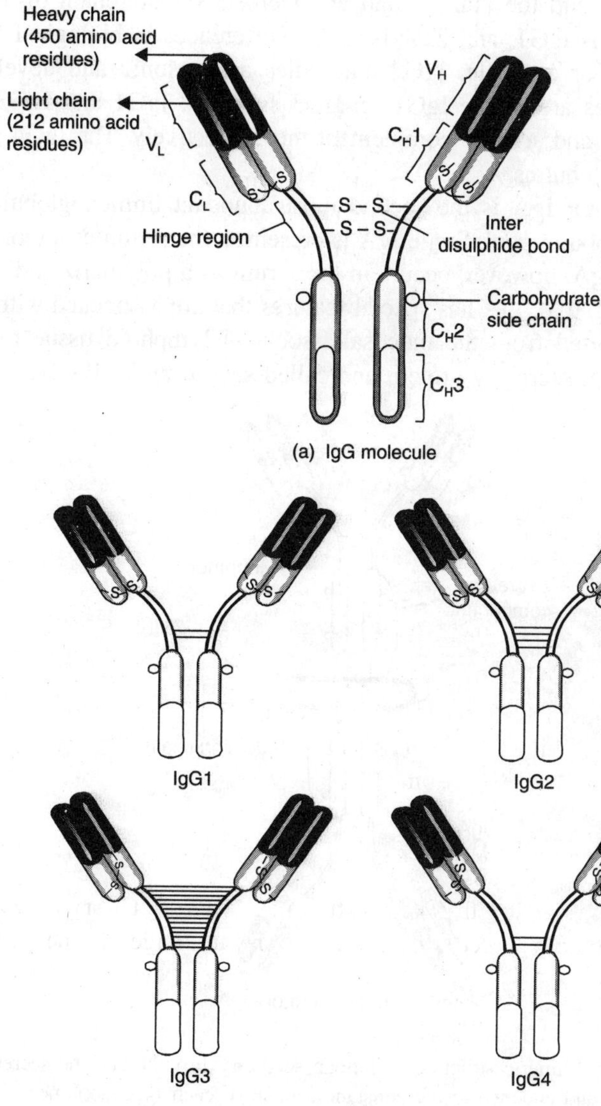

Fig. 9.2. Immunoglobulin G. (a) The basic structure of human IgG. (b) The structure of the four human IgG subclasses IgG1, IgG2, IgG3 and IgG3. Note the arrangement and numbers of disulphide bonds (shown as thin black lines). Carbohydrate side chain is represented by circles

Immunoglobulin G (IgG): IgG is a monomer unit (Fig. 9.2). This is the major immunoglobulin (most abundant) in human serum and makes up 70 to 75% of the total immunoglobulin pool. IgG is present in blood plasma and tissue fluids. The IgG class acts against bacteria and viruses by opsonizing the invaders and neutralizing toxins. It is also one of the two immunoglobulin classes that activated complement by the classical pathway. IgG is the only immunoglobulin molecule able to cross the placenta (hence termed as **maternal antibody**) and provide naturally acquired immunity for the new-born and persists for several months.

Totally there are four IgG subclasses (IgG1, IgG2, IgG3, and IgG4) that vary chemically in their (γ heavy) chain composition and the number and arrangement of interchain disulfide bonds. About 65% of the total serum IgG is IgG1, and 23% is IgG2. Differences in biological function have been noted in these subclasses. For example, IgG2 antibodies are opsonic and develop in response to antitoxins. Anti-Rh antibodies are of the IgG1 or IgG3 subclass. IgG1 and IgG3 also bind best to monocytes and macrophages and activate complement most effectively. The IgG4 antibodies function as skin sensitizing immunoglobulins.

Immunoglobulin A (IgA): IgA is the second most abundant immunoglobulin and account for about 15% of the immunoglobulin pool. Some IgA is present in the serum as a monomer of two heavy and two light chains. Most IgA, however, occurs in the serum as a polymerized dimer joined together by a polypeptide J chain (Fig. 9.3). IgA has special features that are associated with secretory mucosal surfaces. IgA, when transported from the mucosal-associated lymphoid tissue to mucosal surfaces, acquires a polypeptide termed secretory protein and called **secretory antibody**.

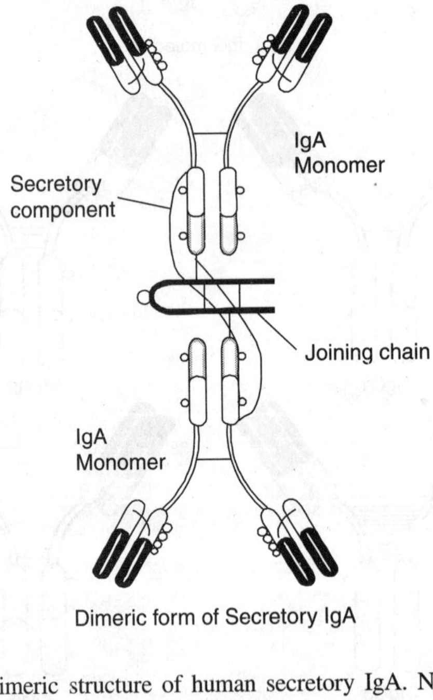

Fig. 9.3. Immunoglobulin A. The dimeric structure of human secretory IgA. Notice the secretory component wound around the IgA dimer and attached to the constant domain of each IgA monomer. Circles indicate carbohydrate side chain

The secretory IgA (sIgA), as the modified molecule is now called the primary immunoglobulin of the secretory immune system. This system is found in the gastrointestinal tract, upper and lower respiratory tracts, and genitourinary system. Secretory IgA is also found in saliva, tears, and breast milk (*colostrum*). In these fluid and related body areas, sIgA by the formation of an immune barrier plays a major role in protecting surface tissues against infectious microorganisms. For example, in breast milk sIgA (maternal antibodies) helps to protect nursing newborns. In the intestine, sIgA attaches to viruses, bacteria, and protozoan parasites such as *Entamoeba histolytica*. This prevents pathogen adherence to mucosal surfaces and invasion of host tissues, a phenomenon known as **immune exclusion**. In addition, sIgA binds to antigens within the mucosal lamina propria, and the antigen-sIgA complexes are excreted through the adjacent epithelium into the gut lumen. This rids the body of locally formed immune complexes and decreases their access to the circulatory system. Secretory IgA may neutralize viruses and other intracellular pathogens that reside within epithelial cells. Secretory IgA also plays a role in the alternate complement pathway.

Immunoglobulin M (IgM): IgM is the third most abundant immunoglobulin and accounts for about 10% of the immunoglobulin pool. It is usually a polymer of five monomeric units (pentamer) joined by disulfide bond (Fig. 9.4). The monomers are arranged in a pinwheel array with the Fc ends towards the center and consist of special J (joining) chain between two monomers. The heavy chain has fourth constant domain (C_{H4}). IgM is the first immunoglobulin made during B-cell maturation and the first secreted into serum during primary antibody response. Since IgM is so large, it does not leave the bloodstream or cross the placenta. IgM agglutinates bacteria, activates complement by the classical pathway, and enhances the ingestion of pathogens by phagocytic cells. These classes also contain special antibodies such as red blood cell agglutinins and heterophile antibodies.

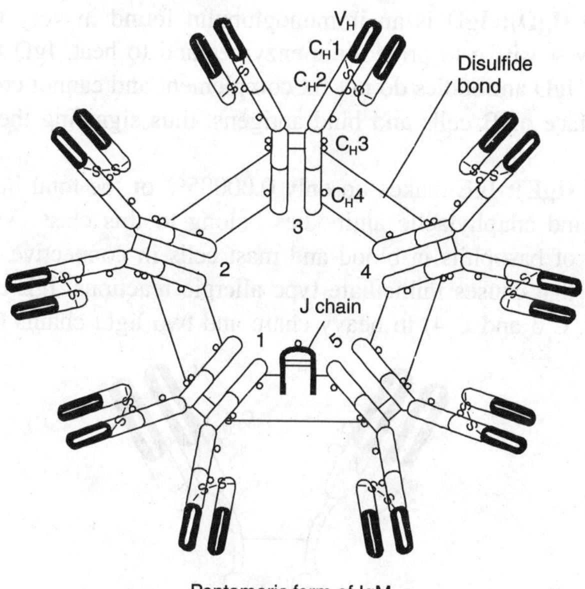

Fig. 9.4. Immunoglobulin M. The pentameric structure of human IgM. The disulphide bonds linking peptide chains are shown by line; carbohydrate side chains are indicated by dots. Note that 10 antigen-binding sites are present

Although most IgM appears to be pentameric, around 5% or less of human serum IgM exists in a hexameric form. This molecule contains six monomeric units but seems to lack a J chain. Hexameric IgM activates complement up to twenty fold more effectively than does the normal pentameric form. It has been suggested that bacterial cell wall antigens such as gram-negative lipopolysaccharides may directly stimulate B cells to form hexameric IgM without a J chain. If this is the case, the immunoglobulins formed during primary immune responses are less homogenous than previously thought.

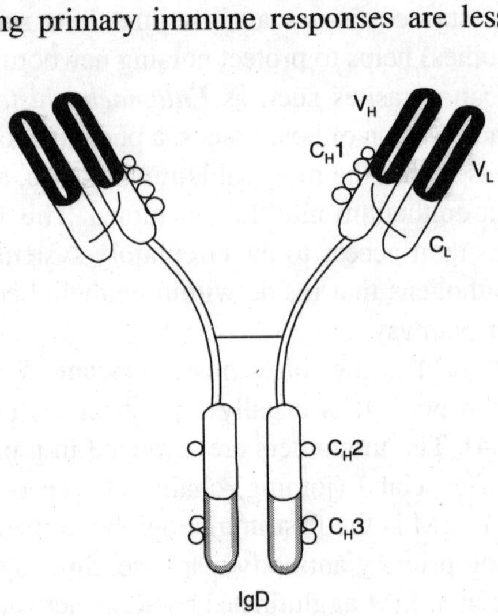

Fig. 9.5. Immunoglobulin D. The structure of human IgD. The disulphide bonds linking protein chains are shown by line; carbohydrate side chains are represented by circles

Immunoglobulin D (IgD): IgD is an immunoglobulin found in very low concentration in the blood serum. It is highly sensitive to proteolytic enzymes and to heat. IgD has a monomer structure similar to IgG (Fig. 9.5). IgD antibodies do not fix complement and cannot cross the placenta, but they are abundant on the surface of B cells and bind antigens, thus signaling the B cell to start antibody production.

Immunoglobulin E (IgE): IgE makes up only 0.00005% of the total immunoglobulin pool. The classic skin-sensitizing and anaphylactic antibodies belong to this class. Most IgE is bound to the cytoplasmic membranes of basophils in blood and mast cells in connective tissue, where it mediates the release of histamine that causes immediate type allergic reactions. IgE molecules have four constant domains ($C_\varepsilon 1$, $C_\varepsilon 2$, $C_\varepsilon 3$ and $C_\varepsilon 4$) in heavy chain and two light chains (Fig. 9.6).

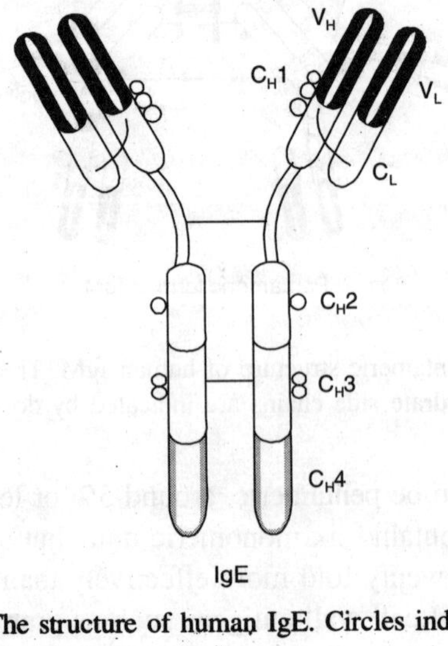

Fig. 9.6. Immunoglobulin E. The structure of human IgE. Circles indicate carbohydrate side chain

Table 9.1: Physicochemical Properties of Human Immunoglobulin Classes

Property	*Immunoglobulin Classes*				
	IgG[a]	*IgM*	*IgA*[b]	*IgD*	*IgE*
Heavy chain	γ_1	μ	α_1	δ	ε
Mean serum concentration (mg/ml)	9	1.5	3.0	0.03	0.00005
Percent of total antibody	80-85	5-10	5-15	<1	0.002-0.05
Valency	2	5 (10)	(2-4)	2	2
Mass of heavy chain (kDa)	51	65	56	70	72
Mass of entire molecule (kDa)	146	970	160[c]	184	188
Placental transfer	+	–	–	–	–
Half-life in serum (day)[d]	23	5	6	3	2
Complemental activation					
Classical pathway	++	+++	–	–	–
Alternative pathway	–	–	+	–	–
Induces mast cell degradation	–	–	–	–	+
Major characteristics	Most abundant Ig in body fluids; neutralises toxins, opsonises bacteria, activates complement, maternal antibody	First to appear after antigen stimulation; very effective agglutinator; expressed as membrane-bound antibody on B cells	Secretory antibody; protects external surfaces	Present on B-cell surface; B-cell recognition of antigen	Anaphylactic-mediating antibody: resistance to helminths
% carbohydrate	3	7-10	7	12	11

[a]Properties of IgG subclass 1.

[b]Properties of IgA subclass I.

[c]sIgA = 360 – 400 kDa

[d]Time required for half of the antibodies to disappear.

The Fc portion of the $C_ε4$ chain can bind to special Fc receptors on mast cells and basophils. When two IgE molecules on the surface of these cells are cross-linked by binding to the same antigen, the cells show degranulation. This degranulation releases histamine and other pharmacological mediators of anaphylaxis. It also stimulates eosinophils and gut hypermotility (increased rate of movement of the intestinal contents) that aid in the elimination of helminthic parasites. Thus, though IgE is present in small amounts, this class of antibodies has very potent biological capabilities.

9.5 THE BASIS OF ANTIBODY DIVERSITY

The most remarkable and unique property of antibodies is their diversity. According to current estimation each human or mouse can synthesize more than 10 million different kinds of antibodies. This remarkable diversity is generated due to rearrangement of antibody gene segments, somatic mutations and generation of different codons during antibody gene splicing.

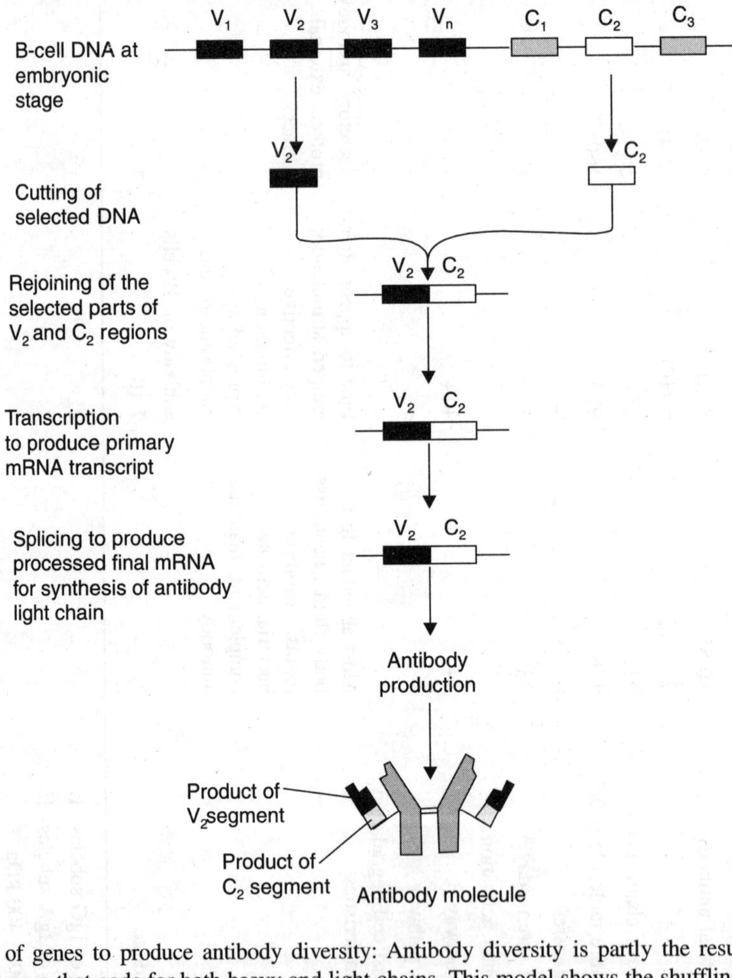

Fig. 9.7. Shuffling of genes to produce antibody diversity: Antibody diversity is partly the result of the shuffling of gene sequences that code for both heavy and light chains. This model shows the shuffling cutting, and splicing process used to produce an assembled light chain of the antibody molecule.

Immunoglobulin genes are split or interrupted genes having many exons. Embryonic B cells contain a small number of exons, close together on the same chromosome, that determine the constant

(C) region of the light chains (Fig. 9.7). Separated from them, but on the same chromosome, is a larger cluster of exons that determines the variable (V) region of the light chains. During B-cell differentiation, one exon for the constant region is spliced exon for the variable region. This splicing produces a complete light-chain antibody gene. A similar splicing mechanism also occurs to join the constant and variable exons of the heavy chains.

As the light chain genes actually consist of three parts and the heavy chain genes consist of four parts, formation of a final antibody molecule is slightly more complicated than previously mentioned. The germ line DNA for the light chain gene contains multiple coding sequences called V and J (joining) regions. During the differentiation of a B cell, a deletion (which is variable in length) occurs that fuse one V exon with one J exon. This DNA joining process is termed combinatorial joining since it can create many combinations of the V and J regions. When the light chain gene is transcribed, transcription continues through the DNA region that encodes for the constant portion of the gene. RNA splicing subsequently joins the V-J and C regions creating mRNA.

The formation of a heavy chain gene through combinatorial joining occurs by means of DNA splicing of the heavy chain counterparts of V and J along with a third set of D (diversity) sequences (Fig. 9.8). Initially, all heavy chains have the μ type of constant region. This corresponds to antibody class IgM. Another DNA splice joins the VDJ region with a different constant region that can subsequently change the class of antibody produced by the B cell.

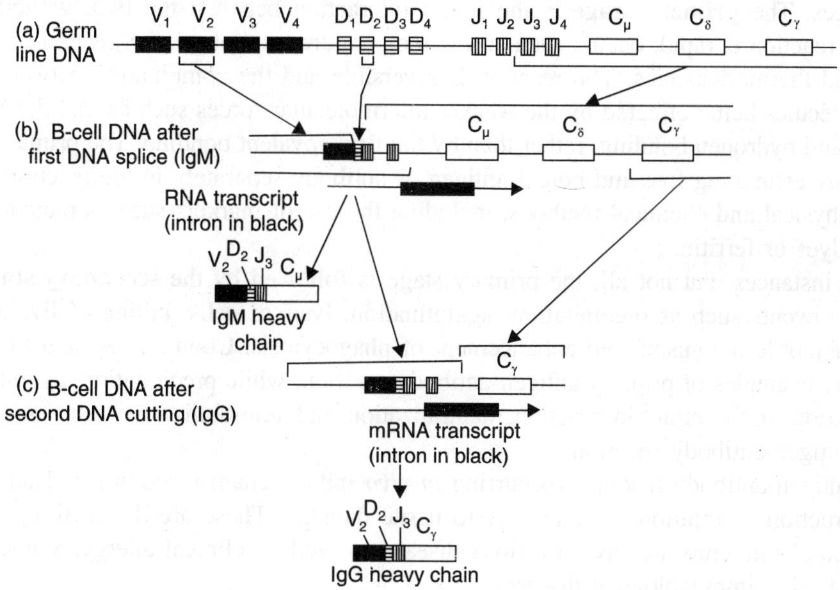

Fig. 9.8. The formation of a gene for the heavy chain of an antibody molecule

The kappa (κ) light chains in mouse are formed from combinations of about 250-350 V_κ and 4 J_κ regions giving a maximum of approximately 1,400 different κ chains. The O chains have their own V_O and J_O regions but smaller in number than their κ counterparts (6 different O chains). The heavy chains have approximately 250-1,000 V_H, 10-30 D, and 4 J_H regions giving maximum 120,000 different combinations. Because any light chain can combine with any heavy, there will be a maximum of 2×10^8 possible κ chain antibody types.

The value 2×10^8 different antibodies is actually an under estimate because antibody diversity is further augmented by two processes:

1. The V regions of germ line DNA are susceptible to high rate of somatic mutation during B cell development in the bone marrow. These mutations allow B cell clones to produce different polypeptide sequences.
2. The junction for either VJ or VDJ splicing in combinatorial joining can occur between different nucleotides and thus one VJ splicing event can join the V sequence CCTCCC with the J sequence TGGTGG in two ways:

CCTCCC + TGGTGG = CCGTGG,

which codes for the amino acids proline and tryptophan; and

CCTCCC + TGGTGG = CCTCGG,

which codes for proline and arginine. Thus the same VJ joining could produce polypeptides differing in a single amino acid.

9.6 ANTIGEN-ANTIBODY INTERACTIONS

The antigen-antibody interaction is a bimolecular association similar to an enzyme-substrate interaction, with an important distinction: it does not lead to an irreversible chemical alteration in either the antibody or the antigen. It therefore is, reversible. The reactions between antigens and antibodies occur in three stages. The **primary stage** is the initial interaction between the two, without any visible effects. This reaction is rapid, occurs even at low temperatures and obeys the general laws of physical chemistry and thermodynamics. The reaction is reversible and the combination between antigen and antibody molecules being effected by the weaker intermolecular forces such as Van der Waal's forces, ionic bonds and hydrogen bonding, rather than by the firm covalent bonding. The primary reaction can be detected by estimating free and bound antigen or antibody separately in the reaction mixture by a number of physical and chemical methods, including the use of markers such as radioactive isotopes, fluorescent dyes or ferritin.

In most instances, but not all, the primary stage is followed by the **secondary stage** leading to demonstrable events such as precipitation, agglutination, lysis of cells, killing of live antigens, neutralization of motile organisms and enhancement of phagocytosis. Coomb's test and fluorescent antibody tests are examples of primary antigen-antibody reaction, while precipitation, agglutination, complement fixation, toxin-antitoxin reaction, neutralization and immobilization tests are dependent on secondary antigen-antibody reaction.

Some antigen-antibody reactions occurring *in vivo* initiate chain reaction that lead to neutralization or destruction of injurious antigens, or to tissue damage. These are the **tertiary reactions** and include humoral immunity against infectious disease as well as clinical allergy, xenotransplantation rejection and other immunological diseases.

The exquisite specificity of antigen-antibody interactions has led to the development of a variety of immunologic assays. These assays can be used to detect the presence of either antibody or antigen and have played vital roles in diagnosing diseases, monitoring the level of the humoral immune response and identifying molecules of biological or medical interest. These assays differ in their speed and sensitivity, some are qualitative and others are quantitative. In the body, reactions between antigens and antibodies form the basis of antibody-mediated immunity to infectious diseases, or to tissue injury in some types of hypersensitivity and autoimmune diseases. In the laboratory, they help in the diagnosis of infections, in epidemiological surveys, in the identification of infectious agents and of non-infectious antigens such as enzymes. In general, these reactions can be used for the detection and quantification of either antigens or antibodies. Antigen-antibody reactions *in vitro* are known as serological reactions.

General Characteristics of Antigen-antibody Reactions

1. The reaction is specific, and antigen combine only with its homologous antibody and vice versa. The specificity, however, is not absolute and 'cross reactions' may occur due to antigenic similarity or relatedness.
2. An entire molecule reacts and not fragments. When an antigenic determinant present in a large molecule or on a 'carrier' particle reacts with its antibody, whole molecules or particles are agglutinated.
3. There is no denaturation of the antigen or the antibody during the reaction.
4. During combination only the surface antigens participate.
5. Combination is firm, but reversible. The affinity and avidity of the reaction influence the firmness of the union. Affinity refers to the intensity of attraction between antigen and antibody molecules. Avidity is the strength of the bond after the formation of antigen-antibody complexes.
6. Both antigens and antibodies participate in the formation of agglutinates or precipitates.
7. Antigens and antibodies can combine in varying proportions, unlike chemicals with fixed valencies. Both antigens and antibodies are multivalent. Antibodies are generally bivalent, though IgM molecules may have 5 or 10 combining sites. Antigens may have valencies up to hundreds.

Strength of Antigen-antibody Interactions

The noncovalent interactions that form the basis of antigen-antibody (Ag-Ab) binding include hydrogen bonds, ionic bonds-hydrophobic interactions and Van der Waals interactions.

Because the strength of each of these interactions is weak (compared with that of a covalent bond) a large number of such interactions are required to form a strong antigen-antibody interaction. Furthermore; each of these noncovalent interactions operated over a very small distance, generally about 1×10^{-7} mm (1 A) or a little less; consequently, a strong antigen-antibody interaction depends on a very close fit between the antigen and antibody. Such fits require a high degree of complementarity between antigen and antibody, and this requirement is the basis of the exquisite specificity that characterizes antigen-antibody interactions.

Antibody Affinity

The strength of the total noncovalent interactions between a single antigen-binding site on an antibody and a single epitope is the 'affinity' of the antibody for that epitope. Low affinity antibodies bind antigen weakly and tend to dissociate readily, whereas high-affinity antibodies bind antigen more tightly and remain bound longer. The association between a binding site on an antibody (Ab) with monovalent antigen (Ag) can be described by the equation

$$Ag + Ab \underset{k_{-1}}{\overset{k_1}{\rightleftharpoons}} Ag - Ab$$

where k_1 is the forward (association) rate constant
k_{-1} is the reverse (dissociation) rate constant

The ratio of k_1/k_{-1}, is the association constant K_a (i.e., $k_1/k_{-1} = K_a$), a measure of affinity.

Because K_a is the equilibrium constant for the above reaction, it can be calculated from the ratio of the molar concentration of bound antigen-antibody complex to the molar concentrations of unbound antigen and antibody at equilibrium as follows:

$$K_a = \frac{\{Ag - Ab\}}{\{Ab\}\,\{Ag\}}$$

The value of K_a varies for different antigen-antibody complexes and depends upon both k_1, which is expressed in units of liters/mol/sec (L/mol/s), and k_1, which is expressed in units of 1/second. For small haptens, the forward rate constant can be extremely high; in some cases, k_1 can be as high as 4×10^8 L/mol/s. For larger protein antigens, however, k_1 is smaller, with values in the range of 10^5 L/mol/s.

Antibodies Avidity

The affinity at one binding site does not always reflect the true strength of the antibody-antigen interaction. When complex antigens containing multiple, repeating antigenic determinants are mixed with antibodies containing multiple binding sites, the interaction of an antibody molecule with an antigen molecule at one site will increase the probability of reaction between these two molecules at a second site. The strength of such multiple interactions between a multivalent antibody and antigen is called the 'avidity'.

The avidity of an antibody is a better measure of its binding capacity within biological systems (e.g., the reaction of an antibody with antigenic determinants on a virus or bacterial cell) than the affinity of its individual binding sites. High avidity can compensate for low affinity. For example, secreted pentameric IgM often has a lower affinity than IgG, but the high avidity of IgM, resulting from its higher valence, enables it to bind antigen effectively.

9.7 PRECIPITATION REACTIONS

When a soluble antigen combines with its specific antibody in the presence of electrolytes (NaCl) at a suitable temperature and pH, the antigen-antibody complex forms an insoluble precipitate. This reaction is called precipitation. The precipitates thus formed by antibodies are known as precipitins. When the precipitate remains suspended as floccules, the reaction is called flocculation. Precipitation reaction may occur in liquid medium or in semisolid medium like agar gels, agarose or polyacrylamide.

Quantitative Precipitation in Liquid Medium

Antigen and antibody combine in varying proportions. When same amount of antiserum is mixed with increasing amounts of corresponding pure soluble antigen in test tubes, precipitation occurs most rapidly and abundantly in one of the middle tubes, in which the Antigen and the antibody are present in optimal or equivalent proportion. In the preceding tube, in which the antibody is in excess, and in the later tubes, in which the antigen is in excess, the precipitation will be weak or even absent. For a given antigen-antibody system, the optimal or equivalent ratio will be constant, irrespective of the quantity of the reactants. If the amounts of precipitate in the different tubes are plotted on a graph, the resulting curve will have three phases (Fig. 9.9):

(i) An ascending part (Prozone or Zone of antibody excess) (A), in which the uncombined antibody is present.

(ii) A peak (Zone of equivalence) (B), in which both antigen and antibody are completely precipitated and no uncombined antigen and antibody are present.

(iii) A descending part (post zone or zone of antigen excess) (C), in which all antibody has combined with antigen and some uncombined antigen is present.

This is called the zone phenomenon. The prozone is of importance in clinical serology, as sometimes sera rich in antibody may give a false negative precipitation of agglutination result, unless several dilutions are tested.

Mechanism of Precipitation

Marrack (1934) proposed the lattice hypothesis to explain the mechanism of precipitation. The mechanism is supported by considerable experimental evidence and is now widely accepted. According to this concept, multivalent antigens combine with bivalent antibodies in varying proportions, depending on the antigen-antibody ratio in the reaction mixture. Precipitation results when a large lattice is formed consisting of alternating antigen and antibody molecules. This is possible only in the zone of equivalence. In the zones of antigen or antibody excess, the lattice does not enlarge, as the valencies of the antibody and the antigen, respectively, are fully satisfied.

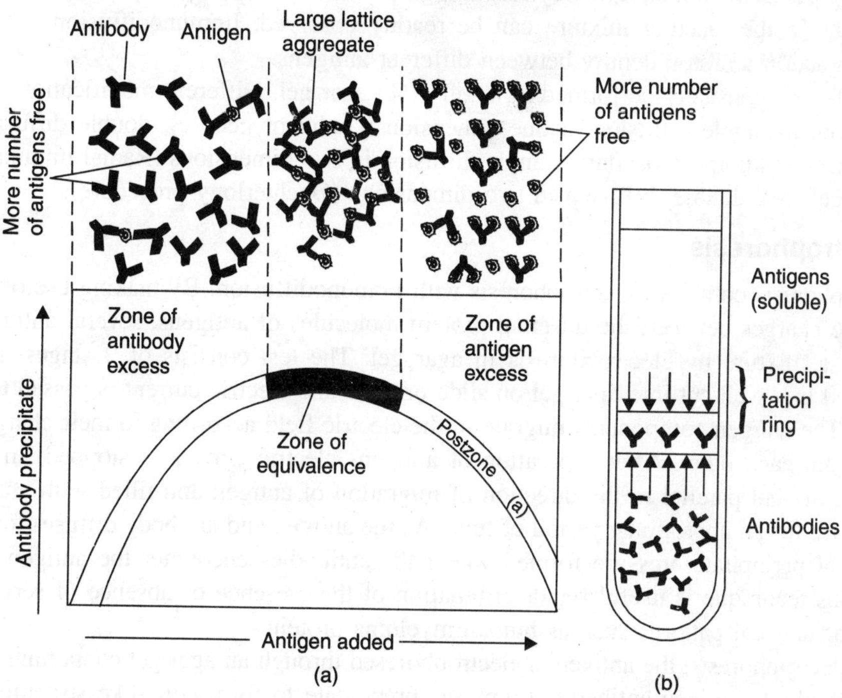

Fig. 9.9. Immunoprecipitation reaction. (a) Graph indicating a precipitation curve is based on the ratio of antigen to antibody. The zone of equivalence represents the optimal ratio for precipitation. (b) A precipitation ring test. Antibodies and antigens diffuse toward each other in a test tube. A precipitation ring is formed at the zone of equivalence

Applications of Precipitation Reactions

Precipitation reactions are very sensitive and can detect as little as 1 μg protein antigen. The precipitation test may be carried out either as a qualitative or quantitative test. The qualitative test is much more widely used to detect the presence of antigen and is of considerable value in:

1. Identification of bacteria, e.g., detection of group specific poly-saccharide of *Streptococcus* (Lancefields's groupings).
2. Identification of bacterial components in infected tissues, e.g., *B. anthracis* (Ascoli's thermoprecipitin test).
3. Detection of unknown antibody e.g., VDRL and KT in syphilis.
4. Forensic application in identification of human blood and seminal stains.
5. Standardization of toxins and antitoxins.
6. Testing for food adulteration.

9.8 IMMUNODIFFUSION (PRECIPITATION IN GEL)

Immunoprecipitation can form not only in solution but also in an agar matrix. When antigen and antibody diffuse towards one another in agar, or when antibody is incorporated into the agar and antigen diffuses into the antibody containing matrix, a visible line of precipitation will form. The reaction is visible as a distinct band of precipitation, which is stable and can be stained for preservation, if necessary. As each antigen-antibody reaction gives rise to a line of precipitation, the number of different antigens in the reacting mixture can be readily observed. Immunodiffusion also indicates identity, cross reaction and nonidentity between different antigens.

Immunodiffusion is usually performed in a soft (1%) agar gel. Different modifications of the test are available such as single diffusion in one dimension (Oudin procedure), double diffusion in one dimension (Oakley-Fulthorpe procedure), single diffusion in two dimensions (Radial immunodiffusion or Mancini procedure), double diffusion in two dimensions (Ouchterlony procedure).

Immunoelectrophoresis

Immunoelectrophoresis combines electrophoresis with immunodiffusion. By making use of the differences in surface charges between the different protein molecules of antigens, several antigens can be separated from a mixture by electrophoresis in agar gel. The test consists of 2 stages. In stage 1, antigen is placed in a well cut into agar gel on slide or plate and electric current is passed through the agar (1-2 hrs). The antigen components migrate in the electric field according to their charge and size and separate from each other. After separation of antigens electric current is stopped. In stage 2, a trough is cut in the gel parallel to the direction of migration of antigen and filled with antibody, and allowed to diffuse for an appropriate period of time. As the antigen and antibody diffuses towards each other, a series of precipitate arcs are formed where the antibodies encounter the antigen in optimal proportions. This technique is useful for determination of the presence or absence of serum proteins and detection of unusual patterns such as human myeloma protein.

In rocket electrophoresis, the antigen is electrophoresed through an agar gel containing Ab. This is a quantitative method. Antigen-antibody complexes precipitate to form cone-like structures (appearance of a rocket), the length of which corresponds to the concentration of antigen.

Counter Immunoelectrophoresis (CIE): It relies on movement of antigen and antibody in opposite direction. The antigen carrying a negative charge migrates towards the positive pole and the positive charged antibody towards the negative pole and vice versa. Antigen (specimen) and the homologous antibody are placed in separate wells in agar gel and through which electric current is passed. Line of precipitation appears at a point where the two meet in 30 to 60 min This test is faster and 10 times more sensitive than double diffusion technique. CIE is useful in detection of hepatitis B Antigen and antibodies, DNA Abs in SIE etc.

9.9 AGGLUTINATION REACTIONS

When an antibody combines with a particulate antigen in presence of electrolytes at optimal pH and temperature, it results in visible clumping of particles. This is called agglutination. Antibodies that produce such reactions are called agglutinins.

Agglutination is more sensitive than precipitation for the detection of antibodies. Agglutination reactions are similar in principle to precipitation reactions; they depend on the cross linking of polyvalent antigens. Agglutination reaction occurs when antigen and antibody are mixed in optimal proportions. Just as an excess of antibody inhibits precipitation reactions, such excess can also inhibit agglutination reactions; this inhibition is called the prozone effect.

Because prozone effects can be encountered in many types of immunoassays, understanding the basis of this phenomenon is of general importance. Several mechanisms can cause the prozone effect. First, at high antibody concentrations, the number of antibody binding sites may greatly exceed the number of epitopes. As a result, antibodies can bind antigen only univalently instead of multivalently. Antibodies (Abs) that bind univalently cannot cross link one antigen to another. Prozone effects are readily diagnosed by performing the assay at a variety of antibody (or antigen) concentrations.

When one is using polyclonal Abs, the prozone effect can also occur for another reason. The antiserum may contain high concentrations of Abs that bind to the antigen but do not induce agglutination; these antibodies, called 'incomplete' or 'monovalent' antibodies are often of the IgG class. At high concentrations of IgG, incomplete Abs may occupy all of the antigenic sites, thus blocking access by IgM, which is a good agglutinin. This effect is not seen with agglutinating monoclonal Abs.

Slide Agglutination Test

This is a widely used method to detect unknown antigen (bacteria, red blood cells etc.). A fine suspension of antigen is made in saline on a slide. Small drop of antiserum is added and mixed properly and allowed to react for a minute. If specific antigens are present then it forms clumps within a few minutes. This method is routinely used in blood grouping, cross matching and bacterial identification.

Tube Agglutination Test

Serum is serially diluted in saline in a series of tubes. An equal volume of antigen suspension (e.g. 0.5 ml of cell suspension) is added to all the tubes and incubated for known period. The highest dilution of serum at which agglutination occurs is recorded as antibody titer. This method is used for determination of antibody titer. It is routinely used for serological diagnosis of typhoid, paratyphoid, brucellosis and typhus fever.

Radioimmunoassay (RIA)

In this method generally antibodies are labeled with radioactive substances such as Iodine (I^{125} or I^{131}). This technique was first developed by two endocrinologists Berson, S.A. and Rosalyn Yallow (1960) for the determination of insulin and anti-insulin complexes in diabetics. They introduced RIA in clinical medicine in 1961 and since then it has been successfully utilized for quantitative assay of all hormones, drugs and Hbs antigen. For this pioneering work Yallow was awarded Nobel Prize in 1977. This is one of the most sensitive techniques used for detecting antigen and antibody in the sample.

Immunofluorescence

Coons and his coworkers (1942) showed that fluorescent dyes can be conjugated to antibodies and such labeled antibodies can be used to locate and identify antigens in tissues. Generally used fluorescence compounds are fluorescein isothiocyanate that shows blue green colour and lissamine rhodamine that impart orange red fluorescence.

Enyzme Linked Immunosorbent Assay (ELISA)

It is developed by Engvall and Perlmann and van Weeman and Schuurs independently in 1971. ELISA is a highly sensitive test used to detect pathogens, toxins, hormones or any antigenic agents at picogram levels in the sample. First antigen is attached to micro titre plate through antibody bounded on plate surface. Then enzyme conjugated (linked) antibodies specific for antigen was added. This specifically binds with the respective antigen. The most commonly used enzymes are horse radish peroxidase, alkaline phosphatase and β-galactosidase. Later substrate (colourless) solution was added. If antigen present in the sample enzyme linked antibodies binds tightly and will not be washed out in the washing step. After incubation period enzyme acts on colorless substrate and produces coloured products that can be visualized and measured.

EXERCISE

1. Define antibody and write a note on structure of antibody molecule.
2. Mention different types of antibodies.
3. Describe antigen-antibody interactions.
4. What do you mean by affinity and avidity?
5. Define precipitation and explain precipitation reaction.
6. What is meant by agglutination? Mention its types. (BU Oct. 2002)
7. What is meant by radio-immunoassay (RIA)? (BU Oct. 2001)
8. Write a note on immunoelectrophoresis. (BU Oct. 2001)
9. Explain properties of different immunoglobulins. Add a note on their significance. (BU April 2001)
10. Give an account of any two antigen-antibody interactions. (BU Oct. 2001) (BU April 2003)
11. Write a note on epitopes. (BU April 2001) (BU April 2003)
12. Mention effect of adjuvants. (BU Oct. 2002)
13. Briefly write a note on zone phenomenon. (BU Oct. 2002)
14. What are the factors that influence a compound to be an ideal antigen? (BU Oct. 2001)
15. Describe epitope in brief. (BU April 2001) (BU April 2003)
16. Write a short note on ELISA.

Chapter

10

Humoral (Antibody-Mediated) Immunity

10.1 INTRODUCTION

Humoral immunity depends first on the ability of B cells (B lymphocytes) to recognize specific antigens and second on their ability to initiate responses that protect the body against foreign agents (antigens). In most cases the antigens are present on surfaces of infecting microorganisms. The most common response is the production of antibodies that will inactivate an antigen and lead to destruction of infecting microorganisms.

All B cells contain specific antibody/receptor on its membrane and can bind immediately to a specific antigen. This antigen binding activates or sensitizes the parent B cell and results in division of the cell to form more number of cells. Most of the resulted cells act as **plasma cells** and a few cells act as **memory cells** (Fig. 10.1). Plasma cells are large B lymphocytes that synthesize and release antibodies specific for a particular antigen. The activated plasma cells can produce as many as 2000 antibodies per second. The plasma cells have short life span of five to seven days. The memory cells can remain active for years or decades to instantly stimulate antibody production. These memory cells help in response to second and subsequent exposures to an antigen much more rapidly than to the first exposure.

10.2 ACTIVATION OF B CELLS

The activity of B cells can be influenced by certain T lymphocytes. If a B cell encounters large antigens with many epitopes, B cells generate plasma cells and antibodies without the help of T cells (Fig. 10.2). But most of the infective organisms have only few epitopes. In such cases T helper cells assist B cells by processing of antigens, so that B cells can respond to them. Macrophages also present antigens to B cells. The suppressor T cells block the activity of T helper cells and thus stop the antibodies production by the B cells.

10.3 MOLECULAR EVENTS IN ACTIVATION OF B CELLS

The B cells directly produce antibodies when exposed to large number of antigens. In this method bacterial lipopolysaccharides (TI-1; **Thymus-independent antigen**-1) or bacterial flagellin/cell wall

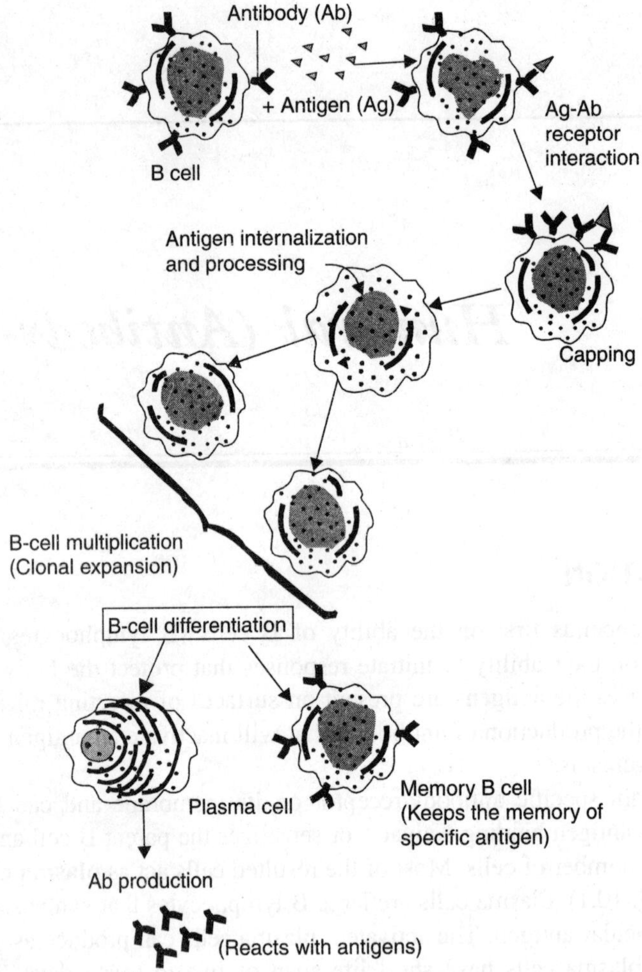

Fig. 10.1. The immune system can respond specifically to a myriad of possible antigens entering into body, whether they are individual molecules or are attached to pathogens and abnormal cells such as cancer cells. B cells or B lymphocytes constantly roam the body, particularly the blood and lymphoid tissues. Each B cell synthesises only one of the millions of possible antibodies and displays this antibody on its surface. When the antigen meets a B cell having a surface antibody of the proper specificity (top left), it complexes with the antibody and capping occurs. (Capping is the regional aggregation of antibodies on the surface of the cell following Ag-Ab interaction). The antigen is then internalised; the B cell swells and begins to divide rapidly, producing a B-cell clone. The activated B-cell clone differentiates into plasma cells and memory cells. Plasma cells form the specific antibody that immediately attacks the antigen that provoked its formation. Memory B cells persist in the body and boost the immune system's readiness to eliminate the same antigen if it presents itself in the future. Most antibody responses also require signals from T helper cells (not shown).

polysaccharides (TI-2) acts as an activator for B cells and both the signals are produced (Fig. 10.3). Here T cells will not help the B cells hence these antigens are referred as thymus-independent antigens. In case of **thymus dependent** activation first signal produced by cross linking of surface immunoglobulins on surface of B cells by thymus dependent (TD) antigen and second one by the interaction with T helper cells through CD40 (on the B cell) and CD40L (on T_H cells) with assistance by cytokines (Fig. 10.4).

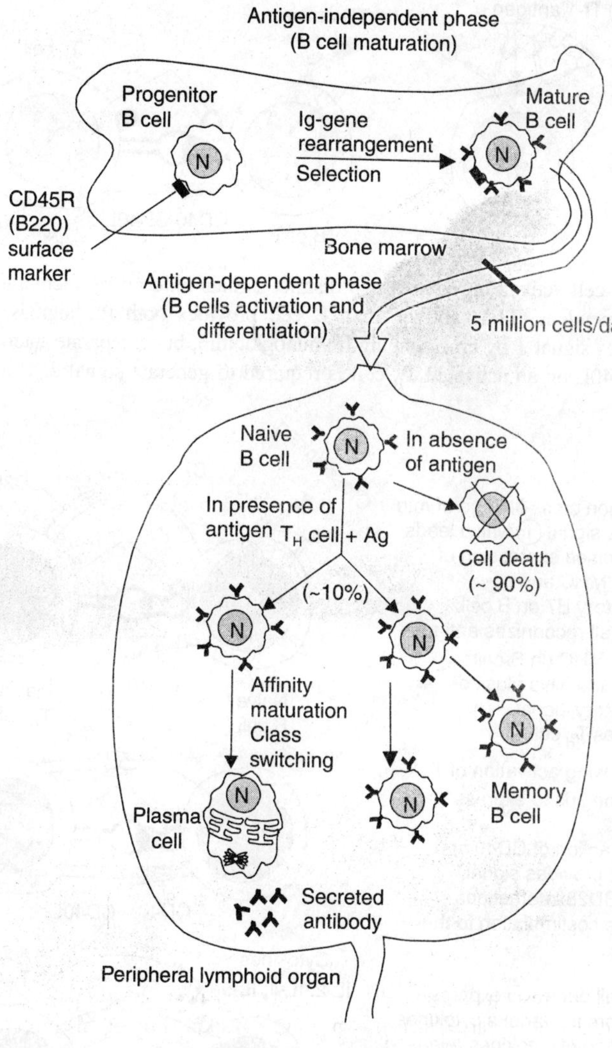

Fig. 10.2. An overview of B-cell development in immune system. During the antigen-independent maturation phase, immunocompetent B cells expressing membrane IgM and IgD are generated in the bone marrow. Only about 10% of the potential B cells reach maturity and exit the bone marrow. In the absence of antigen-induced activation, naive B cells in the periphery die within a few days. In the presence of soluble protein antigen and activated T_H cells, B cells are activated and proliferate within secondary lymphoid organs. Those bearing high affinity membrane-bound immunoglobulin differentiate into plasma cells and memory B cells, which may express different isotypes because of class switching. The numbers cited refer to B-cell development in the mouse, but the overall principles apply to humans as well.

10.4 PRIMARY AND SECONDARY RESPONSES

The initial exposure of antigen to B cells results in production of antibodies by plasma cells and within few days antibodies begin to appear in the blood plasma. This is known as **primary response** and antibodies concentration increases over a period of 1 to 10 weeks (Fig. 10.5). The first antibodies produced are IgM molecules, which can attack antigen directly. Then concentration of IgG production

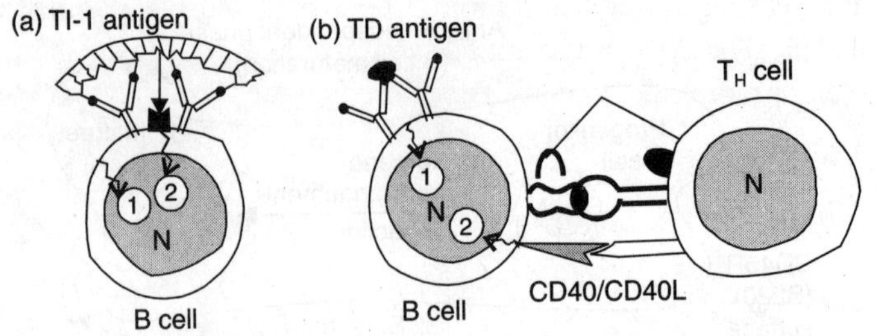

Fig. 10.3. An effective B-cell activation requires two distinct signals induced by membrane events. Binding of a type 1 thymus-independent (TI-1) antigen to a B cell provides both the signals. A thymus-dependent (TD) antigen provides signal 1 by cross linking immunoglobulin, but a separate interaction between CD40 on the B cell and CD40L on an activated T_H cell is required to generate signal 2.

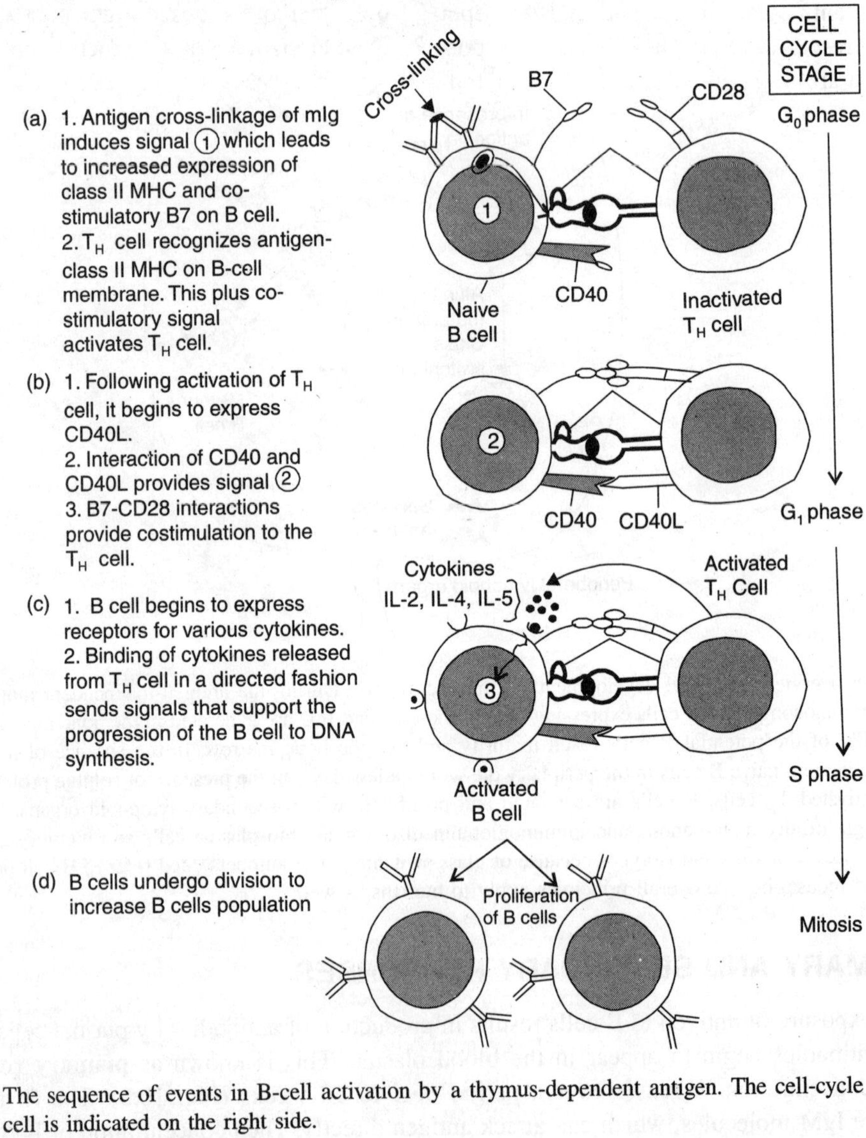

Fig. 10.4. The sequence of events in B-cell activation by a thymus-dependent antigen. The cell-cycle phase of the B cell is indicated on the right side.

increases by B cells. Ultimately concentration of both IgM and IgG can become so less that cannot be detected in the plasma. The type of the antibody molecule produced by the plasma cells depends upon the type of the cytokines binds to the cells. Depending on the type of cytokines plasma cells differentiate to produce different class of antibodies (Fig. 10.5). The concentration of antibodies reaches highest around 10-17 days. But memory cells remain in lymphoid tissues. They generally do not participate in the primary response, whereas they retain their ability to recognize a particular antigen. These cells survive without dividing for many months to many years.

When the same antigen recognized by the memory cells in later stage results in the antibody production at a rapid rate. This is known as **secondary response or anamnestic response.** The presence of memory cells makes the secondary response much faster than the primary response. During this some memory cells divide rapidly to produce plasma cells and others remain as memory cells. Plasma cells synthesize and release rapidly large quantities of antibodies. Similar to primary response, IgM are the first antibodies produced and later replaced by larger quantities of IgG molecules than in the primary response. Thus, the secondary response is characterized by a rapid increase in antibodies most of which are IgG (Fig. 10.5). The concentration of antibodies reaches highest in around 2 to 7 days.

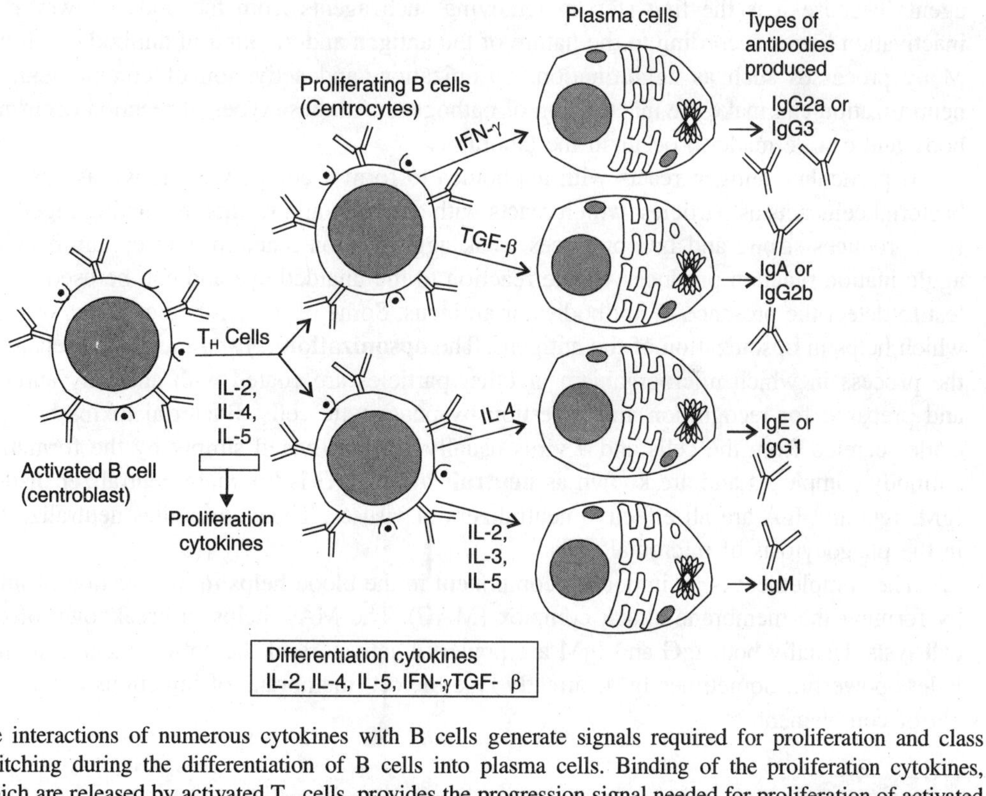

Fig. 10.5. The interactions of numerous cytokines with B cells generate signals required for proliferation and class switching during the differentiation of B cells into plasma cells. Binding of the proliferation cytokines, which are released by activated T_H cells, provides the progression signal needed for proliferation of activated B cells. The effects of the indicated cytokines have been demonstrated; however, similar or identical effects may be mediated by other cytokines. Class switching in the response to thymus-dependent antigens also requires the CD40-CD40L interaction, which is not shown here

Humoral immunity based on antigen-antibody reactions is most useful in giving the protection against bacterial infections. It is also known to neutralize toxins and viruses that have not yet invaded host cells. The defensive mechanism of humoral immunity depends on recognizing antigens associated with pathogens. For the bacteria to colonize on surfaces or for viruses to infect host cells, these pathogenic agents must first attach to surfaces. The IgA antibodies in tears, nasal secretions, saliva and other body fluids react with antigens on the pathogenic microorganisms. They coat bacteria and viruses and prevent them from adhering to mucosal surfaces.

Microorganisms that escape IgA invade tissues and encounter IgE in lymphnodes and mucosal tissues. Gut-associated lymphoid tissues (GALT) release large quantities of IgE, which bind to mast cells to release histamine and other substances that initiate and accelerate the inflammatory process. Inflammatory process means accumulation of large quantities of blood, complement factors, macrophages/phagocytes and antibodies to the site of infection, which results in swelling, redness, pain and heat.

The microbes that have reached lymphoid tissue without being recognized by B cells are acted on by macrophages and presented to B cells. Thus, macrophage acts as antigen presenting cells. Then B cells bind the antigens and produce antibodies with the help of T helper cells. Antibodies binding with antigen on the surfaces of microbes form antigen-antibody complexes.

Antigen-antibody complexes formation is an important component of the inactivation of infectious agents because it is the first step in removing such agents from the body. However, the means of inactivation varies according to the nature of the antigen and the kind of antibody with which it reacts. Many processes such as agglutination, opsonization, and activation of complement, cell lysis and neutralization can make the inactivation of pathogens. All these types of reaction occur naturally in the body and can be made to occur in the laboratory.

If particulate antigen reacts with antibodies to form a complex is known as **agglutination**. Here bacterial cells acts as particles, which reacts with antibody and results in sticking together of microbes. IgM produces strong and IgG produces weak agglutination reaction with certain bacterial cells. This agglutination reaction produces visible reaction to the unaided eye and can be used for the laboratory test to detect the presence of antibodies or antigens. Some of the antibodies act as opsonins or agents, which helps in opsonization of the antigens. The **opsonization** (*opson* means to prepare victims for) is the process in which microorganism or other particles are coated with antibody and/or complement and prepared for recognition and ingestion by phagocytic cells. Bacterial toxins being a small molecule secreted from the cells and a virus usually are inactivated simply by the formation of antigen-antibody complexes and are known as **neutralization**. IgG is the main neutralizer of bacterial toxins. IgM, IgG and IgA are all effective neutralizers of viruses. Thus, antibodies neutralize toxins or helps in the phagocytosis of microbial cells.

The complement is an important component in the blood helps in inactivation of infectious agents by forming the membrane attack complex (MAC). The MAC helps in breakdown of cells known as **cell lysis**. Usually both IgG and IgM are powerful activators of the complement system, whereas IgA is less powerful. Sometimes IgM, directly lyse the cell membrane of infectious organisms without the aid of complement.

EXERCISE

1. What is meant by primary response and secondary response?
2. Write a note on B lymphocytes.
3. What do you mean by humoral immunity?
4. How are B cells activated by antigens for antibody production?
5. Define opsonization.

6. Give an account on neutralization.
7. Give an account of 'B' immuno competent cells. (BU Oct. 2001)
8. How do cytokines influence the class switching in B lymphocytes?
9. Explain mechanism of plasma cells production.
10. Describe molecular mechanism of B cells activation.

Chapter

11

Cellular (Cell-Mediated) Immunity

11.1 INTRODUCTION

In this type of immunity, directly immune cells are involved in protection of body, hence, it is referred as cellular immunity. In the immunity process antibodies alone are not enough to defense in certain circumstances, especially virus and other pathogens that have invaded host cells. This type of immunity is also involved in elimination of tumor (cancer) cells or transplantation rejection. Thus, cell-mediated immunity generally is helpful in killing intracellular pathogens, destruction of tumor cells, rejection of graft tissue, and hypersensitivity reaction to certain antigens (poison, drugs).

Cell-mediated immune response involves the differentiation and action of different T cells for killer activity and the production of chemical mediators called lymphokines. The major cells involved in lymphokines production are T helper ($\mathbf{T_H}$) cells. Cells involved in killing activity include cytotoxic T ($\mathbf{T_C}$) cells, activated macrophages, natural killer cells, antibody dependent killer cells.

The cell-mediated immune response begins usually with processing of an antigen by antigen presenting cells (macrophages, dendritic cells and B cells). These cells scavenge foreign materials (antigens or pathogens) from the blood, tissues and phagocytize such materials. These cells digest them in the cell interior and small pieces of the antigens bind to a special class of proteins called major histocompatibility complex (MHC) class II molecules. The MHC-antigen complex is then transported to the surface of the antigen presenting cells and displayed to circulating T helper cells. Before understanding T cells action and T cells-macrophages interaction some knowledge of major histocompatibility complex molecules.

11.2 MAJOR HISTOCOMPATIBILITY COMPLEX (MHC)

Major histocompatibility (*histo* means tissue; *compatibility* means ability to get along) complex is special group of proteins present on the cell surface. These proteins derived from collection of genes on chromosome 6 in humans (chromosome 17 in mice) and these surface antigens are referred as human leukocyte antigen (HLA) complex (H-2 complex in mice). Almost all human tissue cells contain MHC molecules on their plasma membranes. The MHC molecules can be divided into three classes: class I molecules are found on almost all types of nucleated body cells; class II molecules are present only on those leukocytes involved in T helper cell mediated immune response (macrophages, antigen presenting cells and B cells); class III molecules include various secreted proteins that have immune functions (complement component C2, C4a, factor B and tumor necrosis factor). The class I proteins are obtained from A, B, C locus of MHC, class II by DP, DQ, DR locus and class III by C4, BF, C2 locus on chromosome 6.

Class I MHC molecules contain a large glycoprotein **α** chain (45 kDa) associated noncovalently with smaller proteinaceous **β_2-microglobulin** (12 kDa) molecule (Fig. 11.1). The α chain consists of three external domain (α1, α2 and α3) and β_2-microglobulin contains only one external domain that is associated with α3 domains. This α chain (external domains) is anchored in the plasma membrane by its hydrophobic transmembrane segment and hydrophilic cytoplasmic tail. The α1 and α2 domains exposed outside and form the antigen binding pocket.

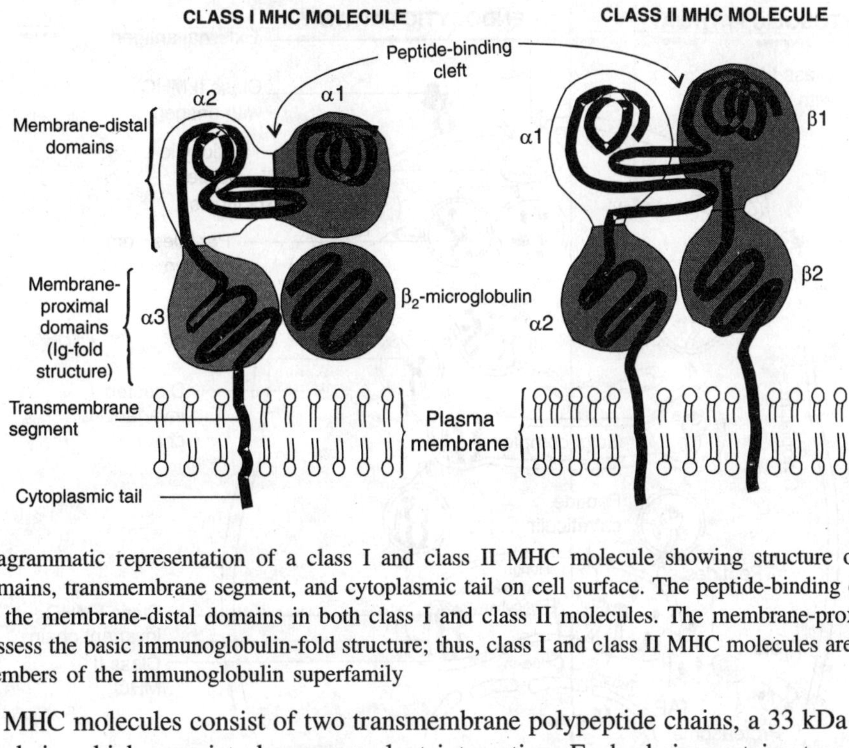

Fig. 11.1. Diagrammatic representation of a class I and class II MHC molecule showing structure of the external domains, transmembrane segment, and cytoplasmic tail on cell surface. The peptide-binding cleft is formed by the membrane-distal domains in both class I and class II molecules. The membrane-proximal domains possess the basic immunoglobulin-fold structure; thus, class I and class II MHC molecules are considered as members of the immunoglobulin superfamily

Class II MHC molecules consist of two transmembrane polypeptide chains, a 33 kDa α chain and a 28 kDa β chain which associate by noncovalent interaction. Each chain contains two external domains in both α chain (α1, α2) and β chain (β1, β2). Both chains have transmembrane segment and cytoplasmic segment. The α1 and β1 domains exposed outside and form the antigen binding part.

11.3 ANTIGEN PROCESSING AND ANTIGEN PRESENTATION

Recognition of foreign antigen by a T cell requires that peptides derived from the antigen should display with MHC molecules on the membrane of cell. The formation of antigenic peptide from antigen by sequence of degradation events (by proteasome-multifunctional protease complex) in the cell is known as **antigen processing**. The degraded peptides then associate with MHC molecules within the cell interior and the peptide-MHC complexes are transported to the membrane surface, and they are displayed on the outer membrane surface known as **antigen presentation**.

Class I and class II MHC molecules bind peptides that have been processed in different intracellular compartments. Class I MHC molecules bind peptides derived from endogenous antigens that have been processed within the cytoplasm of the cell (normal cellular proteins, tumor proteins or viral and bacterial proteins produced within infected cells). The protein antigen is degraded into peptides by proteasome and transported into rough endoplasmic reticulum lumen by a transport protein called TAP (transporters associated with antigen processing). The calnexin associated class I MHC α chain binds to β_2-microglobulin (calnexin, calreticulin and tapasin are molecular chaperones, which helps in maintaining protein stability). The β_2-microglobulin binding dissociates calnexin and binds to calreticulin and tapasin. This complex captures antigenic peptide, which allows dissociation of the MHC molecule from calreticulin and tapasin. At the end class I MHC-peptide complex is transported from the rough endoplasmic reticulum lumen through the Golgi complex to the plasma membrane surface. This processes is known as **cytosolic pathway of antigen presentation** (Fig. 11.2).

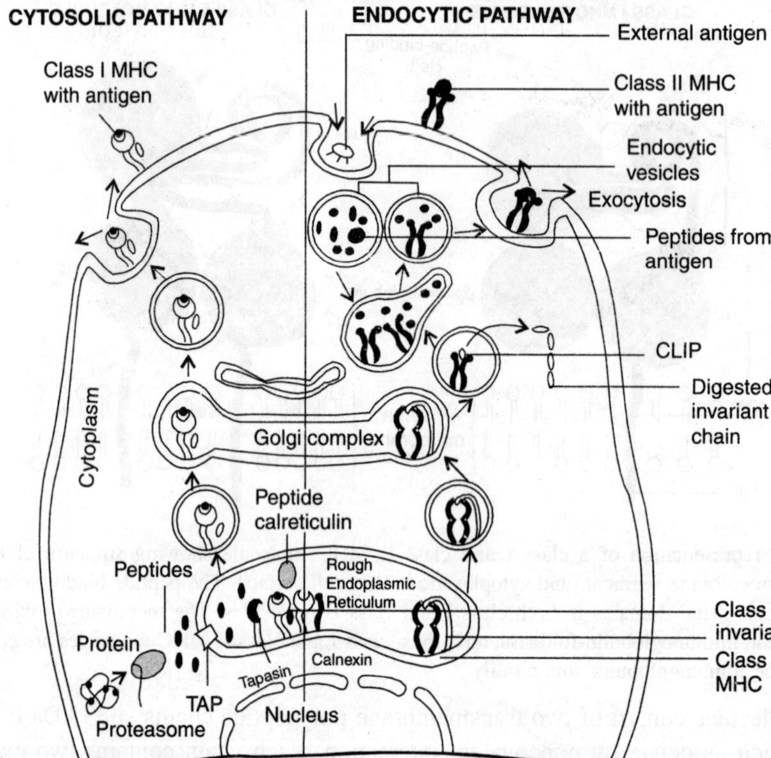

Fig. 11.2. Scheme showing separate antigen-presenting pathways for cytosolic (endogenous) and endocytic (exogenous) antigens. The mode of antigen entry into cells and the site of antigen processing appear to determine whether antigenic peptides associate with class I MHC molecules in the rough endoplasmic reticulum or with class II molecules in endocytic compartments

Class II MHC molecules bind peptides derived from exogenous antigens that are taken inside the cell by phagocytosis, endocytosis or both. The internalized antigen is degraded into peptides in the phagocytic vesicles by various hydrolytic enzymes with various pH and forms endosomes/phagosomes and endolysosomes. The class II MHC molecules synthesized in rough endoplasmic reticulum, is associated with another protein called the invariant chain. This prevents from binding of other internal peptide to the class II MHC molecules. This is transported to endolysomes through Golgi complex. During this transport invariant chain undergoes degradation leaving only CLIP (class II associated invariant chain peptide) in the cleft of class II MHC. The CLIP is later exchanged for the peptide through the help of another protein called HLA-DM (similar to MHC molecules). Once the peptide has bound, the class II MHC-peptide complex is transported to the plasma membrane surface. This processes is known as **endocytic pathway of antigen presentation** (Fig. 11.2).

The non-proteinaceous materials such as lipids, glycolipids are also known to have immunogenic capacity. A special type of receptors called CD1 receptors generally present these materials on the plasma membrane. The exact mechanism is not yet understood.

The T cells are known to bind to the antigen peptide, which are processed and present on the self-cells (antigen presenting cells). There are three subtypes identified in the T cells, namely, T helper (T_H) cells, T cytotoxic (T_C) cells and T suppressor (T_S) cells. T helper (T_H) cells and T suppressor (T_S) cells are involved in regulation of specific immune cells and hence called regulatory T cells. T cytotoxic (T_C) cells generally cause cytolysis and cells death of target cells (antigen containing cells), hence these cells are called effector T cells.

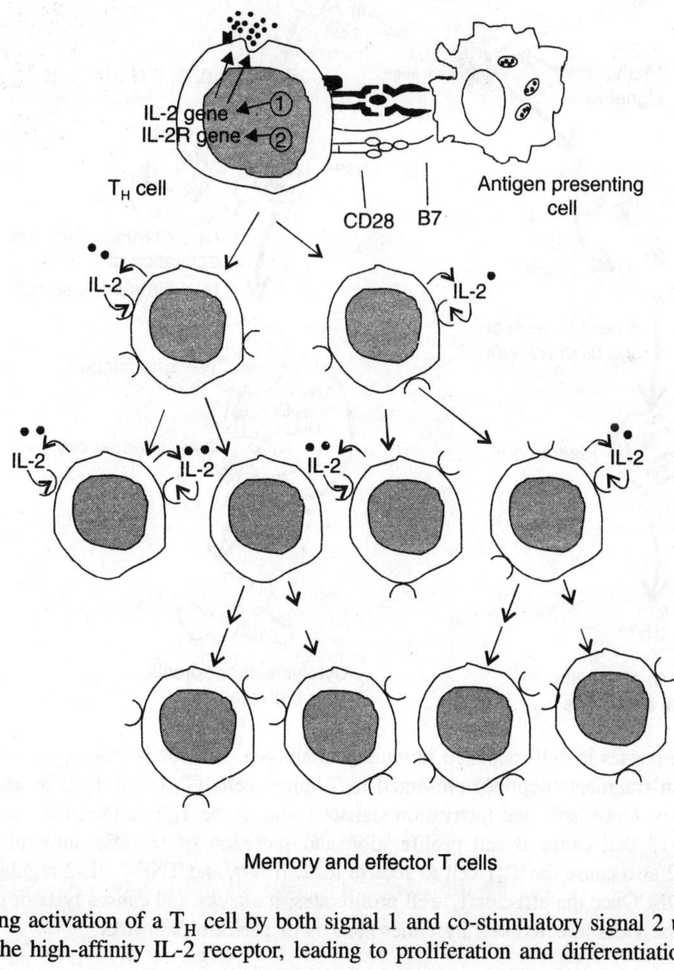

Fig. 11.3. Model showing activation of a T_H cell by both signal 1 and co-stimulatory signal 2 up-regulates expression of IL-2 and the high-affinity IL-2 receptor, leading to proliferation and differentiation

11.4 ACTIVATION OF T CELLS

The naive T cells are activated by the antigen-MHC complex, which generates memory T cells, regulatory T cells and effector T cells (T_H, T_C, and T_S cells) (Fig. 11.3). The T_H cells are also known as $CD4^+$ cells and two subtypes are noticed i.e., T_H 1 and T_H 2 cells. These subtypes cells produce and secrete a specific mixture of cytokines or lymphokines (Table 10.1). T_H 1 cells produce IL-2, IFN-γ and tumor necrosis factor that are involved in cellular immunity, especially macrophage activation (stimulate to destroy the pathogens by phagocytosis). The T_H 2 cells produce IL-4, IL-5, IL-10 and IL-13 that are involved in humoral immunity. T_H 2 cells interact with B cells (helpers for B cells) to make antibodies against antigen (Fig. 11.4).

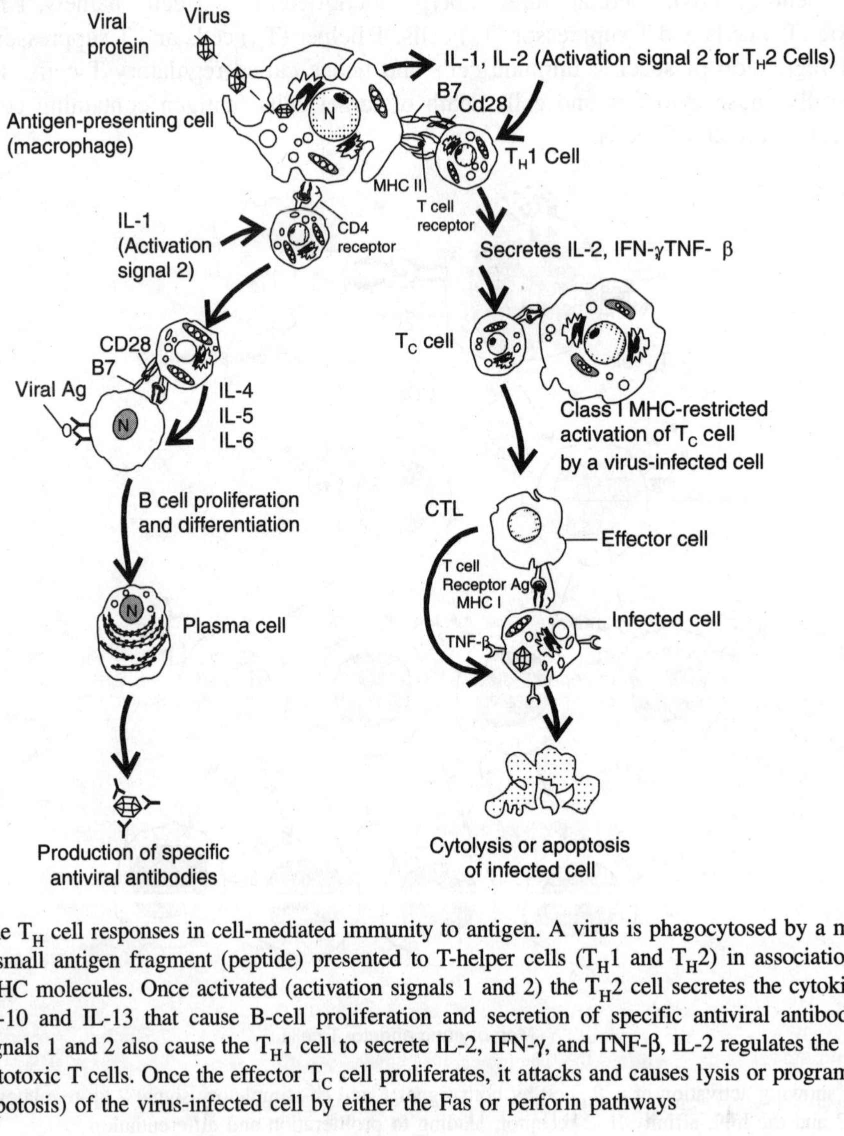

Fig. 11.4. The T_H cell responses in cell-mediated immunity to antigen. A virus is phagocytosed by a macrophage and a small antigen fragment (peptide) presented to T-helper cells (T_H1 and T_H2) in association with class II MHC molecules. Once activated (activation signals 1 and 2) the T_H2 cell secretes the cytokines IL-4, IL-5, IL-10 and IL-13 that cause B-cell proliferation and secretion of specific antiviral antibodies. Activation signals 1 and 2 also cause the T_H1 cell to secrete IL-2, IFN-γ, and TNF-β, IL-2 regulates the proliferation of cytotoxic T cells. Once the effector T_C cell proliferates, it attacks and causes lysis or programmed cell death (apoptosis) of the virus-infected cell by either the Fas or perforin pathways

11.5 MOLECULAR MECHANISM OF T_H1 CELLS ACTIVATION

The T helper (T_H1) cells require two signals (costimulation) to be stimulated by an antigen. The first signal is through interaction of TCR (T cell receptor)-CD3 complex (on T cells) and antigen bound to class II MHC on antigen presenting cells. The CD4 is also involved in the activation. The second signal for activation involves the binding of CD28 on T_H1 cell with a protein B7 (CD80) on macrophage (an APC). Both the activation signal goes to cytoplasm of T_H1 cells (Fig. 11.5) by following a signal transduction pathway in a cascade manner to show effector mechanism.

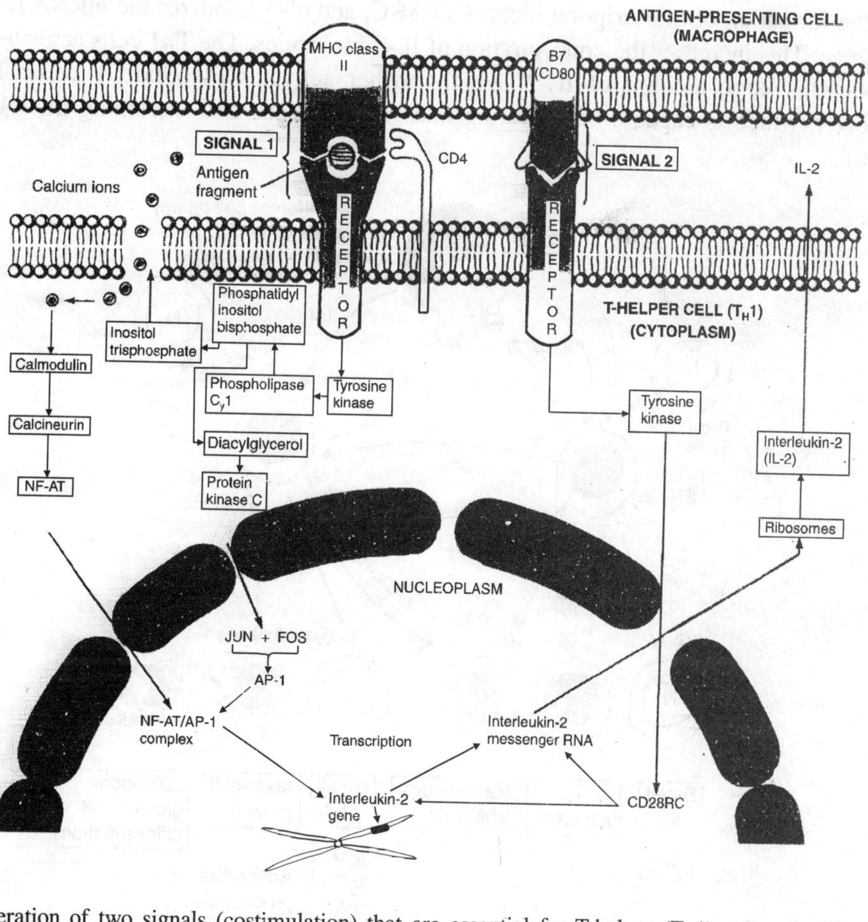

Fig. 11.5. Generation of two signals (costimulation) that are essential for T-helper (T_H1) cell activation. The signal (1) is the presentation of the antigen fragment by a macrophage or other antigen-presenting cell along with the MHC class II molecule to the T-helper cell receptor and CD4 protein. The signal (2) initiates when the macrophage presents the B7 (CD80) protein to the T-helper cell with its CD28 protein receptor. Both signals send information into the cytoplasm of the T-helper cell. The signal (1) causes interleukin-2 mRNA to be produced. The signal (2) boosts the production of interleukin-2 to effective concentrations. It is now known that the gene for interleukin-2 is under very tight regulation. It cannot be transcribed unless the NF-AT/AP-I complex are activated when the T-helper cell is stimulated through its antigen-specific receptor.

First signal activates tyrosine kinase and it adds phosphate to phospholipase Cγ1 which cleaves phosphatidylinositol biphosphate to two products, diaceylglycerol and inositol triphosphate. The diaceylglycerol further activates the protein kinase C then it moves to nucleus where it catalyses the formation of a protein complex called AP-1 from separate components. The other product inositol triphosphate causes a calcium channel to open, thus calcium ions rush into the cytoplasm leading to further enzymatic activity and the activation of calmodulin, calcineurin and nuclear factor of activated T_H1 cells (NF-AT). Then NF-AT migrates into the nucleoplasm where it binds to the newly formed AP-1, forming complex NF-AT/AP-1 (transcription factor). This complex binds to specific gene in the DNA causing IL-2 mRNA to be transcribed. This messenger RNA moves out of the nucleus for protein (IL-2) synthesis.

Second signal mediated by CD28 receptor and B7 molecule activates another tyrosine kinase result in the formation of transcription factor CD28RC, and also stabilizes the mRNA transcribed from the IL-2 gene. This increases the concentration of IL-2 molecules. The T_H1 cells activated by these two signals secretes large amount of IL-2, which further activates cytotoxic T cells. T_H1 cells also secrete IFN-γ, which is capable of activating macrophages cells and increasing microbicidal activity (Fig. 11.6).

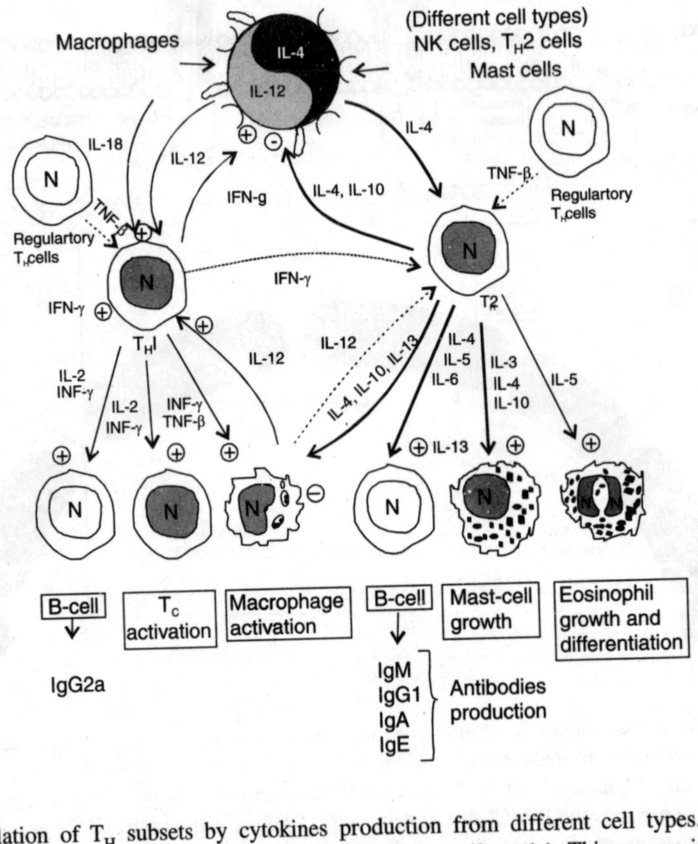

Fig. 11.6. The regulation of T_H subsets by cytokines production from different cell types. Solid arrows indicate stimulatory effects (+); dashed arrows indicate inhibitory effects (–). Thin arrows indicate the cytokines of the T_H1 pathway; thick arrows, the T_H2 pathway.

The T helper (T_H2) cells also require two signals (costimulation) **to** be stimulated by an antigen. The first signal is through interaction of TCR (T cell receptor)-CD3 complex (on T cells) and antigen bound to class II MHC on antigen presenting cells along with CD4. This activated T_H2 cells express CD40L and reacts with CD40 on B cell and provides the second signal to the B cells for expression of cytokines receptors. Binding of B cell receptor (B7) and T cell receptor (CD28) also stimulates the activation. After binding to cytokines (IL-2, IL4 and IL-5) released from T_H2 cells, B cells divides to plasma cells and memory cells. Thus, B cells get activated by the T_H2 cell for the increased antibody production.

11.6 T CYTOTOXIC (T_C) CELLS

These cells are responsible for elimination of host cells harboring viruses and other pathogens or tumor cells. The T_C cells are also known as $CD8^+$ cells. These are activated by a complex process that involves at least three signals. The T_C cell receptor bind to class I MHC-antigen complex on the surface of virus infected cells or tumor cells (target cells) and provides first signal (Fig. 11.7). A CD28 (T_C cell) and B7 (target cell) complex formation acts as costimulation signal. Finally the T_C cells require exposure to IL-2 from T_H 1 cells. The cells then proliferate and differentiate to form active

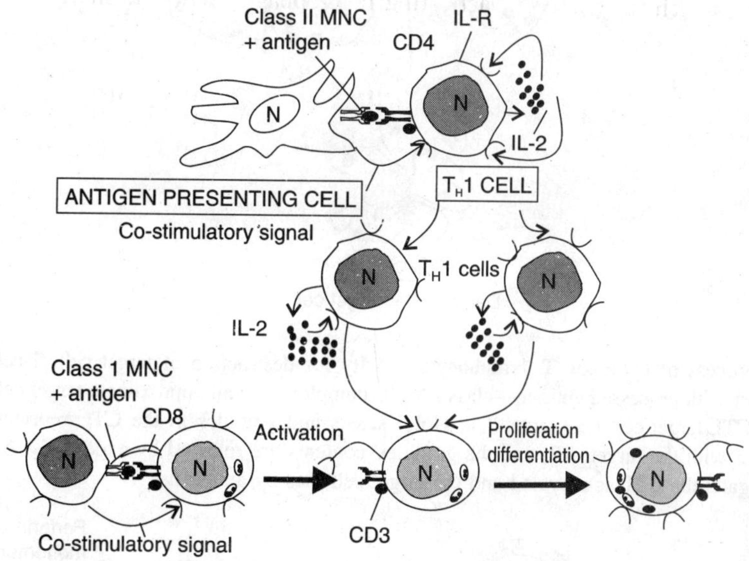

Target cell	Cytotoxic T-Lymphocyte progenitor	Ag-Activated CTL-P	CTL
Cell state		Activity	
IL-2R expression	–	+	+
IL-2 expression	–	±	±
Proliferation	–	–	+
Effector cytotoxic function	–	–	+

Fig. 11.7. Model showing generation of effector cytotoxic T-lymphocytes (CTLs). After interaction with antigen—class I MHC complexes on appropriate target cells, CTL-Progenitors (CTL-Ps) begin to express IL-2 receptors (IL-2R) and lesser amounts of IL-2. Proliferation and differentiation of antigen-activated CTL-Ps generally require additional IL-2 secreted by T_H1 cells resulting from antigen activation and proliferation of $CD4^+$ T cells. In the subsequent effector phase, CTLs destroy specific target cells.

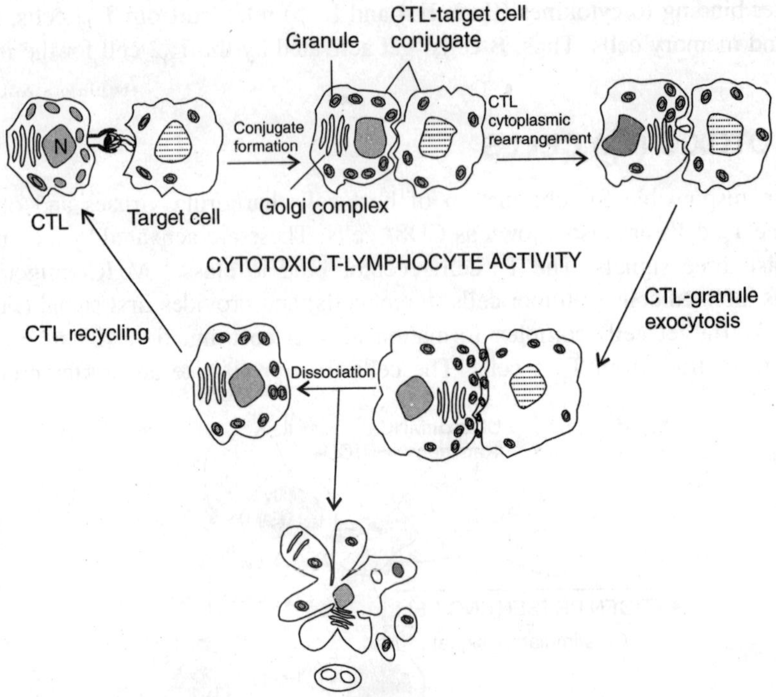

Fig. 11.8. The process of cytotoxic T. lymphocyte activity for destruction of target cell. T-cell receptors on a CTL interact with processed antigen—class I MHC complexes on an approximate target cell, leading to formation of a CTL/target-cell conjugate. The Golgi stacks and granules in the CTL reorient towards the point of contact with the target cell, and the granules' contents are released by exocytosis. After dissociation of the conjugate, the CTL is recycled and the target cell dies by apoptosis.

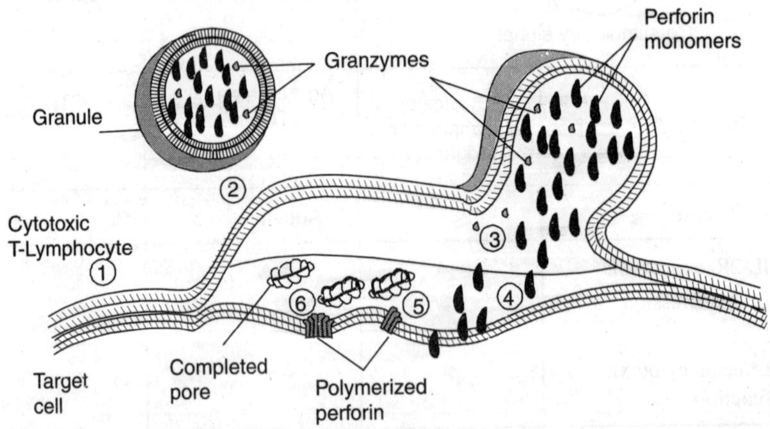

Fig. 11.9. CTL-mediated pore formation in target cell membrane. The rise in intracellular Ca^{2+} triggered CTL-target cell interaction, (1) induces exocytosis in which the granules fuse with the CTL cell membrane, (2) and release monomeric perforin into the small space between the two cells, (3) The released perforin monomers undergo a Ca^{2+}-induced conformational change that allows them to insert into the target-cell membrane, (4) In the presence of Ca^{2+}, the monomers polymerise within the membrane, (5) forming cylindrical trans-membrane pores.

cytotoxic T lymphocytes (CTL) from precursor CTLs (CTL-Ps), which can attack target cell. CTL and T_H 1 cells release cytokines such as gamma interferon, whereas both macrophages and T cells release tumor necrosis factor-α. These cytokines inhibits virus reproduction, meanwhile they also activate macrophages and other phagocytes that can kill the infected cell.

The cytotoxic T lymphocytes can destroy target cells in at least two ways. The CD95 pathway triggers apoptosis or programmed cell death. The perforin pathway is probably more often used by CTL through perforin and granzymes to induce both osmotic cytolysis and apoptosis.

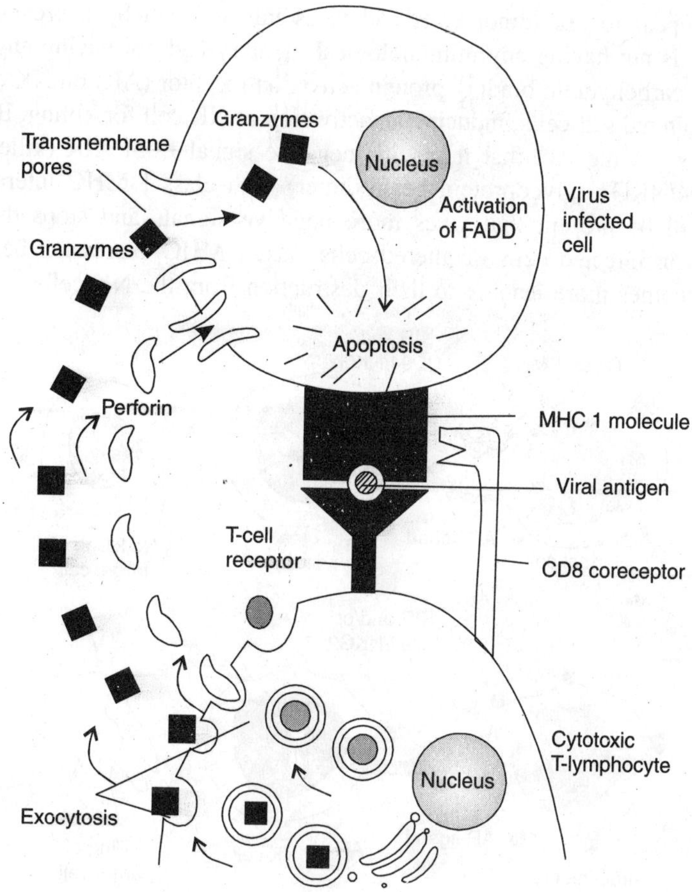

Fig. 11.10. How an effector cytotoxic T cell destroys a virus-infected target cell. T-cell cytotoxicity often involves the perforin pathway and leads to apoptosis and cytolysis. A cytotoxic T cell (down) contacts a target cell (top). The T cell secretes perforin that forms transmembrane pores in the target cells' plasma membrane. These pores allow the contents of the target cell to leak out (cytolysis). The granzyme activates the FADD (Fas associated death domain) that starts events of apoptosis.

After the activation of CTL, it follows conjugation, membrane attack, CTL dissociation and target cell destruction. During conjugation (fusion) with target cell, the membrane integrin leukocyte functional antigen 1 (LFA-1) (leukocyte adhesion molecule) binds to intercellular cell-adhesion molecules (ICAM) on target cells. Then CTLs cytoplasmic changes occur by reorientation of Golgi complex and storage granules towards target cell junction (Fig. 11.8). The granules are vacuoles that contain monomers of pore forming protein **perforin** and several serine proteases called **granzymes** (also called fragmentins).

These granules release perforin and granzyme into juction space by exocytosis. The perforin forms a large number of cylindrical pore of 5-20 nm size in the membrane by polymerization in presence of Ca^{2+} (Fig. 11.9 and 11.10). This permits the entry of granzymes into target cell interior and initiates the events of apoptosis by fragmenting the DNA. CTLs can also activates Fas ligand by binding of FADD protein (Fas associated Death Domain) on target cells that initiate events of apoptosis.

11.7 NATURAL KILLER (NK) CELLS

Natural Killer cells appear to kill tumor cells and virus infected cell by a process similar to that employed by CTLs. It is not having any immunological memory and not having any antigen specific receptors. The general carbohydrate binding protein activation receptor (AR) on NK cells interact with ligand on normal and altered self cells, inducing an activation of NK cell for killing. But, it does killing activity or not is based on the fact that it obtains negative signal from KIR (killer cells inhibitory receptor) and/or CD94/NKG2 (glycoprotein heterodimer) with class I MHC interaction. In normal cells class I MHC will be present that gives more negative signals and stops the killing activity (Fig. 11.11). Whereas in infected cells or altered cells class I MHC expression becomes decreased, thus killing signals becomes more leading to their destruction from the NK cells.

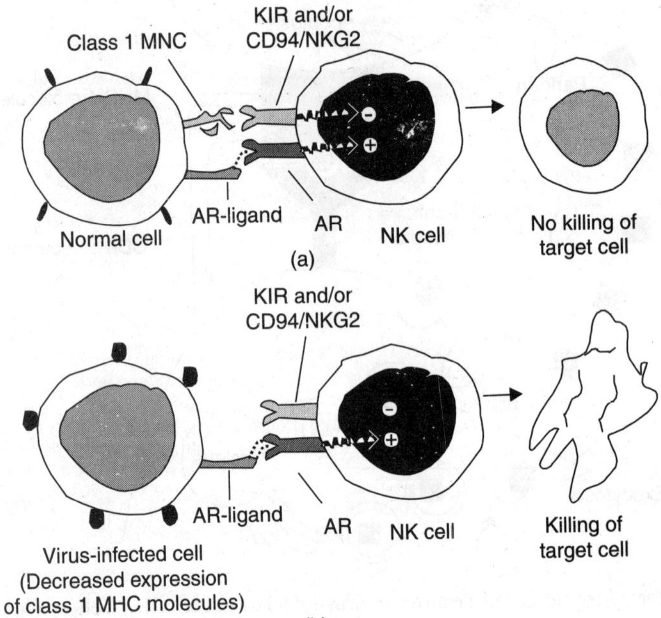

Fig. 11.11. Molecular events in opposing-signals model of how cytotoxic activity of NK cells is restricted to only altered target self-cells. An activation receptor (AR) on NK cells interacts with its ligand on normal and altered self-cells, inducing an activation signal that results in killing. However, engagement of inhibitory NK cell receptors such as KIR and CD94/NKG2 by class I MHC molecules delivers an inhibitory signal that counteracts the activation signal. Expression of class I molecules on normal cells thus prevents their destruction by NK cells. Because class I expression is often decreased on altered self-cells, the killing signal predominates, leading to their destruction.

T suppressor (T_S) cells are known to suppress the functions of B and T cells in immune response.

EXERCISE

1. What is meant by cellular immunity? (BU Oct. 2002)
2. Describe various types of T cells.
3. Give an account on major histo-compatibility complex. (BU April 2001)
4. Explain antigen presentation.
5. Describe activation of T_H cells.
6. Illustrate the cytotoxic activity by T_C cells.
7. Write a note on Natural Killer cells. (BU April 2001)
8. Define cellular immunity. Briefly explain the cluster of differentiation. (BU Oct. 2002)
9. Give an account of 'T'-Immuno competent cells. (BU Oct. 2001)
10. What is MHC? Explain its role in immunity. (BU April 2001, Oct. 2001)
11. Briefly explain cytosolic and endocytic pathway of antigen presentation.
12. Describe effect of cytokines produced by the T cells on different cells of immune system.

Chapter

12

Effector Mechanisms

12.1 INTRODUCTION

Recognition of antigen by the immune system induces a variety of effector mechanisms associated with the humoral and cell-mediated responses. The cytokines are small proteins secreted by activated lymphocytes, macrophages and other immune cells that assist in regulating the development of immune effector cells and/or possess direct effector functions. The antibodies produced and complement proteins are other effectors of humoral immunity. Both specific cells (T_H and T_C cells) and non-specific cells (natural killer cells, macrophages, neutrophils, eosinophils) are effectors of cell-mediated immunity. In humoral immunity, system primarily functions to eliminate extracellular bacteria and bacterial products, whereas the cell-mediated immunity functions to eliminate altered self-cells (infected cells and tumor cells) and foreign grafts cells.

12.2 EFFECTOR MECHANISM IN HUMORAL IMMUNITY

Cytokines

Cytokines are low molecular weight regulatory proteins or glycoproteins secreted by white blood cells and various other cells in the body in response to a number of stimuli. Cytokines help in regulating the development of immune effector cells and some cytokines act directly and function as effector molecules. Cytokines bind to receptors on the membrane of the same cell that secreted it (**autocrine** action) or other target cell (**paracrine** action), which are very near by or target cells which are far away placed in body (**endocrine** action). They regulate the intensity and duration of immune response by stimulating or inhibiting the activation, proliferation, and/or differentiation of various cells (Fig. 12.1). They also regulate secretion of antibodies and other cytokines.

Cytokines produced by lymphocytes were earlier known as **lymphokines**. Cytokines produced from monocytes and macrophages were formerly known as **monokines**. Cytokines secreted from some leukocytes and acting on other leukocytes are called **interleukins**. Cytokines that induce an antiviral condition in other cells are called **interferons**. Cytokines that help in the chemical attraction of cells (chemotaxis) are called **chemokines**. Cytokines, toxic to tumor cells, promote inflammation (a localized tissue response to injury characterized by pain, heat, redness and swelling resulted due to changed blood flow, movement of phagocytes or other immune cells) or programmed cell death is known **as**

tumor necrosis factor (TNF). Cytokines help in blood cells growth and differentiation (hematopoiesis) are called **hematopoietins.**

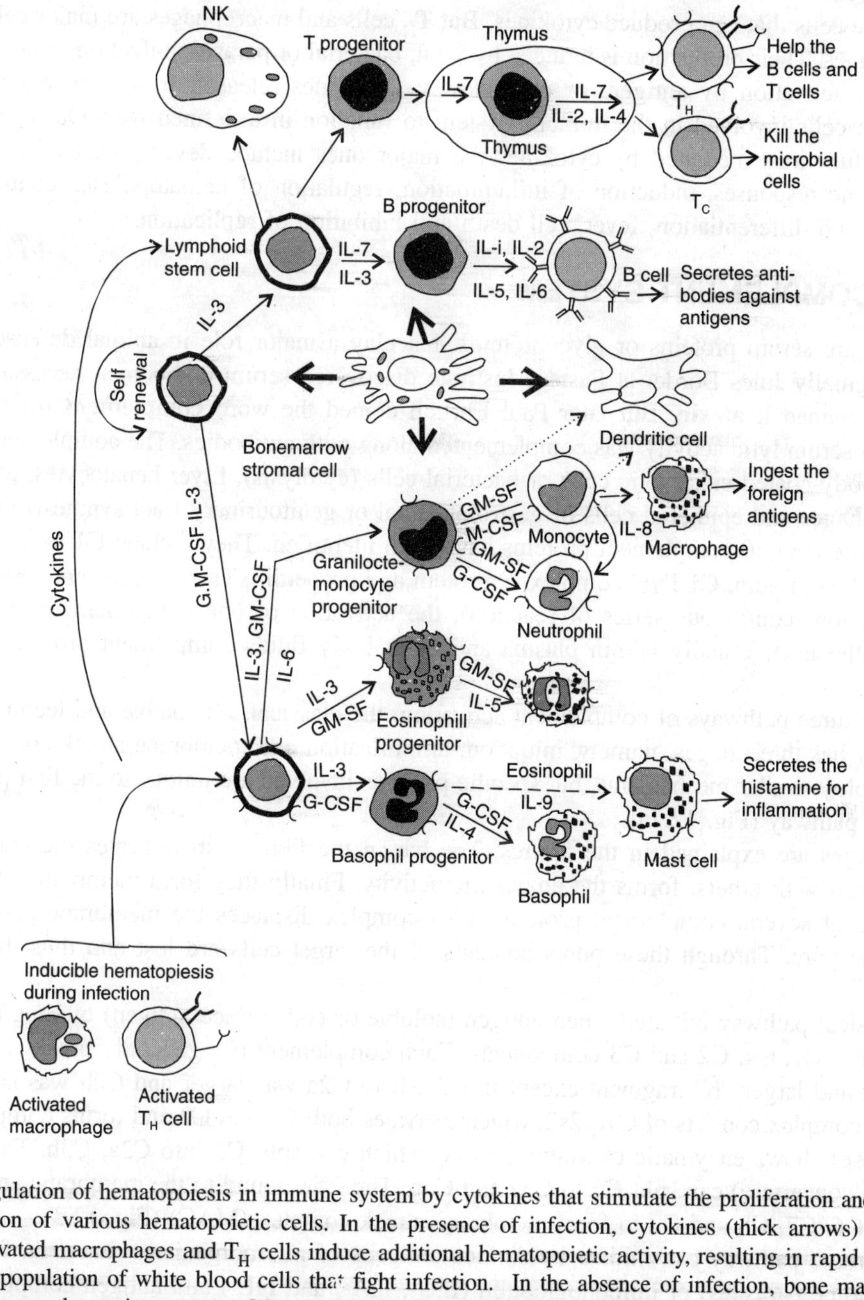

Fig. 12.1. Regulation of hematopoiesis in immune system by cytokines that stimulate the proliferation and/or differentiation of various hematopoietic cells. In the presence of infection, cytokines (thick arrows) produced by activated macrophages and T_H cells induce additional hematopoietic activity, resulting in rapid expansion of the population of white blood cells that fight infection. In the absence of infection, bone marrow stromal cells are the major source of hematopoietic cytokines (thick arrows). IL = interleukin: CSF = colony-stimulating factor; NK = natural killer cell. Stem cells are shown in thick outline; progenitor and immature cells in dark shades; stromal cell in without any shade; and mature differentiated cells in thin outline. Dashed lines indicate likely, but still hypothetical pathways.

All these cytokines function as intercellular messenger molecules that induce particular biological activities after binding to a receptor on target cells. Monocytes, macrophages, T helper cells, B cells, dendritic cells, mast cells, natural killer cells, bone marrow cells, endothelial cells and other leukocytes are some of the cells that can produce cytokines. But T_H cells and macrophages are major cells, which produce cytokines. Their production is induced by viral, bacterial or parasitic infection, cancer, inflammation, T cell activation by antigen, or chemokines. Cytokines released from the parent cell can activate all the cells involved in the immune system to function in a defined way. Out of numerous physiological functions mediated by cytokines the major ones include development of cellular and humoral immune responses, induction of inflammation, regulation of hematopoiesis, control of cell multiplication and differentiation, fever, cell death and inhibition of replication.

12.3 THE COMPLEMENT SYSTEM

Complements are serum proteins or glycoproteins that play a major role in animal defense immune response. Originally Jules Bordet at Pasteur Institute discovered serum component necessary for the cell lysis and named it **alexin**. But later Paul Ehrlich coined the word **complement** for the alexin, because blood serum lytic activity was complemented along with antibodies. The complement proteins can lyse antibody-coated eukaryotic cells or bacterial cells (cytolysis). Liver hepatocytes, Monocytes, tissue macrophages and epithelial cells of gastrointestinal or genitourinary tract synthesize these proteins. Now more than 30 complement proteins have been identified. They include C1 to C9, factor D, H, I, C4b binding protein, C1 INH complex, S protein and properdin. The complement system acts in a cascade fashion (continuous series of reaction), the activation of one component resulting in the activation of the next. Usually within plasma and other body fluids complement proteins are in an inactive state.

There are three pathways of complement activation: the classical, alternative and lectin pathways. Each pathway has three stages, namely, initiation, amplification and membrane attack complex. They generally employ similar mechanisms, but specific proteins involved are unique to the first part (initiation) of each pathway (Fig. 12.2a).

These stages are explained in the figures. The bar in the Fig. 12.2b indicates the complements when associated with others, forms the enzymatic activity. Finally they form membrane attack complex made up of several complement proteins. This complex displaces the membrane phospholipids and forms the pore. Through these pores contents of the target cells are lost and thus the viability (Fig. 12.2b).

The classical pathway initiated when antigen (soluble or cell surface antigen) binding to the IgM or IgG, initiates C1, C4, C2 and C3 components. Each complement (C) generally breaks into smaller "a" fragment and larger "b" fragment except in C2 where C2a was larger and C2b was smaller. The activated C1 complex consists of C1qr2s2, which activates both C4 and C2 and forms complex C4a2b (C3 convertase) shows enzymatic cleavage activity, which can split C3 into C3a, C3b. The C4a2b3b complex (C5 convertase) can split C5 to C5a and C5b. The C5b bound to the membrane and activates sequentially C6, C7, C8 and C9 to form membrane attack complex (MAC) (Fig. 12.3).

The alternate pathway can initiate by surface antigens of microorganisms (bacteria, fungi, some viruses and some parasites) or immunoglobulin (IgG-, IgA-, and IgE-) and antigen complexes. These activate C3 then factor B, factor D and properdin get activated and form the complex C3bBb (C3 convertase) and which then forms complex C3bBb3b (C5 convertase). This further forms C5a and C5b from the C5. The bound C5b forms membrane attack complex similar to classical pathway.

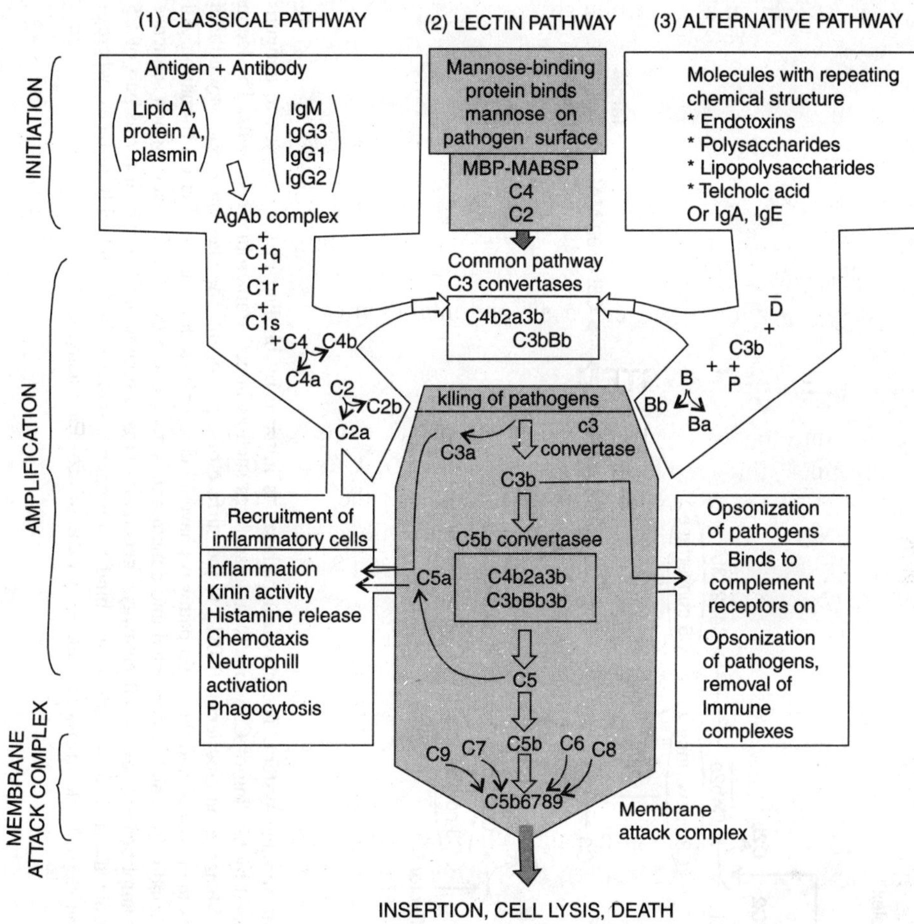

Fig. 12.2 (a) Complement activation. Within each pathway the components are arranged in order of their activation and aligned opposite their functional and structural analogs in the opposite pathway. There are three pathways of complement activation: (1) the classical pathway (left), which is mediated by antibody; (2) the lectin-mediated pathway (middle), which is triggered by a serum protein that binds to mannose residues of bacteria; and (3) the alternative pathway (right), which is triggered directly on pathogen surfaces. All three generate a crucial enzymatic activity (common pathway of C3 covertases) that in turn generates the effector activity of complement. The three main consequences of complement activation are recruitment of inflammatory cells (left box), direct killing of pathogens by the membrane attack complex (middle box), and opsonization of pathogens (right box).

The lectin pathway can initiate by binding of a serum protein MBL (mannose binding lectin) to carbohydrate residues on infective pathogenic microorganisms. This reacts with MASP (MBL associated serine protease) and forms complex similar to C1 complex. This complex activates C3, C4 and C2 similar to the classical pathway. The C1 similar complex gives raise to complex C4a2b (C3 convertase), which can split C3 into C3a, C3b. Then C4a2b3b complex (C5 convertase) can split C5 to C5a and C5b. The C5b bound to the membrane and activates sequentially C6, C7, C8 and C9 to form membrane attack complex (MAC). The MAC is a tubular transmembrane pore that helps in lysis of the target cells by out flow of cellular solutes and electrolytes across the cell membrane.

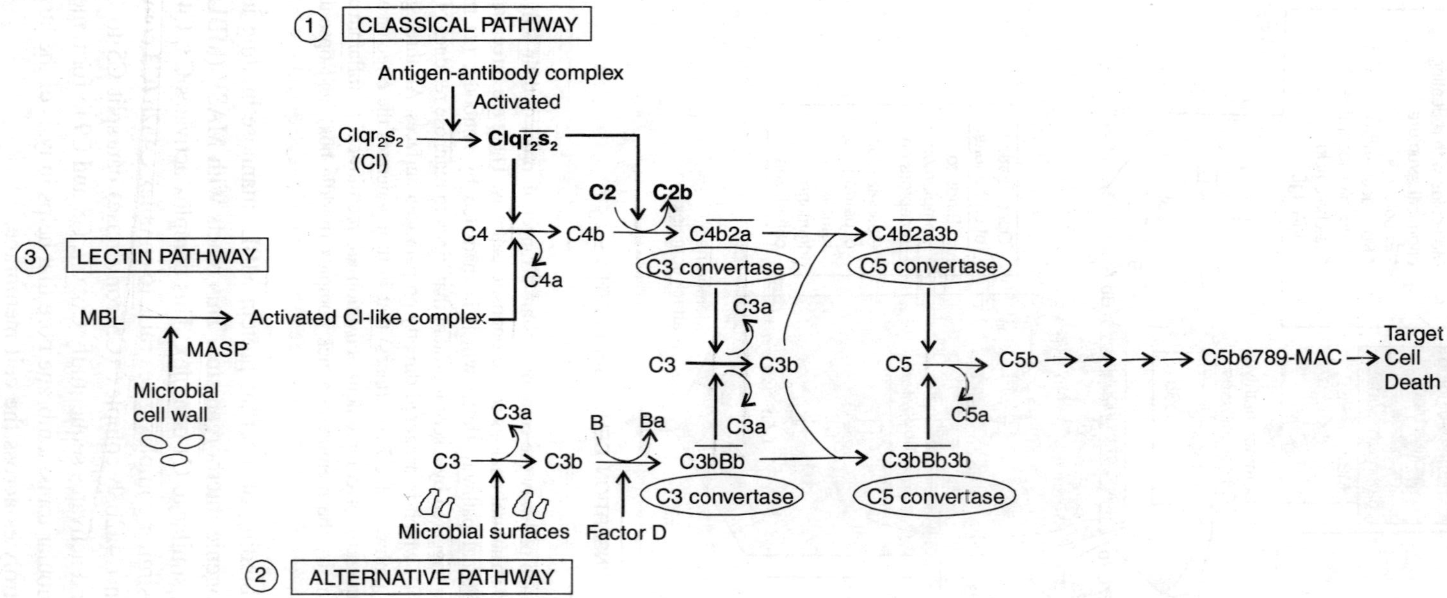

Fig. 12.2 (b) Model of the complement activation pathways in immune system. The classical pathway is initiated by binding of C1 complex to antigen-antibody complexes. The alternative pathway is initiated by binding of C3b to activating surfaces such as infective microbial cell walls. The lectin pathway is initiated by binding of the serum protein Mannose binding lectin (MBL) to the surface of a pathogen. The alternative pathway utilises some components (or factors) that are unique, whereas the classical and the lectin pathways have all components in common after C1 formation. All three pathways generate C3 and C5 convertases and bound C5b, which is converted into a membrane-attack complex (MAC) by a common sequence of terminal reactions. Breakdown of C3 is the major amplification step in all pathways, generating large amounts of C3b, which forms part of C5 convertase. C3b also has capacity to diffuse away from the activating surface and bind to immune complexes or foreign cell surfaces, where it functions as an opsonin. Thin arrows indicate reaction steps; thick arrows indicate enzymatic or activation activity.

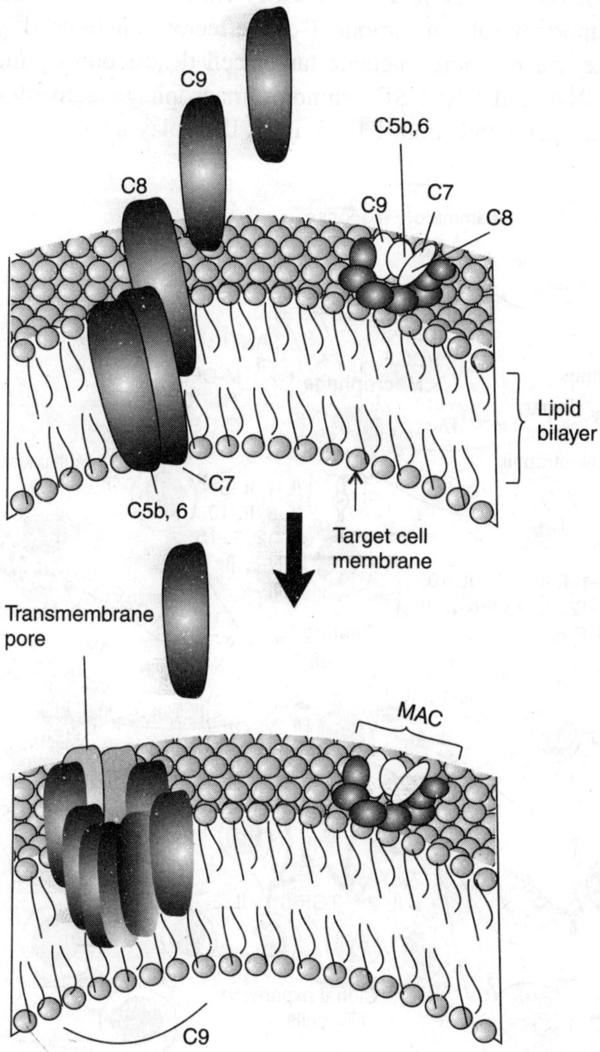

Fig. 12.3. The membrane attack complex in target cell membrane. The membrane attack complex (MAC) is a tubular structure that forms a transmembrane pore in the target cells' plasma membrane that results in cell death. This diagrammatic representation is the subunit architecture of the membrane attack complex. The transmembrane channel is formed by a C5b768 complex and 10 to 16 polymerised molecules of C9.

12.4 EFFECTOR MECHANISM IN CELLULAR IMMUNITY

The effector T cells produce various effector molecules, which may be membrane bound or soluble compounds. The membrane bound molecules include Fas ligand (FAS) on CD8$^+$ cells (CTLs), TNF-β on T_H1 cells and CD40 ligand on T_H2 cells. Each of these T cell also produces soluble effector molecules. The CTLs secrete cytotoxins (perforins and granzymes) as well as two cytokines IFN-α and TNF-β. T_H1 cells produce IL-2, IL-3, TNF-β, IFN-γ, high GM-CSF and T_H2 cells produce IL-3,

IL-4, IL-5, IL-6, IL-10, IL-13, IFN-γ, low GM-CSF. Each of these membrane-bound and secreted molecules plays an important role in various T-cell effector functions (Fig. 12.4). The Fas ligand, perforins and granzyme, for example, mediate target cell destruction by the CTL; membrane-bound TNF-β and soluble IFN-γ and GM-CSF promotes macrophage activation by the T_H1 cells; and membrane-bound CD40 ligand and soluble IL-4, IL-5, IL-6 play a role in B cell activation by the T_H2 cells.

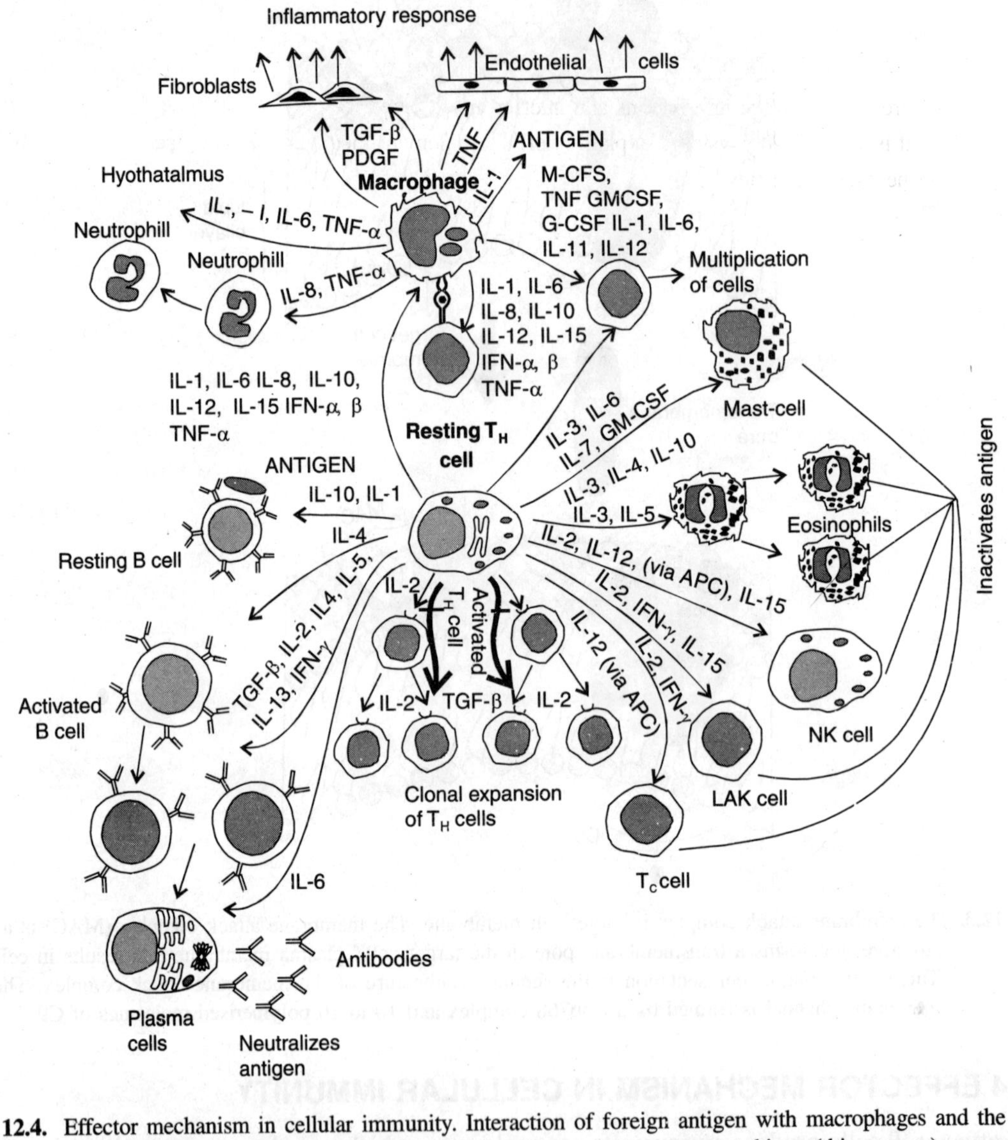

Fig. 12.4. Effector mechanism in cellular immunity. Interaction of foreign antigen with macrophages and the subsequent activation of resting T_H cells leads to release of numerous cytokines (thin arrows), generating a complex network of interacting cells in the immune response. A small subset of T_H cells secretes TGF-β, a cytokine that can inhibit the proliferation of T cells (see thick arrows). All the resulting cells work in their own way to fight against the foreign antigen.

EXERCISE

1. What is complement system and mention its importance in immune system? (BU April 2001)
2. Define cytokines.
3. Mention types of action of cytokines.
4. List various types of cytokines.
5. Elucidate in detail complement activation.
6. Describe formation of membrane attack complex.
7. Give the outlines of classical pathway of complement activation.
8. Differentiate between interleukins and interferons.
9. What is complement system? Explain briefly complement-mediated Ag-Ab reactions. (BU Oct. 2002)
10. Define hematopoietins.

Chapter

13

Immunity to Infectious Diseases

13.1 INTRODUCTION

The term **host** means any organism that is used by another organism for shelter and food. Any organisms that obtain their food and shelter from the host organism are known as **parasites**. The parasite or parasitic organisms take entry into the host organisms by crossing all the barriers of the host and the process of entry is known as **infection**. Infection may or may not result in disease. When the host is visibly or sensibly injured by the parasite, **disease** will occur. The parasite that causes the disease is known as **pathogen** (*pathos* means sadness or pain). For example, the mosquito can be called parasite but not pathogen, whereas *Plasmodium vivox* can be called pathogen. The capacity to cause disease is known as **virulent** pathogen i.e., those organisms able to produce serious disease or able to initiate an infection with very few cells (as low as 1 to 10 cells). An organism that cannot infect or cause disease is known as **avirulent (attenuated)**.

For a particular disease to occur pathogenic organisms have to be virulent. The virulent pathogen, to induce disease in the host, have to follow three stages; infectivity, invasiveness and pathogenicity. **Infectivity** means ability to cause infection and establishing inside the host. **Invasiveness** means capacity of the pathogen to leave from initial site of infection and grow in other tissue of the body. **Pathogenicity** is the ability of the pathogen to injure the host to the maximum extent and production of disease is known as pathogenesis. All the three phenomena include complex events in the host organisms. The varieties of infection processes may occur in host organisms but infection occurs normally by portal of entry (means by which the pathogen enters). These include wound on skin surface, mucous membranes (eye, nose, throat, lungs, genitourinary tract or gastrointestinal tract) (*Clostridium tetani, Shigella, Neisseria gonorrhoeae*). Some enters through intact skin surfaces in the area of sweat glands or hair follicle (*Staphylococcus aureus*). Invasiveness occurs by multiplication of pathogen and then they gain entry to blood which moves in the entire body. Pathogenicity occurs by the increased pathogen number or by the production of poisons called toxins.

The animals are in continuous contact with the various virus, bacteria, fungi and parasitic organisms. If an organism to be pathogenic it should establish infection in a susceptible host and a series of host defense mechanisms has to be overcome by the pathogen. The host immunity includes both nonspecific (innate) and specific (adaptive) defenses. The host innate immunity acts as first line of defense and consists of physical, chemical and biological barriers. This innate or non-specific immunity lacks immunological memory, that is, these responses occur to the same extent each time a pathogen or antigen is encountered. The specific immunity mechanisms are detailed in the earlier chapters with respect to cellular immunity, humoral immunity and effector mechanisms. In this chapter the emphasis is given to the non-specific or innate immune mechanisms.

13.2 PHYSICAL AND CHEMICAL BARRIERS IN NON-SPECIFIC IMMUNITY

Physical or mechanical barriers, along with the host's secretions (flushing mechanisms) act as first line of resistance against the infectious microorganisms. The intact skin, and mucous membranes along with chemicals greatly help in non-specific host defense.

Skin consists of thick keratin rich cells that are resistant to pathogenic degradation, mild acidic condition (pH 5-6), dryness, sebum (oil) from sebaceous glands and shedding (of outer epithelial cells) activity are known to act as a barrier to the pathogenic invasion or entry. Sometimes pathogens gain access to the tissue under the skin. Here they encounter defense by specialized cells called Langerhan's cell (specialized dendritic cell that can phagocytose antigens/pathogens). These cells are associated with skin associated lymphoid tissue (SALT) or cutaneous associated lymphoid tissue (CALT) and act as antigen presenting cells which activate the T cells for destruction of pathogen. Some of the intraepidermal lymphocytes (cells of a SALT) can directly act as T cells to destroy the pathogen.

The mucous membranes of the eye (conjunctiva), respiratory, digestive and urinogenital systems consist of the stratified squamous epithelium that resists the penetration and also traps the pathogens for removal. These epithelial systems also known to contain several types of mucosal associated lymphoid tissue (MALT) that includes gut associated lymphoid tissue (GALT) functions in digestive tract and tonsils, and bronchial associated lymphoid tissue (BALT) that functions in respiratory system. They are having secretory antibodies and macrophages for the protection.

The chemical barrier includes antimicrobial substances produced by the animals especially lysozyme (muramidase), lactoferrin, gastric juices, salivary glycoproteins, sebeum oleic acid on the skin, urea and low pH in urine are known to have microbicidal effects when they come in contact with these compounds. In addition, blood, lymph and other body fluids contain many defensive chemical compounds such as bacteriocins, beta-lysin and other polypeptides.

Lysozyme is an enzyme present in sweat, tears and mucous that degrades the cell wall peptidoglycan material of bacterial organisms. As a result bacteria loses viability leads to the death. **Lactoferrin** is an iron chelating compound present in plasma thus makes iron unavailable for the pathogenic multiplication inside. The gastric juices pH will be around 2-3 in which most of the organisms engulfed are killed. **Sebum**, fluids secreted by the oil glands contain triglycerides degraded to form free unsaturated fatty acids (oleic acid) that are antimicrobial in nature. **Bacteriocins** are toxic proteins produced by normal microbiota on the mucous surfaces. These toxic proteins are able to cause the cell lysis of target bacterial organisms. **Beta-lysin** is a cationic polypeptide released from blood platelets, which can kill some gram-positive bacteria by disrupting their plasma membranes. Other examples for such polypeptides having lethal effect are leukins, plakins, cecropins and phagocytin.

The biological barriers include the saprophytic microbial population which generally associated with the healthy individuals is known as normal microbiota. These microorganisms by their activity provide an environment that is not suitable for the pathogens to survive in that environment. Some of the organisms such as *Bacillus, Bacteroids, Enterococcus, Lactobacillus, Propinobacterium, Prepto streptococcus, Pseudomonas, Staphylococcus* and *Streptococcus* are members of the normal microbiota of humans.

13.3 INFLAMMATION

Inflammation (*inflammation* means to set on fire) is an important non-specific defense reaction to tissue injury caused by a pathogen or wound. It was characterized by cardinal signs or symptoms such as *calor* (increased temperature), *rubor* (redness), *tumor* (swelling), *dolor* (pain) and inflation of immune cells. The inflammation may be acute or chronic types.

The acute inflammatory response starts when injured tissue cells release chemical signals (inflammatory mediators-histamines, prostaglandins and leukotrienes) activates endothelial cells which lining the blood capillaries to produce P-selectins and E-selectins (cell adhesion molecules). These molecules attract and attach neutrophils to endothelial cells and neutrophils activated by inflammatory mediators. This activation occurs through integrins (cell adhesion receptor), which binds neutrophils tightly to endothelial cells. This results in neutrophils to change shape, squeeze through the endothelial wall (the process is known as **extravasation**) (Fig. 13.1) into the interstitial tissue fluid, then migrate to the site of injury and attack the pathogen or other cause of the tissue damage. The chemotactic factors released by pathogen or mast cells attract the neutrophils first, then lymphocytes, monocytes and macrophages depending on the severity. Increased blood flow to the injured sites results in swelling, redness and pain.

The chronic inflammation is a slow process and occurs when neutrophils and macrophages are unable to protect host from the tissue damage. The body attempt to separate injured site and isolate from the surrounding tissues by the formation of granuloma (*granulum* means a small particle and *oma* means to form). This results in formation of new connective tissue and causes permanent tissue damage.

13.4 INTERFERONS (IFNs)

Interferons are cytokines produced by many eukaryotic cells in response to inducers such as virus infection, double stranded RNA, endotoxins and pathogens. These are proteins or glycoproteins known to inhibit viral multiplication inside the cells.

Several classes of interferons are identified based on physicochemical and antigenic properties such as IFN-α, IFN-β and IFN-γ. The IFN-α and IFN-β and generally degrade the mRNA synthesis in the virally infected cells. They are also known to activate the macrophages, T cells, NK cells and neutrophils that are involved in specific immunity.

13.5 FEVER

During the viral or bacterial infections, pyrogenic (heat generating) compounds such as IL-1, IL-6 and tissue necrosis factor are produced by the macrophages. These compounds circulate to the hypothalamus and induce neurons to secrete prostaglandins. Prostaglandins reset the thermostat mechanisms of the body to higher temperatures from normal 37°C/98.6°F (rectum 37.5°C/99.5°F). The higher temperature stimulates leukocytes for the destruction of pathogen, enhances the specific immune activity and increases microbiostasis (pathogen growth inhibition) by decreasing available iron to the pathogen.

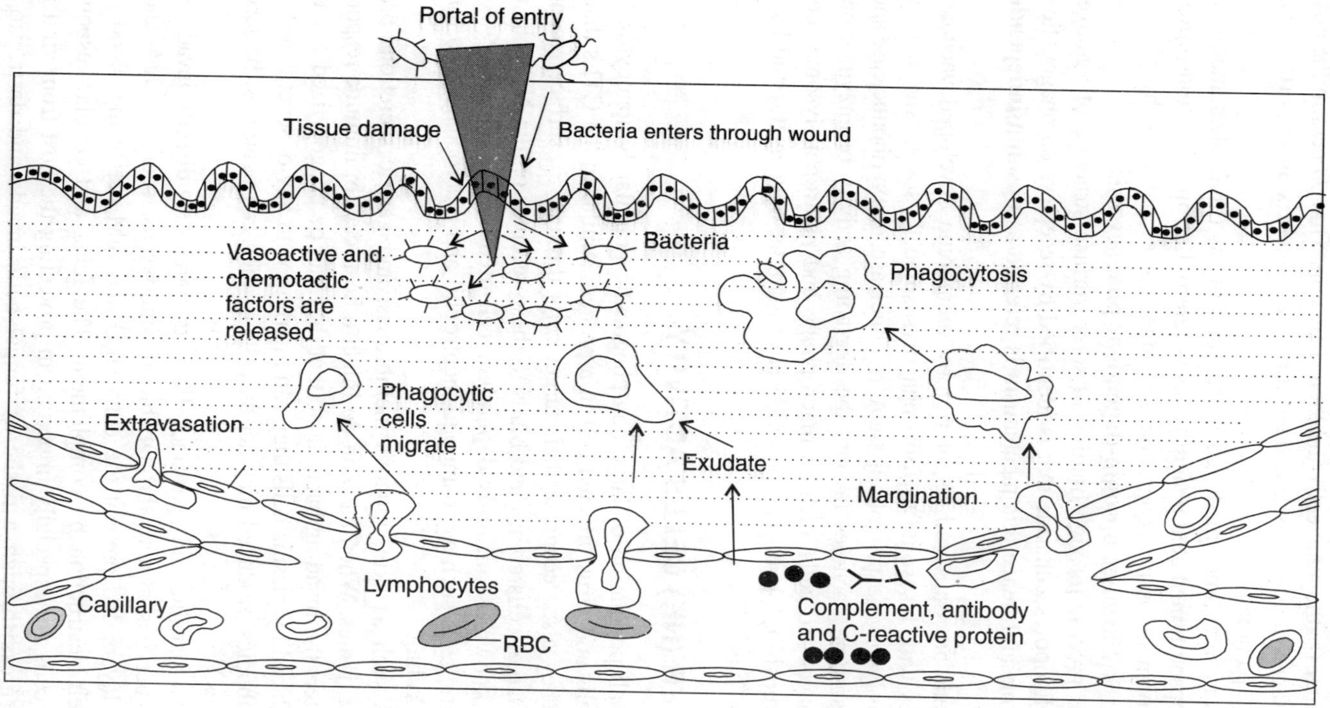

Fig. 13.1. Major events occurring in the inflammatory response. A bacterial infection causes tissue damage with production of various vasoactive and chemotactic factors. These factors induce increased blood flow to the damaged area, increased capillary permeability, and an influx of white blood cells, including phagocytes and lymphocytes, from the blood into the affected tissues. The serum proteins contained in the exudate have antibacterial properties, and the phagocytes begin to engulf the bacteria.

13.6 PHAGOCYTOSIS

Phagocytosis (*phagein* means to eat, *cyte* means cell and *osis* means process) is the process of engulfment of solid particulate matter (pathogenic cell) and their digestion. This process in the host is carried out by monocytes, macrophages, neutrophils and dendritic cells which is an important early defense mechanism against invading pathogens.

The phagocytic process is sometimes mediated by serum proteins (antibodies and/or complement C3b) called **opsonin**. This opsonin mediated phagocytosis is referred as **opsonization** (*opson* means to prepare victim). These opsonin helps in recognition and ingestion of pathogen by phagocytic cells. This phagocytic mechanism is known as **opsonin-dependent phagocytosis**.

Phagocytic cells do this process through binding of some surface structure of pathogens such as lectins, proteins, mannose, lipopolysaccharide that is identified by receptors on phagocytic cells. Here opsonin will not play any role and hence this mechanism is referred as **opsonin-independent phagocytosis** (Fig. 13.2).

Once ingested pathogens are delivered to interior by membrane enveloped vacuole known as phagosome. This later fuses with lysosome to form phagolysosome. Lysosomes contain a variety of hydrolysing enzymes such as lysozyme, phospholipase A, ribonuclease, deoxyribonuclease and proteases. The acidic pH of phagolysosome vacuoles favours the degrading activity of enzymes on entrapped pathogen. Some of the phagocytic cells such as macrophage and neutrophil lysomes contain toxic compounds (superoxide radical-O_2^-, hydrogen peroxide-H_2O_2, singlet oxygen-$^1O_2$ and hydroxyl radical-OH·) that kills the pathogens.

13.7 NATURAL KILLER (NK) CELLS ACTIVITY

Natural Killer cells are small population of large non-phagocytic granular lymphocytes. Their major function is to kill tumor cells and virus infected cell in a process similar to the CTLs that employing perforin and granzymes. They also recognize their target cells through F_C surface receptor for IgG antibody. The IgG binded target cells comes and binds to these receptor on the NK cells and favours the killing process and is known as **antibody-dependent cell-mediated cytotoxicity** (ADCC) (Fig. 13.3). The other non-specific cells such as eosinophil, neutrophil and macrophage also follow similar mechanism for target cell killing.

Non-specific defenses such as skin, mucous membranes, and gastric secretions prevent many bacteria from entering host tissues. When bacteria do infect a host, certain immune responses such as antibodies, complement alters the invading bacteria so that they can be phagocytized and undergo cell lysis. Once antibodies produced by plasma cells they can interfere with any of several steps in bacterial invasion (Table 13.1). Antibodies produced may act on cells or they can neutralize bacterial toxins or inactivate bacterial enzymes.

Viruses first attack cells that line body passages. From this area they directly invade target organs such as lungs or liver or nervous system through blood. Interferons, secretory IgA and some IgG molecules act at mucous layers and prevent the entry. IgG and IgM act in the blood to neutralize viruses or to promote their destruction by complement. The fever and cellular responses such as cytotoxic T cells and NK cells are especially important in protecting the host from viral infection.

Fungal infections progress through a tissue as fungal cells invade and destroy one cell after another. Fungal skin infections probably are protected by IgA antibodies and helper T cells, which release certain cytokines that activate macrophages. The macrophages in turn, engulf and digest fungi.

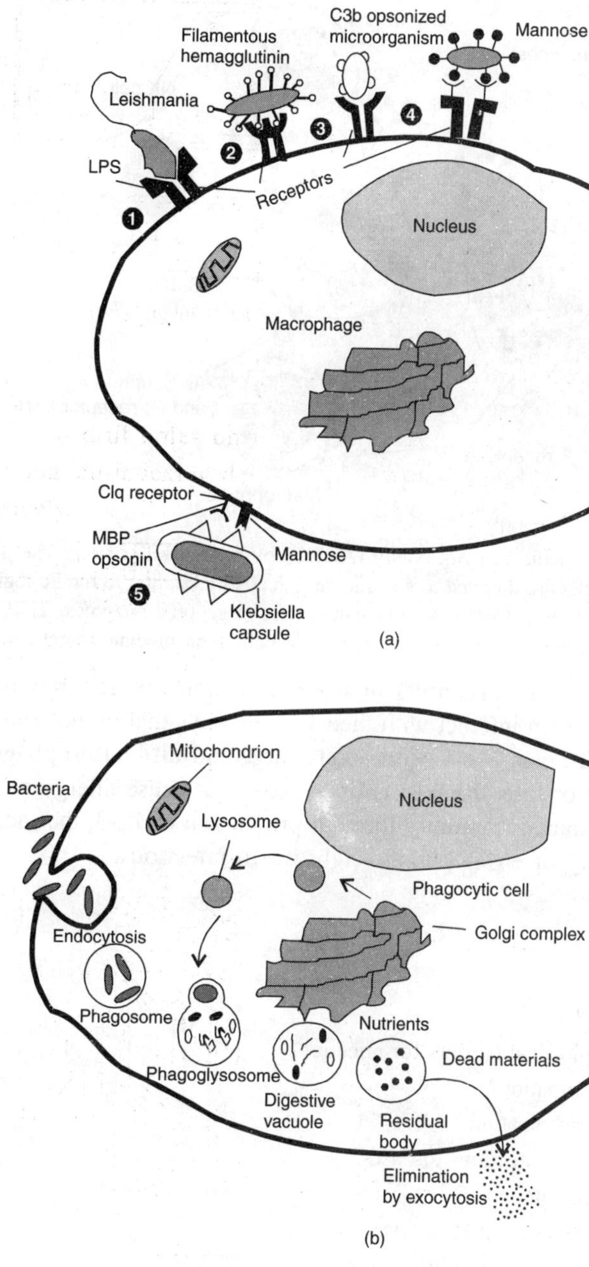

Fig. 13.2. Phagocytosis by macrophage. (a) Drawing shows receptors on a phagocytic cell, such as a macrophage, and the corresponding adhesions on foreign pathogenic microbial surfaces participating in phagocytosis. (1) The LPS binding site for lipophosphoglycan of *Leishmania* spp. (2) filamentous hemagglutinin of *Bordetella pertussis;* (3) binding to C3b on an opsonised cell; (4) the mannose-containing oligosaccharide side chain for lectin phagocytosis mediated by type 1 fimbriated; and (5) the participation of capsular polysaccharide of *Klebsiella pneumoniae* in nonpsonic binding mediated by the mannose receptor. (b) Drawing of phagocytosis, showing ingestion, intracellular digestion, and exocytosis.

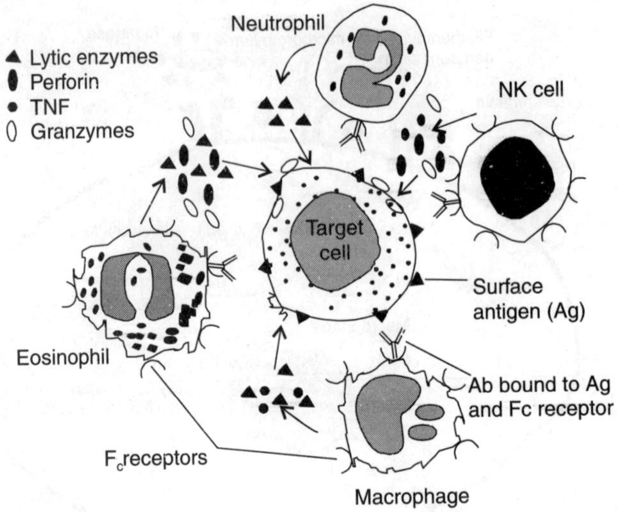

Fig. 13.3. Antibody-dependent cell-mediated cytotoxicity (ADCC) by natural killer cells. The natural killer cells and non-specific cytotoxic cells are directed to specific target cells by binding to the Fc region of antibody bound to surface antigens on the target cells. Various substances (e.g., lytic enzymes, TNF, perforin, granzymes) secreted by the natural killer and non-specific cytotoxic cells then mediate target-cell destruction.

Protozoa and helminths are quite dissimilar in size and complexity, but may use similar methods of invading the body. Many of them interact with their host in ways that do not endanger host survival and thereby ensure parasitic survival. Most of these are large and difficult to phagocytize, some may change the antigenic character or kills the host cells by toxins or cause allergy or develop cysts. The host combats these by cell-mediated immunity through production of IL-3, that activate macrophages or IL-5, that enhance the ability of eosinophils to fight worm infections.

EXERCISE

1. Define host and parasite.
2. What is meant by pathogen?
3. Mention different types of physical barriers for infection.
4. What do you mean by inflammation?
5. Explain phagocytosis and opsonization. (BU Oct. 2002)
6. Describe in detail immunity to infectious diseases.
7. Give an account of inflammation.
8. Write a short note on protective function of fever.
9. How do interferons protect cells by viral infection?
10. Briefly explain antibody-dependent cell-mediated cytotoxicity.

Chapter

14

Vaccines

14.1 INTRODUCTION

Vaccines are the preparations from an infectious microorganisms that is injected to humans or other animals to induce protective immunity. The complete protection shown by the host towards infectious disease is called immunity. Vaccines may consist of a preparation of killed or living weakened (attenuated) microorganisms, inactivated bacterial toxins (toxoids), purified macromolecules, recombinant vectors or DNA vaccines.

The administration of vaccines to animals is known as **vaccination**. For the first time Edward Jenner and Louis Pasteur advocated vaccination for small pox and Rabies diseases respectively. Now vaccines are developed against most of the epidemic diseases (anthrax, cholera, diphtheria, German measles, hepatitis A and B, measles, mumps, polio and whooping cough). The immunity can be active immunity or passive immunity. The **active immunity** means the host organisms are active in the process of obtaining immunity. This can be obtained by natural infections or vaccines administration. The **passive immunity** means the host organisms obtaining immunity without being involved in the immunity process. This can be obtained from the prepared antibodies which comes from the different organisms such as child obtaining immunity from maternal antibodies or antitoxin injections given to snake bitten persons.

The successful vaccine must be able to stimulate immune responses that are similar to those resulting from natural exposure to the pathogenic microorganisms. At present many vaccines are given to humans starting from birth to adolescent stages (Fig. 14.1). Many efforts are being made to produce successful vaccines for the most dangerous such as AIDS, herpes, hepatitis C, malaria and tuberculosis, but not completely successful result is obtained.

14.2 TYPES OF VACCINES

There are many types of vaccines available for the various bacterial, viral, and Protozoan diseases. Based on the type of the materials used for the preparation vaccines are classified.

Whole-organism Vaccines

The preparations consist of whole organisms that are killed (inactivated) or attenuated (avirulent live) are known as whole-organisms vaccines. Many of the common vaccines currently used for the bacterial

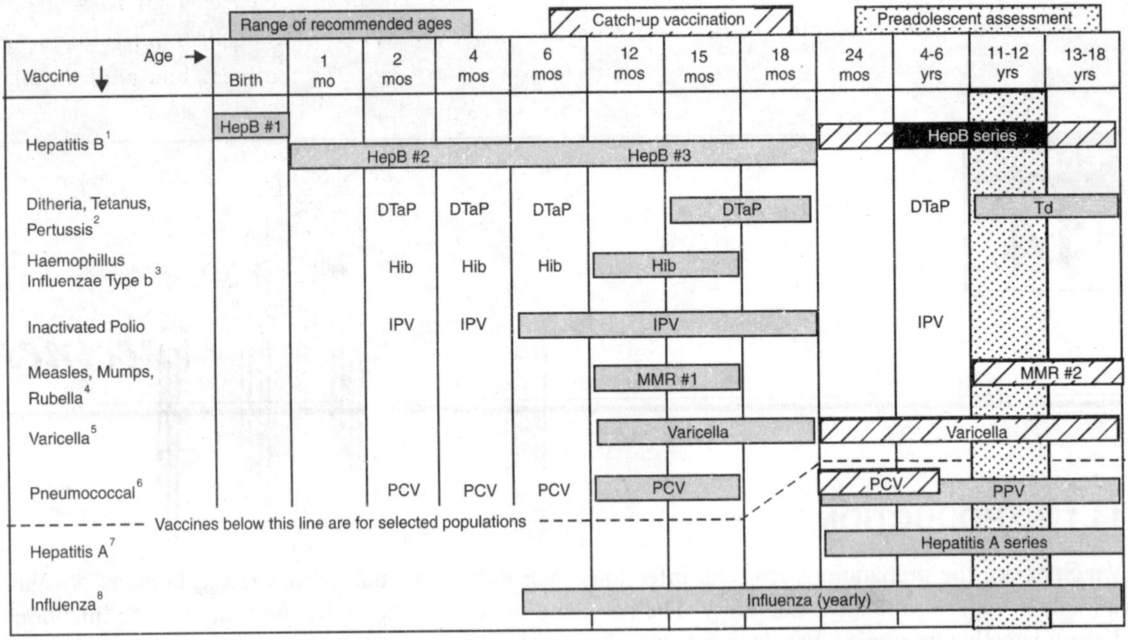

Fig. 14.1. Generally recommended childhood and adolescent immunization schedule

or viral diseases come under this type. The anthrax vaccine, BCG vaccine, cholera vaccine, hepatitis A vaccine, measles vaccine, polio vaccine and yellow fever vaccine are example for whole-organism vaccines. The inactivated vaccines are effective, but require several booster doses and they do not stimulate cell-mediate immunity and secretory IgA production properly. Usually attenuated vaccines work effectively in a single dose and stimulate both humoral and cell-mediated immunity.

Purified Macromolecular Vaccines

The whole-organisms vaccines has some disadvantages such as some fail to give immunity, reverts to virulent form by mutation, causes unknown effect or acts as allergenic agents in the host. To overcome this purified specific macromolecules derived from the pathogenic microorganisms are used in the preparation of vaccines. The pure capsular polysaccharide (*Streptococcus* pneumonia vaccine), denatured toxins called toxoid (tetanus vaccine) and recombinantly produced surface antigen (Hepatitis B vaccine) are used for the vaccine preparation.

Recombinant-vector Vaccines

The genes that encode major antigens of dangerous pathogenic organisms are introduced into non-pathogenic or attenuated microorganisms (bacteria or virus), which act as vector. This attenuated microorganisms are introduced into the host, where these vector microorganisms expressing antigenic proteins during their growth. These antigenic compound induce both cellular and humoral immunity in the host organisms. Adenovirus, canarypox virus, vaccinia virus, attenuated polio virus, attenuated *Salmonella* and *Mycobacterium bovis* have been used as vector vaccines.

DNA Vaccines

The recombinant plasmid DNA (Fig. 14.2) encoding antigenic proteins are injected directly to the muscle cells of the recipient. Here muscle cells take up the DNA and express the antigenic proteins

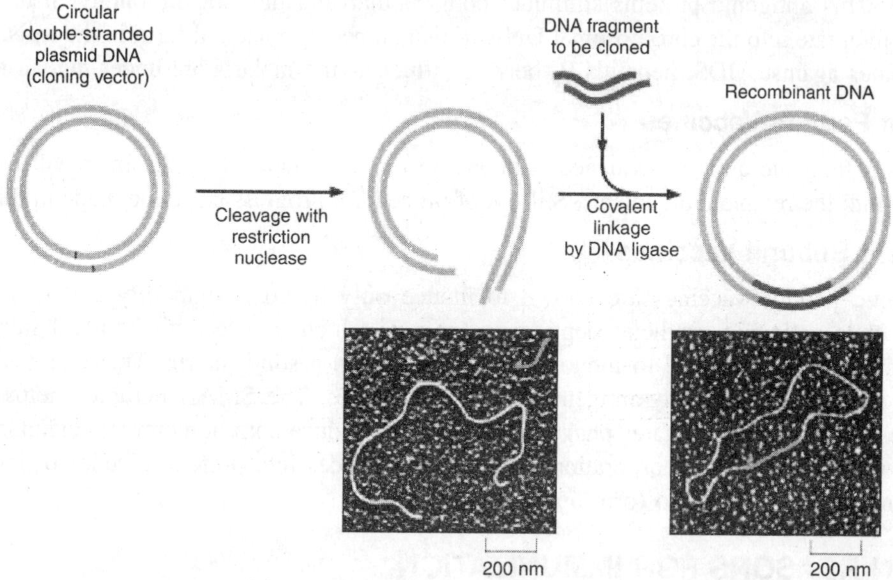

Fig. 14.2. Production of recombinant DNA plasmid that can be used in DNA vaccines

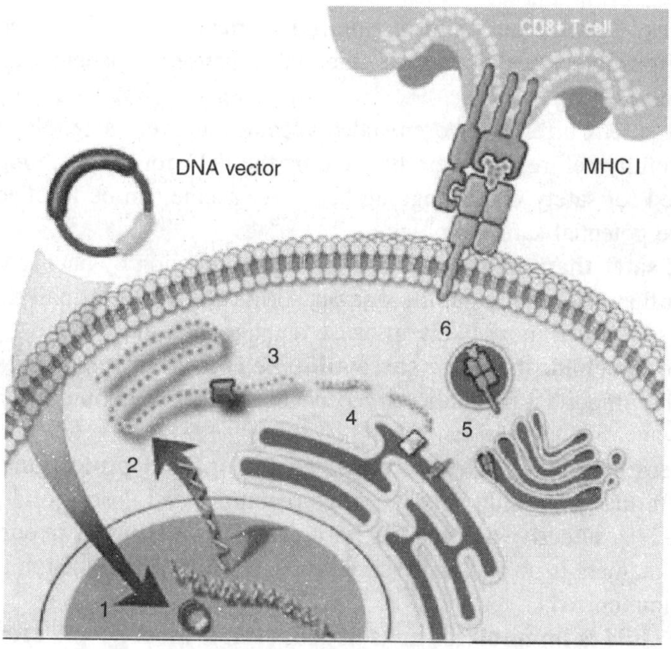

Fig. 14.3. DNA vaccines or genetic vaccines role inside the host muscle cell for the production of required antigen to get exposed in to immune system. 1. The plasmid vector is taken up into cells and transcribed in the nucleus. 2. The single stranded mRNA is translated into protein in the cytoplasm. The DNA vaccine-derived protein antigen. 3. is then degraded by proteosomes into intracellular peptides. 4. The vaccine derived-peptide binds MHC class I molecules. 5. Peptide antigen/MHC I complexes are presented on the cell surface. 6. binding cytotoxic CD 8+ lymphocytes and inducing a cell-mediated immune response.

(Fig. 14.3). This antigenic proteins stimulate both cellular and humoral immunity in the host. The DNA may integrate into the chromosomal DNA or maintained as episomal form in the cells. At present DNA vaccines against AIDS, hepatitis B, herpes, influenza and malaria are under human trials.

Synthetic Peptide Vaccines

As various pathogenic antigen sequence is known efforts are made to synthesize small peptides and introduced into the recipient organisms. Still lot of procedural progress has to be made in this category.

Multivalent Subunit Vaccines

The synthetic peptide vaccines are known to induce only humoral immunity and less of cellular immunity, thus antigenic synthetic peptide compounds which induce both humoral immunity and cellular immunity are attached to monoclonal antibodies on a solid matrix. This matrix approach is known as solid matrix-antibody-antigen (SMAA) complexes. The SMAA complex helps in making particulate matter, which facilitates phagocytosis and thus induce both humoral and cellular immunity. The other method includes incorporation of antigenic peptides into protein micelles or incorporation into lipid membrane vesicles to form liposomes.

14.3 TEN REASONS FOR IMMUNIZATION

There are ten important factors that favours the immunization to be practiced in the people and they are as follows.

1. **Vaccines work.** Having your child immunized on time is one of the very best ways you can protect him from infections having the potential to cause permanent damage and or even kill the individuals.
2. **Vaccines are safe.** Currently recommended vaccines are very safe. Not only was their safety evaluated carefully before they were licensed by the FDA for use in children, but they continue to be reviewed for safety on an ongoing basis. A vaccine can be recalled immediately if it is found to have potential safety problems.
3. **Vaccines are safer than the disease.** Preventing an infection by having your child immunized is much safer than letting him get the disease. Although there are small risks with vaccines, the natural disease is *much* more likely to cause damage than the vaccine.
4. **The younger the child, the more susceptible he is to developing severe infections.** That is why it is vitally important that children receive each of their immunizations as soon as they are due.
5. **Multiple doses of some vaccines are needed for optimal protection.** For that reason, you should be sure that your child gets *all* of his recommended doses—and gets them *on time.*
6. **Nothing else is as effective as vaccines** for preventing the diseases in question. No matter how careful you are, there is no way to be sure that your child will not catch these diseases short of having him immunized.
7. **Unless your child is immunized, he is at risk even if most people around him are already immunized,** People who have *not* been immunized are at increased risk for catching the vaccine-preventable diseases and developing their complications.
8. **If your child is not immunized he can spread disease to others**—including your family and friends. At particular risk are those who haven't been immunized, and the few individuals whose immunizations didn't "take"—in particular people with immune problems. They also are at especially high risk for developing severe disease.

9. **Immunizations help strengthen your child's immune system** by preparing it to defend him against disease.
10. **Diseases will return if people are not immunized.** If children are not immunized, there will be a resurgence of devastating diseases like polio and measles.

EXERCISE

1. What are vaccines?
2. Differentiate between active immunity and passive immunity. (BU April 2001, Oct. 2002)
3. Describe different types of vaccines. (BU Oct. 2001)
4. List disadvantages of whole-organisms vaccines.
5. Define vaccination.
6. What do you mean by DNA vaccines?
7. Give an account of types and properties of polio vaccine. (BU Oct. 2001)
8. Define immunity. Explain active and passive immunity. (BU Oct. 2001)
9. What is meant by SMAA complex?
10. What do you mean by multivalent subunit vaccine?

Chapter

15

Hypersensitivity Reactions

15.1 INTRODUCTION

Immune system works efficiently to remove invading antigens without causing noticeable damage to the host. But under certain circumstances immune system cause damages to the host resulting in tissue loss and/or damage and even death. This type of inappropriate immune response is known as hypersensitivity. Allergic reactions occur when an individual who has produced IgE antibody in response to an innocuous antigen, or **allergen**, subsequently encounters the same allergen. The allergen triggers the activation of IgE-binding mast cells in the exposed tissue, leading to a series of responses that are characteristic of *allergy*. The IgE is involved in protective immunity, especially in response to parasitic worms, which are prevalent in less developed countries. In the industrialized countries, however, IgE responses to innocuous antigens predominate and allergy is an important cause of disease (Table 15.1). Although rarely life-threatening, it causes much distress and lost time from school and work. Because of the medical importance of allergy in industrialized societies, much more is known about the pathophysiology of IgE-mediated responses than about the normal physiological role of IgE.

15.2 ALLERGY AND HYPERSENSITIVITY

The term **allergy** was originally defined by Clemens Von Pirquet as "an altered capacity of the body to react to a foreign substance," which was an extremely broad definition that included all immunological reactions. Allergy is now defined in a much more restricted manner as "disease following a response by the immune system to an otherwise innocuous antigen." Allergy is one of a class of immune system responses that are termed *hypersensitivity reactions.* These are harmful immune responses that produce tissue injury and may cause serious disease. Hypersensitivity reactions were classified into four types by Coombs and Gell (Table 15.2). Allergy is often equated with type I hypersensitivity (immediate-type hypersensitivity reactions mediated by IgE), and will be used in this sense here.

In this chapter we will first consider the mechanisms that favor the production of IgE. We then describe the pathophysiological consequences of the interaction between antigen and IgE that is bound by the high-affinity Fc receptor (FcRI) on mast cells. Finally, we will consider the causes and consequences of other types of immunological hypersensitivity reactions.

Hypersensitivity refers to undesirable (damaging, discomfort-producing and sometimes fatal) reactions produced by the normal immune system. The term hypersensitivity is used to describe immune responses which are damaging rather than helpful to the host. Hypersensitivity reactions require a pre-sensitized (immune) state of the host. Hypersensitivity reactions can be divided into four types: type I, type II, type III and type IV, based on the mechanisms involved and time taken for the reaction. The four-group classification was expounded by **Philip George Houthem Gell** and **Robert Royston Amos ("Robin") Coombs** in 1963. The first 3 are mediated by antibody, the fourth by T cells. Frequently, a particular clinical condition (disease) may involve more than one type of reaction.

Table 15.1: IgE-mediated Reactions to Extrinsic Antigens. All IgE-mediated responses involve mast-cell degranulation, but the symptoms experienced by the patient can be very different depending on whether the allergen is injected, inhaled, or eaten, and depending also on the dose of the allergen.

Syndrome	*Common Allergens*	*Route of Entry*	*Response*
Systemic anaphylaxis	Drugs Serum Venoms Peanuts	Intravenous (either directly or following oral absorption into the blood)	Edema increased vascular permeability Tracheal occluson Circulatory collapse Death
Acute urticaria (wheal-and-flare)	Insect bites Allergy testing	Subcutaneous	Local increase in blood flow and vascular permeability
Allergic rhinitis (hay fever)	Pollens (ragweed, smothy, birch) Dust-mite feces	Inhalation	Edema of nasal mucosa limitation of nasal mucosa
Asthma	Dangers (cat) Pollens Dust-mite feces	Inhalation	Bronchial constriction Increased mucus production Airway inflammation
Food allergy	Tree nuts Peanuts Shellfish Milk Eggs Fish	Oral	Vomiting Diarrhea Prunitis (itching) Urticaria (hives) Anaphylaxis (rarely)

Type I Hypersensitivity

Type I hypersensitivity is also known as **immediate** or **anaphylactic** hypersensitivity is an allergic reaction provoked by reexposure to a specific type of antigen referred to as an allergen. Exposure may be by ingestion, inhalation, injection, or direct contact. The difference between a normal immune response and a type I hypersensitive response is that plasma cells secrete IgE. The reaction may involve skin (urticaria and eczema), eyes (conjunctivitis), nasopharynx (rhinorrhea, rhinitis), bronchopulmonary tissues (asthma) and gastrointestinal tract (gastroenteritis). The reaction may cause a range of symptoms from minor inconvenience to death. The reaction usually takes 15-30 minutes from the time of exposure to the antigen, although sometimes it may have a delayed onset (10-12 hours). Immediate hypersensitivity is mediated by **IgE**. The primary cellular component in this hypersensitivity is the **mast cell** or **basophil**. The reaction is amplified and/or modified by platelets, neutrophils and eosinophils. A biopsy of the reaction site demonstrates mainly **mast cells** and **eosinophils**.

Table 15.2: Comparison of Hypersensitivity Types

Type	*Alternative names*	*Often Mentioned Disorders*	*Mediators*
I	Allergy (immediate)	• Atopy • Anaphylaxis • Asthma	• IgE
II	Cytotoxic, antibody-dependent (ADCC)	• Erythroblastosis fetalis • Goodpasture's syndrome • Autoimmune hemolytic anemia	• IgM or IgG • (Complement)
III	Immune complex disease	• Serum sickness • Arthus reaction • SLE	• IgG • (Complement)
IV	Delayed-type hypersensitivity (DTH) Cell-mediated immune memory response, antibody-independent	• Contact dermatitis • Tuberculin skin test • Chronic transplant rejection	• T-cells

The mechanism of reaction involves preferential production of IgE, in response to certain antigens (**allergens**). IgE has very high affinity for its receptor on mast cells and basophils. A subsequent exposure to the same allergen cross links the cell-bound IgE and triggers the release of various pharmacologically active substances (Table 15.1). Cross-linking of IgE Fc-receptor is important in mast cell triggering. Mast cell degranulation is preceded by increased **Ca^{++} influx**, which is a crucial process; ionophores which increase cytoplasmic Ca^{++} also promote degranulation, whereas, agents which deplete cytoplasmic Ca^{++} suppress degranulation.

The agents released from mast cells and their effects are listed in Table 15.3. Mast cells may be triggered by other stimuli such as exercise, emotional stress, chemicals (*e.g.*, photographic developing medium, calcium ionophores, codeine, *etc.*), anaphylotoxins (*e.g.*, C_{4a}, C_{3a}, C_{5a}, *etc.*). These reactions, mediated by agents without IgE-allergen interaction, **are not hypersensitivity reactions** although they produce the same symptoms.

Table 15.3: Pharmacological Mediators of Immediate Hypersensitivity

Preformed mediators in granules	
Histamine	Bronchoconstriction, mucus secretion, vasodilatation, vascular permeability
Tryptase	Proteolysis
Kininogenase	Kinins and vasodilatation, vascular permeability, edema
ECF-A(tetrapeptides)	Attract eosinophil and neutrophils
Newly formed mediators	
Leukotriene B_4	Basophil attractant
Leukotriene C_4, D_4	Same as histamine but 1000x more potent
Prostaglandins D_2	Edema and pain
PAF	Platelet aggregation and heparin release: microthrombi

The reaction is amplified by PAF (platelet activation factor) which causes platelet aggregation and release of histamine, heparin and vasoactive amines. Eosinophil chemotactic factor of anaphylaxis

(ECF-A) and neutrophil chemotactic factors attract eosinophils and neutrophils, respectively, which release various hydrolytic enzymes that cause necrosis. Eosinophils may also control the local reaction by releasing arylsulphatase, histaminase, phospholipase-D and prostaglandin-E, although this role of eosinophils is now in question.

Cyclic nucleotides appear to play a significant role in the modulation of immediate hypersensitivity reaction, although their exact function is ill understood. Substances which alter cAMP and cGMP levels significantly alter the allergic symptoms. Thus, substances that increase intracellular cAMP seem to relieve allergic symptoms, particularly broncho-pulmonary ones, and are used therapeutically (Table 15.4). Conversely, agents which decrease cAMP or stimulate cGMP aggravate these allergic conditions.

Table 15.4: Relationship between Allergic Symptoms and Cyclic-Nucleotides

Lowering of Cyclic-AMP	*Elevation of Cyclic-AMP*
Stimulation of α-adrenergic receptor (nor-epinephrin, phenyl-epinephrin)	Stimulation of β-adrenergic receptor (epinephrine, isoproterenol)
or	Blocking of α-adrenergic receptor (phenoxybenzamine)
Blocking of β-adrenergic receptor (propanolol)	Inhibition of phosphodiesterase (theophylline)
	Binding of histamine-2 or PGE to their receptors
Elevation of cyclic-GMP	
Stimulation of γ-cholinergic receptor (acetyl choline, carbacol)	
WORSENING OF SYMPTOMS	**IMPROVEMENT OF SYMPTOMS**

Diagnostic tests for immediate hypersensitivity include **skin** (prick and intradermal) **tests** (Fig. 15.1), measurement of total IgE and specific IgE antibodies against the suspected allergens. Total

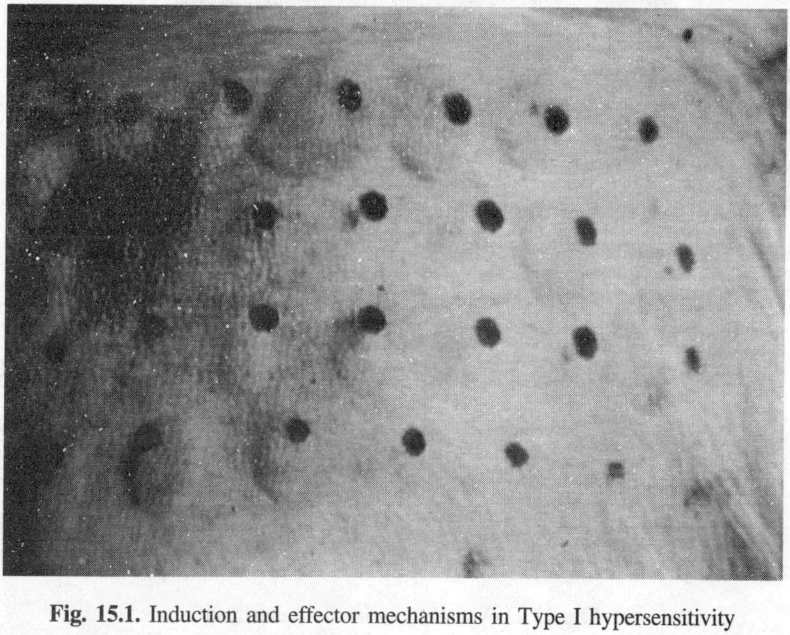

Fig. 15.1. Induction and effector mechanisms in Type I hypersensitivity

IgE and specific IgE antibodies are measured by a modification of enzyme immunoassay (ELISA). Increased IgE levels are indicative of an atopic condition, although IgE may be elevated in some non-atopic diseases (*e.g.*, myelomas, helminthic infection, *etc.*).

Symptomatic treatment is achieved with antihistamines which block histamine receptors. Chromolyn sodium inhibits mast cell degranulation, probably, by inhibiting Ca^{++} influx. Late onset allergic symptoms, particularly bronchoconstriction which is mediated by leukotrienes, are treated with leukotriene receptor blockers (Singulair, Accolate) or inhibitors of the cyclooxygenase pathway (Zileutoin). Symptomatic, although short term, relief from bronchoconstriction is provided by bronchodilators (inhalants) such as isoproterenol derivatives (Terbutaline, Albuterol). Thophylline elevates cAMP by inhibiting cAMP-phosphodiesterase and inhibits intracellular Ca^{++} release is also used to relieve bronchopulmonary symptoms.

There appears to be a genetic predisposition for atopic diseases and there is evidence for HLA (A2) association.

Hyposensitization (immunotherapy or desensitization) is another treatment modality which is successful in a number of allergies, particularly to insect venoms and, to some extent, pollens. The mechanism is not clear, but there is a correlation between appearance of IgG (blocking) antibodies and relief from symptoms. Suppressor T cells that specifically inhibit IgE antibodies may play a role.

Type II Hypersensitivity

Type II hypersensitivity is also known as cytotoxic hypersensitivity and may affect a variety of organs and tissues. The antigens are normally endogenous, although exogenous chemicals (haptens) which can attach to cell membranes can also lead to type II hypersensitivity. Drug-induced hemolytic anemia, granulocytopenia and thrombocytopenia are such examples. The reaction time is minutes to hours. Type II hypersensitivity is primarily mediated by antibodies of the IgM or IgG classes and complement (Fig. 15.2). Phagocytes and K cells may also play a role in ADCC.

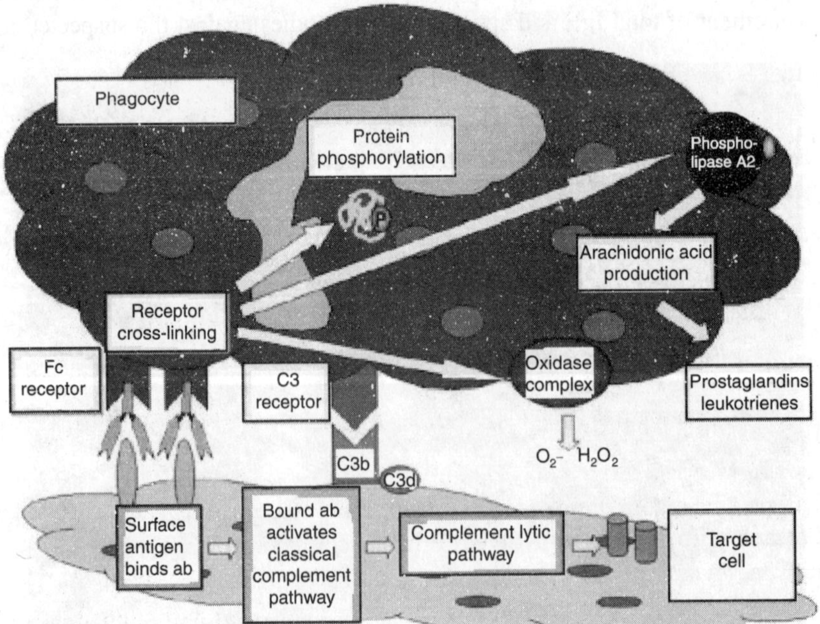

Fig. 15.2. Type II cytotoxicity mechanism

The lesion contains antibody, complement and neutrophils. Diagnostic tests include detection of circulating antibody against the tissues involved and the presence of antibody and complement in the lesion (biopsy) by immunofluorescence. The staining pattern is normally smooth and linear, such as that seen in Goodpasture's nephritis (renal and lung basement membrane) (Fig. 15.3a) and pemphigus (skin intercellular protein, desmosome) (Fig. 15.3b).

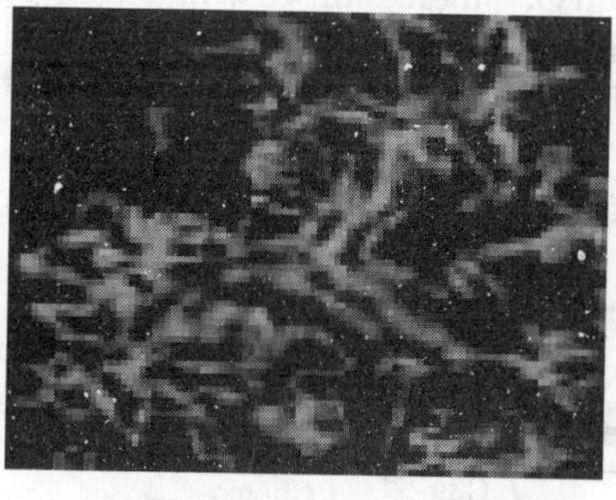

Fig. 15.3 (a) Immunofluorescent stain of immunoglobulin G (IgG) showing linear pattern in Goodpasture's syndrome

Fig. 15.3 (b) Pemphigus vulgaris—Immunofluorescence Treatment involves anti-inflammatory and immunosuppressive agents

Type III Hypersensitivity

Type III hypersensitivity is also known as immune complex hypersensitivity. The reaction may be general (*e.g.*, serum sickness) or may involve individual organs including skin (*e.g.*, systemic lupus erythematosus, Arthus reaction), kidneys (*e.g.*, lupus nephritis), lungs (*e.g.*, aspergillosis), blood vessels (*e.g.*, polyarteritis), joints (*e.g.*, rheumatoid arthritis) or other organs. This reaction may be the pathogenic mechanism of diseases caused by many microorganisms.

The reaction may take 3 - 10 hours after exposure to the antigen (as in Arthus reaction). It is mediated by soluble immune complexes. They are mostly of the IgG class, although IgM may also be involved. The antigen may be exogenous (chronic bacterial, viral or parasitic infections), or endogenous (non-organ specific autoimmunity: *e.g.*, systemic lupus erythematosus, SLE). The antigen is soluble and **not attached** to the organ involved. Primary components are soluble immune complexes and complement (C3a, 4a and 5a). The damage is caused by platelets and neutrophils (Fig. 15.4). The lesion contains primarily neutrophils and deposits of immune complexes and complement. Macrophages infiltrating in later stages may be involved in the healing process.

The affinity of antibody and size of immune complexes are important in production of disease and determining the tissue involved. Diagnosis involves examination of tissue biopsies for deposits of Ig and complement by immunofluorescence. The immunofluorescent staining in type III hypersensitivity is granular (as opposed to linear in type II such as seen in Goodpasture's syndrome). The presence of immune complexes in serum and depletion in the level of complement are also diagnostic. Polyethylene glycol-mediated turbidity (nephelometry), binding of C1q and Raji cell test are utilized to detect immune complexes. Treatment includes anti-inflammatory agents.

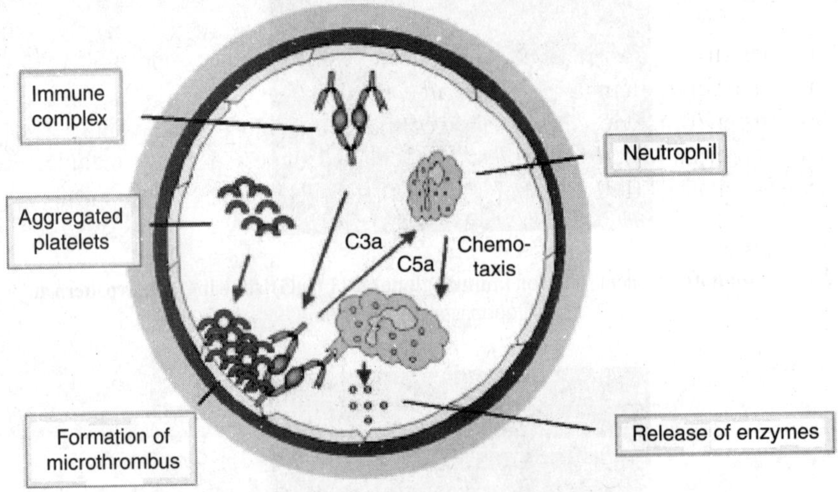

Fig. 15.4. Mechanism of damage in immune complex hypersensitivity

Type IV Hypersensitivity

Type IV hypersensitivity is also known as **cell mediated** or **delayed type hypersensitivity**. The classical example of this hypersensitivity is tuberculin (Montoux) reaction (Fig. 15.5) which peaks 48 hours after the injection of antigen (PPD or old tuberculin). The lesion is characterized by induration and erythema.

Table 15.5: Delayed Hypersensitivity Reactions

Type	*Reaction Time*	*Clinical Appearance*	*Histology*	*Antigen and Site*
Contact	48-72 hr	Eczema	Lymphocytes, followed by macrophages; edema of epidermis	Epidermal (organic chemicals, poison ivy, heavy metals, *etc.*)
Tuberculin	48-72 hr	Local Induration	Lymphocytes, monocytes, macrophages	Intradermal (tuberculin, lepromin, *etc.*)
Granuloma	21-28 days	Hardening	Macrophages, epitheloid and giant cells, fibrosis	Persistent antigen or foreign body presence (tuberculosis, leprosy, *etc.*)

Type IV hypersensitivity is involved in the pathogenesis of many autoimmune and infectious diseases (tuberculosis, leprosy, blastomycosis, histoplasmosis, toxoplasmosis, leishmaniasis, *etc.*) and granulomas due to infections and foreign antigens. Another form of delayed hypersensitivity is contact dermatitis (poison ivy (Fig. 15.6), chemicals, heavy metals, *etc.*) in which the lesions are more popular. Type IV hypersensitivity can be classified into three categories depending on the time of onset and clinical and histological presentation (Table 15.5).

Mechanisms of damage in delayed hypersensitivity include T lymphocytes and monocytes and/or macrophages. Cytotoxic T cells (Tc) cause direct damage whereas helper T (TH1) cells secrete cytokines which activate cytotoxic T cells and recruit and activate monocytes and macrophages, which cause the bulk of the damage (Fig. 15.4). The delayed hypersensitivity lesions mainly contain monocytes and a few T cells.

Major lymphokines involved in delayed hypersensitivity reaction include monocyte chemotactic factor, interleukin-2, interferon-gamma, TNF alpha/beta, *etc.*

Diagnostic tests *in vivo* include delayed cutaneous reaction (*e.g.* Montoux test (Fig. 15.5)) and patch test (for contact dermatitis). *In vitro* tests for delayed hypersensitivity include mitogenic response, lympho-cytotoxicity and IL-2 production. Corticosteroids and other immunosuppressive agents are used in treatment.

Fig. 15.5. Mantoux intradermal tuberculin skin test for tuberculosis

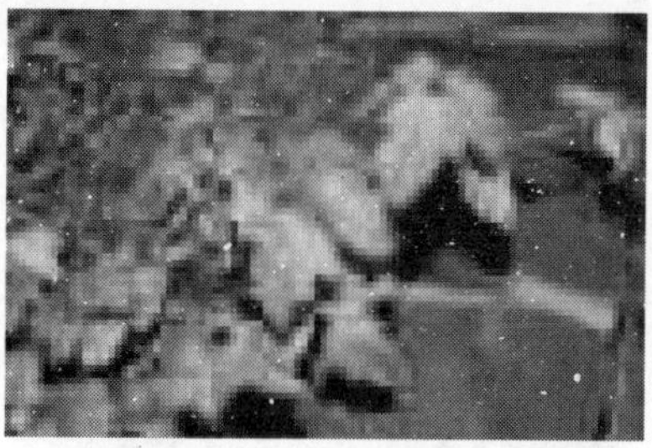

Fig. 15.6. Poison ivy plant

Types I–III are antibody-mediated and are distinguished by the different types of antigens recognized and the different classes of antibody involved (Fig. 15.7 and Table 15.6). Type I responses are mediated by IgE, which induces mast-cell activation, whereas types II and III are mediated by IgG, which can engage Fc-receptor and complement-mediated effector mechanisms to varying degrees, depending on the subclass of IgG and the nature of the antigen involved. Type II responses are directed against cell-surface or matrix antigens, whereas type III responses are directed against soluble antigens, and the tissue damage involved is caused by responses triggered by immune complexes. Type IV hypersensitivity reactions are T cell-mediated and can be subdivided into three groups. In the first group, tissue damage

	Type I	Type II	Type III	Type IV		
Immune reactant	IgE	IgG		T_H1 cells	T_H2 cells	CTL
Antigen	Soluble antigen	Cell- or matrix-associated antigen	Soluble antigen	Soluble antigen	Soluble antigen	Cell-associated antigen
Effector mechanism	Mast-cell activation	FcR^+ cells (phagocytes, NK cells)	FcR^+ cells complement	Macrophage activation	Eosinophil activation	Cytotoxicity
Example of hypersensitivity reaction	Allergic rhinitis, asthma, systemic anaphylaxis	Some drug allergies (e.g. penicilin)	Serum sickness, Arthus reaction	Contact dermatitis, tuberculin reaction	Chronic asthma chronic allergic rhinitis	Contact dermatitis

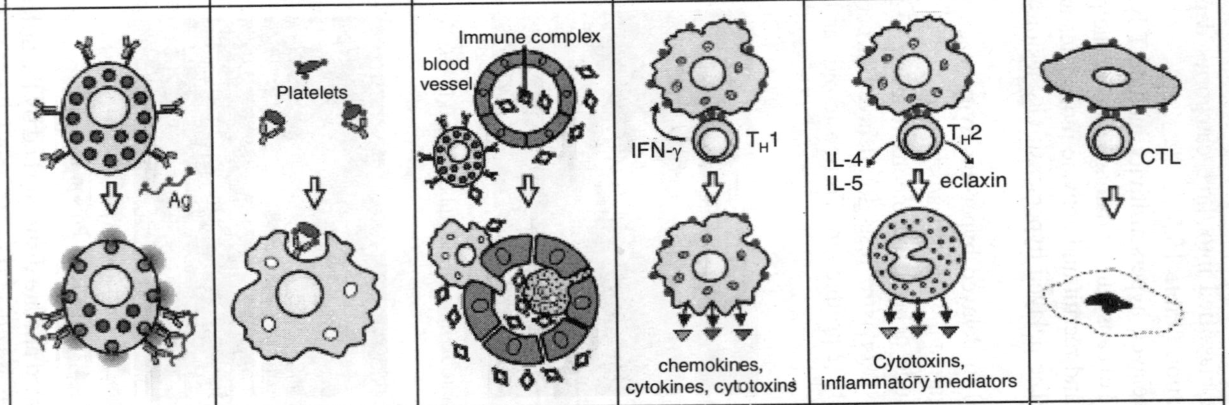

Fig. 15.7. There are four types of hypersensitivity reaction mediated by immunological mechanisms that cause tissue damage

is caused by the activation of macro-phages by T_H1 cells, which results in an inflammatory response. In the second, damage is caused by the activation by T_H2 cells of inflammatory responses in which eosinophils predominate; in the third, damage is caused directly by cytotoxic T cells (CTL).

Table 15.6: Comparison of Different Types of Hypersensitivity

Characteristics	*Type-I (anaphylactic)*	*Type-II (cytotoxic)*	*Type-III (immune complex)*	*Type-IV (delayed type)*
Antibody	IgE	IgG, IgM	IgG, IgM	None
Antigen	Exogenous	Cell surface	Soluble	Tissues & organs
Response time	15-30 minutes	Minutes-hours	3-8 hours	48-72 hours
Appearance	Weal & flare	Lysis and necrosis	Erythema and edema, necrosis	Erythema and induration
Histology	Basophils and eosinophil	Antibody and Complement	Complement and neutrophils	Monocytes and lymphocytes
Transferred with	Antibody	Antibody	Antibody	T-cells
Examples	allergic asthma, hay fever	Erythroblastosis Fetalis, Goodpasture's nephritis	SLE, farmer's lung disease	Tuberculin test, poison ivy, granuloma

15.3 IMMUNE DISORDERS

Many disorders are noticed in the four types of hypersensitivity responses of the immune system which lead to the many complications in humans. The following table 15.7 lists some of the disorders commonly noticed.

Table 15.7: Different Immune Disorders of Hypersensitivity Reactions

I(IgE)	Allergy (Food allergy) · Atopy (Atopic dermatitis) · Anaphylaxis · Urticaria · Hay fever
II(IgM, IgG)	ADCC · Pernicious anemia · Hemolytic disease of the newborn · Penicillin allergy
	autoimmune, cytotoxic: Idiopathic thrombocytopenic purpura · Bullous pemphigoid · Autoimmune hemolytic anemia · Goodpasture's syndrome · Pemphigus vulgaris · Rheumatic fever
	autoimmune, "type 5"/receptor: Graves' disease · Myasthenia gravis
III(Immune complex)	Serum sickness · Arthus reaction · Hypersensitivity vasculitis · Reactive arthritis · Henoch-Schönlein purpura · Farmer's lung · Post-streptococcal glomerulonephritis
	autoimmune: Systemic lupus erythematosus · Subacute bacterial endocarditis
IV(T-cells)	Contact dermatitis · Cell-mediated immunity · Mantoux test · Graft-versus-host disease
	autoimmune: Temporal arteritis · Diabetes mellitus type 1 · Hashimoto's thyroiditis · Guillain-Barré syndrome · Multiple sclerosis · Rheumatoid arthritis · Coeliac disease
Unknown/multiple	Autoimmune hepatitis · Transplant rejection

EXERCISE

1. Define hypersensitivity.
2. Explain why immune system shows hypersensitivity.
3. Describe various classification made in hypersensitivity reactions.
4. Elucidate how hypersensitivity type I develops in the host.
5. Differentiate between Type II and Type III hypersensitivity reactions in the host.
6. What are allergens and how it will result in the hypersensitivity reactions?
7. List out different pharmacological mediators of immediate hypersensitivity.
8. Discuss relationship between allergic symptoms and acyclic-nucleotides in the immune system.
9. Explain how skin-prick test will be performed to detect hypersensitivity Type I reaction.
10. What do you mean by hypersensitivitization and why it will be done?
11. Define antibody dependent cell-mediated cytotoxicity (ADCC).
12. Give an account of immune complex mediated diseases and explain molecular events of disease development.
13. Elucidate the development of delayed type of hypersensitivity in the host organisms.
14. Enumerate the process of contact dermatitis with suitable example.
15. What is meant by Mantoux test and why it is performed?

Chapter

16

Blood Groups, Blood Typing and Blood Transfusions

16.1 INTRODUCTION

Experiments with blood transfusions, the transfer of blood or blood components into a person's blood stream, have been carried out for hundreds of years. Many patients have died and it was not until 1901, when the Austrian Karl Landsteiner (1868-1943) discovered human blood groups, that blood transfusions became safer.

Mixing blood from two individuals can lead to blood clumping or agglutination. The clumped red cells can crack and cause toxic reactions. This can have fatal consequences. Karl Landsteiner discovered that blood clumping was an immunological reaction which occurs when the receiver of a blood transfusion has antibodies against the donor blood cells.

Austrian Karl Landsteiner (1868-1943)

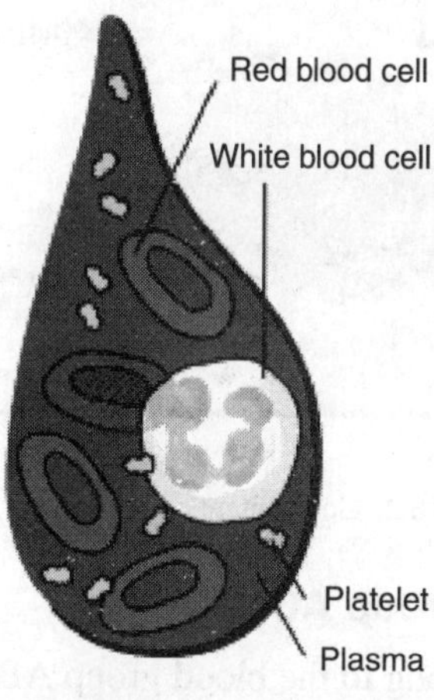

Fig. 16.1. Blood drop showing its components

Karl Landsteiner's work made it possible to determine blood types and thus paved the way for blood transfusions to be carried out safely. For this discovery he was awarded the Nobel Prize in Physiology or Medicine in 1930.

16.2 BLOOD COMPONENTS

An adult human has about 4–6 liters of blood circulating in the body. Among other things, blood transports oxygen to various parts of the body.

Blood consists of several types of cells floating around in a fluid called plasma (Fig. 16.1).

The red blood cells contain hemoglobin, a protein that binds oxygen. Red blood cells transport oxygen to, and remove carbon dioxide from, the body tissues.

The white blood cells fight infection.

The platelets help the blood to clot, if you get a wound for example.

The plasma contains salts and various kinds of proteins.

16.3 ABO BLOOD GROUPING SYSTEM

According to the AB0 blood typing system there are four different kinds of blood types: A, B, AB or O (null).

Blood Group A

If you belong to the blood group A, you have A antigens on the surface of your red blood cells and B antibodies in your blood plasma.

Blood Group B

If you belong to the blood group B, you have B antigens on the surface of your red blood cells and A antibodies in your blood plasma.

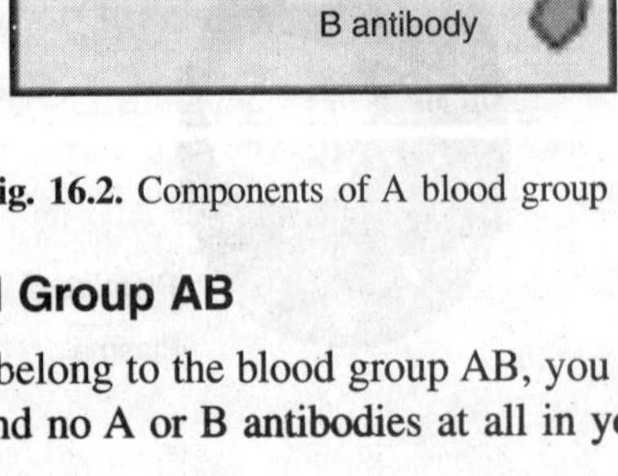

Fig. 16.2. Components of A blood group person

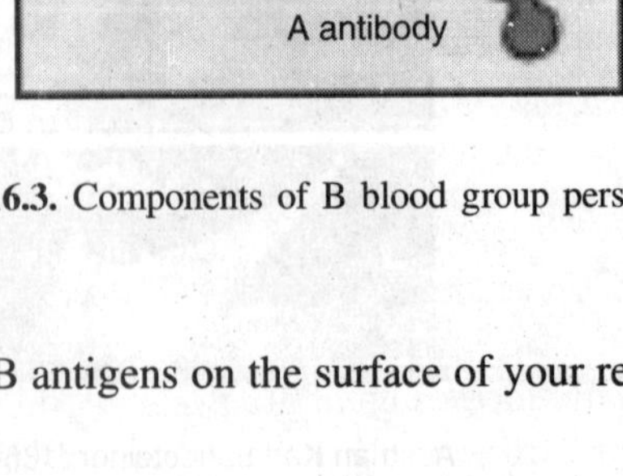

Fig. 16.3. Components of B blood group person

Blood Group AB

If you belong to the blood group AB, you have both A and B antigens on the surface of your red blood cells and no A or B antibodies at all in your blood plasma.

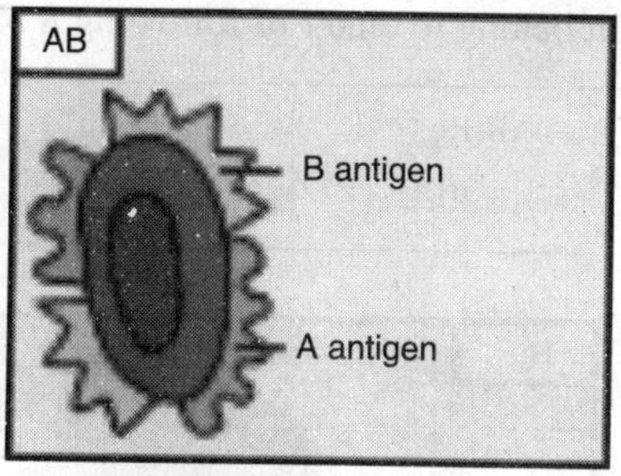

Fig. 16.3. (*a*) Components of AB blood group person

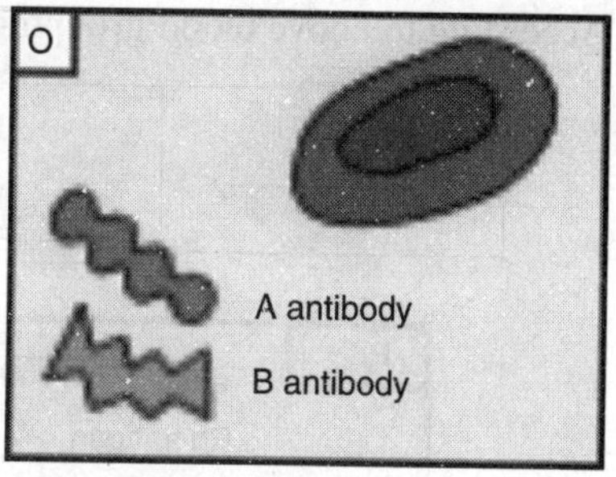

Fig. 16.3. (*b*) Components of O blood group person

Blood Group O

If you belong to the blood group 0 (null), you have neither A or B antigens on the surface of your red blood cells but you have both A and B antibodies in your blood plasma.

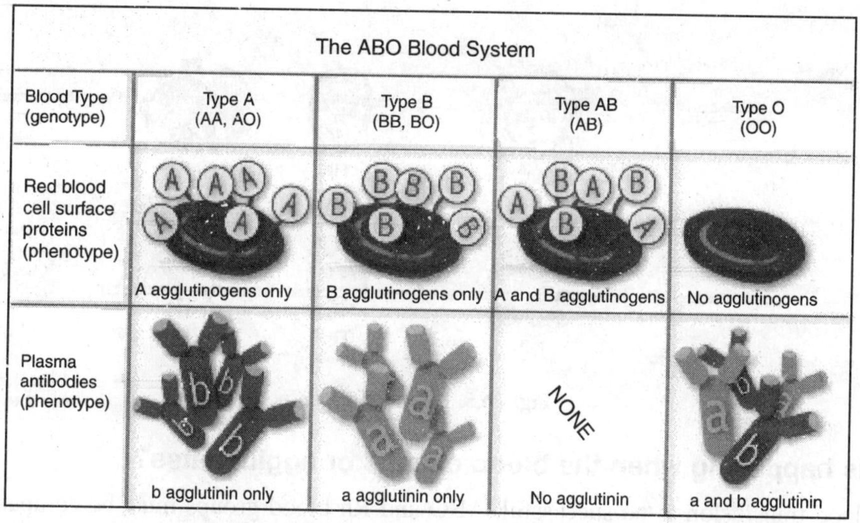

Fig. 16.4. Phenotypes and genotypes of different blood group persons

Agglutinogens are antigens that carry a person's identity. They are embedded in the cell membrane. In the RBC, the antigens can be A or B, both or none.

Plasma antibodies which react with the antigens (anti-A or anti-B). Blood group A with agglutinogen A on RBC has the opposite agglutinin (anti-B) in plasma and can never have or make anti-A.

16.4 Rh FACTOR BLOOD GROUPING SYSTEM

Many people also have a so called Rh factor on the red blood cell's surface. This is also an antigen and those who have it are called Rh^+. Those who haven't are called Rh^-. A person with Rh- blood does not have Rh antibodies naturally in the blood plasma (as one can have A or B antibodies, for instance). But a person with Rh^- blood can *develop* Rh antibodies in the blood plasma if he or she receives blood from a person with Rh^+ blood, whose Rh antigens can trigger the production of Rh antibodies. A person with Rh+ blood can receive blood from a person with Rh^- blood without any problems.

According to above blood grouping systems, you can belong to either of following 8 blood groups:

A Rh^+	B Rh^+	AB Rh^+	O Rh^+
A Rh^-	B Rh^-	AB Rh^-	O Rh^-

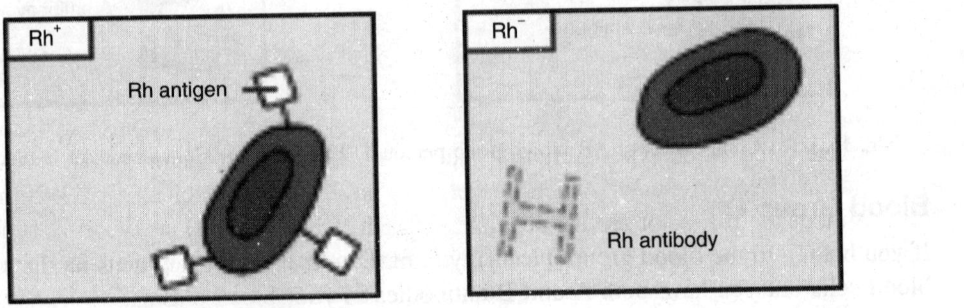

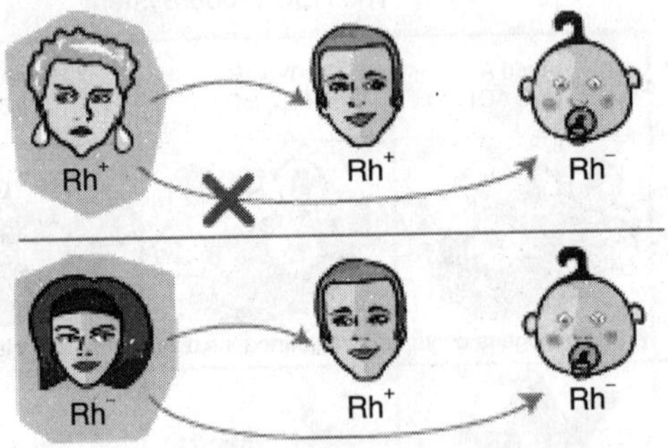

Fig. 16.5. Rh blood group notation

What is happening when the blood clumps or agglutinates?

For a blood transfusion to be successful, ABO and Rh blood groups must be compatible between the donor blood and the patient blood. If they are not, the red blood cells from the donated blood will clump or agglutinate. The agglutinated red cells can clog blood vessels and stop the circulation of the blood to various parts of the body. The agglutinated red blood cells also crack and its contents leak out in the body. The red blood cells contain hemoglobin which becomes toxic when outside the cell. This can have fatal consequences for the patient.

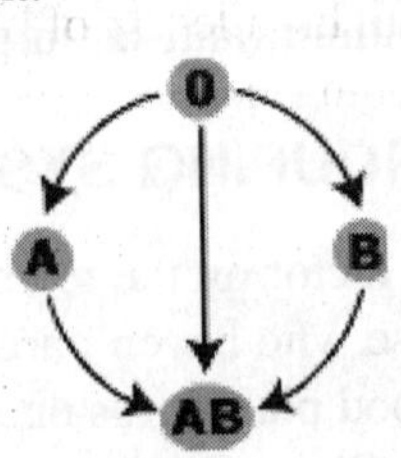

Fig. 16.6. People with blood group O are called "universal donors" and people with blood group AB are called "universal receivers"

The A antigen and the A antibodies can bind to each other in the same way that the B antigens can bind to the B antibodies. This is what would happen if, for instance, a B blood person receives blood from an A blood person. The red blood cells will be linked together, like bunches of grapes, by the antibodies. As mentioned earlier, this clumping could lead to death.

16.5 Rh FACTOR

Rh factor is a protein called an antigen that is found on the red blood cells of most people. (An antigen is any substance that the body considers "foreign" and thus stimulates the body to produce antibodies against it.) Rh factor, like the blood types A, B, and O, is inherited from one's parents. A simple blood test can determine blood type, including the presence of the Rh factor. About 85 per cent of white Americans and 95 per cent of African Americans have the Rh factor and are known as Rh-positive. Those without the Rh factor are Rh-negative. The Rh (**Rh = Rhesus)** status of the blood will be listed as negative (–) or positive (+). People who are RhD-positive have a protein on their red blood cells called D antigen, RhD-negative people do not have this. If you have Rh^- blood, your body may form antibodies against Rh^+ blood and destroy it. In order for this to happen, you must first be exposed to Rh^+ blood (*i.e.*, through a blood tranfusion or carrying an Rh^+ fetus). This can be a problem if you have antibodies against Rh^+ blood and are pregnant with an Rh^+ fetus. However, there is medication that can prevent this reaction from occurring if it is given immediately after you are exposed to Rh^+ blood.

Rh Factor in Pregnancy

Rh factor plays a critical role in some pregnancies. If a woman who is Rh-negative becomes pregnant by a man who is Rh-positive, the fetus may inherit the Rh factor from its father and be Rh-positive. If the blood of the fetus becomes mixed with the mother's Rh-negative blood, a disease called erythroblastosis fetalis can occur in future pregnancies, resulting in destruction of the fetus's red blood cells, brain damage, and even death.

The mixing of blood does not normally occur but may take place before or during birth if a tear in the placenta (the organ through which nutrients pass from the mother to the fetus) allows some fetal blood to enter the mother's circulatory system. If this happens, the fetus's red blood cells bearing the Rh factor stimulate the mother's white blood cells to produce antibodies against the foreign antigen. The mother's blood is now sensitized to the Rh factor.

Once a mother's blood is sensitized, the antibodies her body produces in response to the Rh antigen can cross the placenta and attach to the red blood cells of any Rh-positive fetus that she carries. This results in the rupture of the fetus's red blood cells, causing anemia (a condition marked by weakness and fatigue due to a reduced number of red blood cells). Severe anemia can lead to heart failure and death. The breakdown of red blood cells also causes the overproduction of a reddish-yellow substance called bilirubin. An infant with high levels of bilirubin will develop jaundice (have a yellowish appearance) and may suffer brain damage.

Symptoms

Rh incompatibility can cause symptoms ranging from very mild to fatal. In its mildest form, Rh incompatibility causes destruction of red blood cells.
Symptoms may include:

- Jaundice
- Hypotonia

- Motormental retardation
- Polyhydramnios (before birth)

Exams and Tests

There may be:

- A positive direct Coombs test result
- Higher-than-normal levels of bilirubin in the baby's cord blood
- Signs of red blood cell destruction in the infant's blood

Treatment

Since Rh incompatibility is almost completely preventable with the use of RhoGAM, preventio remains the best treatment. Treatment of the already affected infant depends on the severity of th condition.

Mild Rh incompatibility may be treated with:

- Aggressive hydration
- Phototherapy using bilirubin lights

Outlook (Prognosis)

Full recovery is expected for mild Rh incompatibility.

Possible Complications

Possible complications include:

- Hydrops fetalis
- Kernicterus
- Neurological syndrome with mental deficiency, movement disorder, hearing loss, speech disorde and seizures

When to Contact a Medical Professional

Call your health care provider if you think or know you are pregnant and have not yet seen a docto

Prevention

Rh incompatibility is almost completely preventable. Rh-negative mothers should be followed closel by their obstetricians during pregnancy.

If the father of the infant is Rh-positive, the mother is given a mid-term injection of RhoGAM an a second injection within a few days of delivery.

These injections prevent the development of antibodies against Rh-positive blood. This effectivel prevents the condition.

16.6 ERYTHROBLASTOSIS FETALIS

Among the various forms of congenital dropsy, in which the infants are often prematurely bor erythroblastosis, first described by Schridde and named by Rautmann, is the least understood. Congenit generalized edema may be the result of cardiac anomalies and diseases, portal obstruction, syphilis the liver, fetal peritonitis, abnormality of the D. venosus Arantii, deformities of the intestines ar liseases of the kidneys. Schridde, in 1910, pointed out a form of congenital general dropsy with ıydramnios associated with a pathological blood state.

The disorder is characterized by anasarca and fluid in the cavities, hydramnios and enlargement of he liver and spleen. The latter two organs show the most marked changes, which consist of the ccumulation, both inside and outside of the blood vessels, of large numbers of erythroblasts and a maller number of other marrow cells. The lymph follicles in the spleen are absent and the liver cells re crowded out. Accumulations of lymphoblasts in small numbers may be found in the kidneys, drenals, and lymph glands. Erythroblasts appear in the blood in greatly increased numbers and they how very often mitotic processes. The heart is often hypertrophied.

Because of the presence of hemosiderin in the spleen and liver, Schridde was led to believe that the lisease was due to a severe anemia with compensatory hematopoiesis having no relation to syphilis.)thers have assumed that the extramedullary formation of blood corpuscles was due to some form of nknown toxic action. Chiari described an infant in whom there was no blood pigment in the liver or pleen, and consequently no indications of any antecedent destruction of blood cells. Fischer, in his xamination of the older literature, came to the conclusion that many of the cases described as congenital eukemia were probably instances of erythroblastosis.

revention of Erythroblastosis Fetalis

rythroblastosis fetalis can be prevented by administering a preparation of anti-Rh factor antibodies to n Rh-negative mother whose blood has not yet produced antibodies to the Rh antigen. The preparation, nown as Rh immune globulin, rids the mother's blood of fetal red blood cells before she can become ensitized to them. Rh immune globulin is given whenever there is a possibility of fetal blood mixing ith maternal blood, such as following childbirth, an abortion, a miscarriage, or prenatal testing.

reatment for Erythroblastosis Fetalis

reatment for erythroblastosis fetalis depends to what extent the fetus is affected by the action of its nother's anti-Rh-factor antibodies. A fetus whose red blood cells are being severely destroyed can be eated with transfusions (replacement) of Rh-negative blood while it is still in the uterus. If the fetus ıows signs of illness close to its anticipated birth, the physician may elect to deliver the baby early. he baby's blood is then replaced with Rh-negative blood following birth.

A pregnant woman who is already sensitized to the Rh antigen can have her doctor carefully onitor the level of antibodies in her blood throughout her pregnancy. If the levels rise, the fetus will eed special attention. Unfortunately, once a woman is sensitized, she will always produce antibodies hen exposed to the Rh antigen.

The Rh^+ person cannot make anti-Rh^+. The Rh^- person able to make anti-Rh^+ if exposed to Rh^+.

At birth, there is no Rh agglutinin in the plasma, but a Rh^- person is born with ability to make it. xamples: Rh-gets blood with Rh^+. In 1st transfusion, nothing happens and no physical sign but tibodies anti-Rh^+ are madeNext in the second transfusion, the Rh^+ blood cells are killed by anti-Rh^+ d results in hemolytic reaction. If the baby is Rh^- and mother Rh^+ nothing happens. If the baby is 1^+ and mother Rh^-: 1st child: no physical sigh but the mother makes antibodies anti Rh^+ after the rth of the child. When the second child Rh^+ comes along, the baby's blood cells can be destroyed → molytic reaction (erythroblastosis fetalis). To save the baby, he will need to have a blood transfusion th Rh^- to remove the toxins formed from the hemoglobin degradation and to avoid more hemolysis. prevent the formation of these agglutinins anti Rh^+ by the mother, she is given Rhogam. This drug events mother from producing antibodies. Rhogam is nothing else than the antibodies themselves, anti-Rh^+. They destroy the fetus Rh^+ RBC before the mother's immune system has time to react to them. Can only be given early in pregnancy or after the child is born.

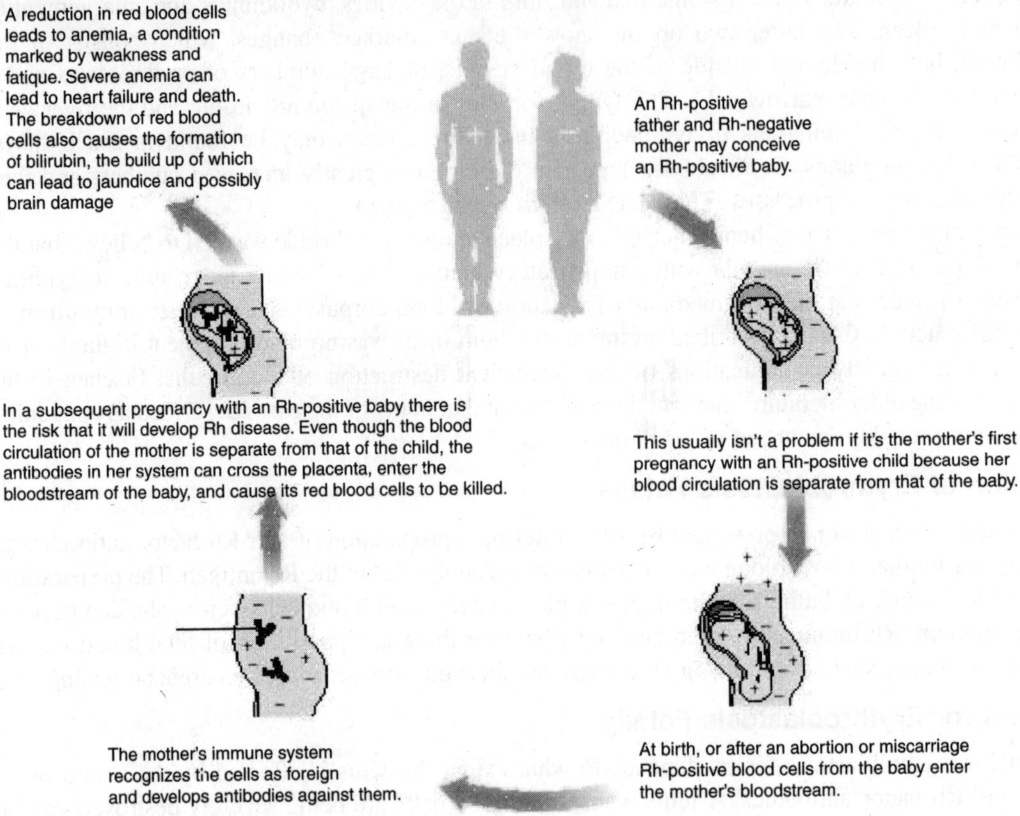

Fig. 16.7. The path of Rh disease

Inheritance of blood type can be complicated. You may have group "A" blood, but have genes for both groups "A" and "O." Because "A" group is stronger or dominant, your blood will be group "A." However, you can give your children genetic material for either group "A" or "O." The same is true of "B" group. If both "A" and "B" genetic material is given to your child, the child will have group "AB" blood. A child with genes for "O" from both parents will be group "O." The chart on the following page shows this pattern.

		Mother's group			
		O	**A**	**B**	**AB**
Father's group	**O**	O	O, A	O, B	A, B
	A	O, A	O, A	O, A, B, AB	A, B, AB
	B	O, B	O, A, B, AB	O, B	A, B, AB
	AB	A, B	A, B, AB	A, B, AB	A, B, AB

Rh status is similiar to blood groups. If you are Rh positive, you may have genes for both positive and negative. If you are Rh negative, you have two genes for Rh negative.

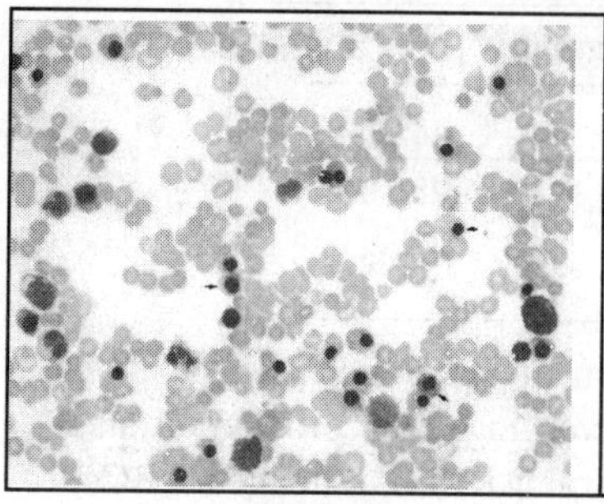

Erythroblastosis fetails

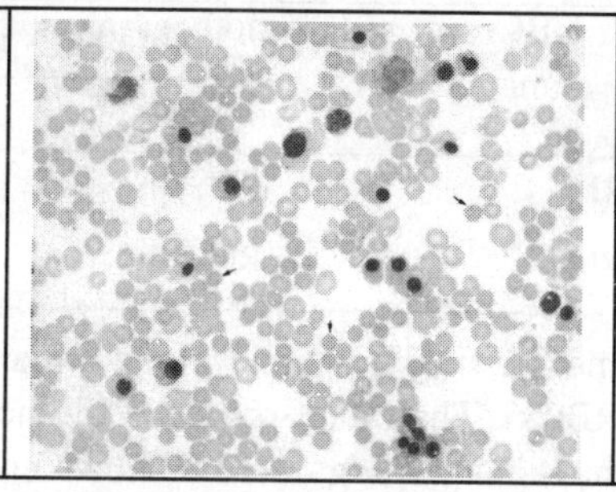

Erythroblastosis fetails

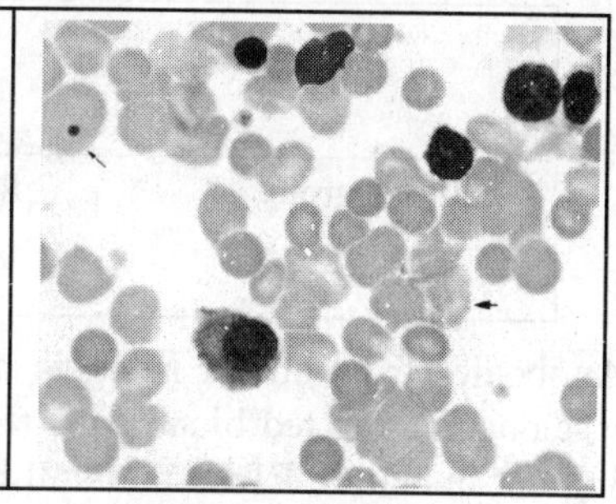

Erythroblastosis fetails

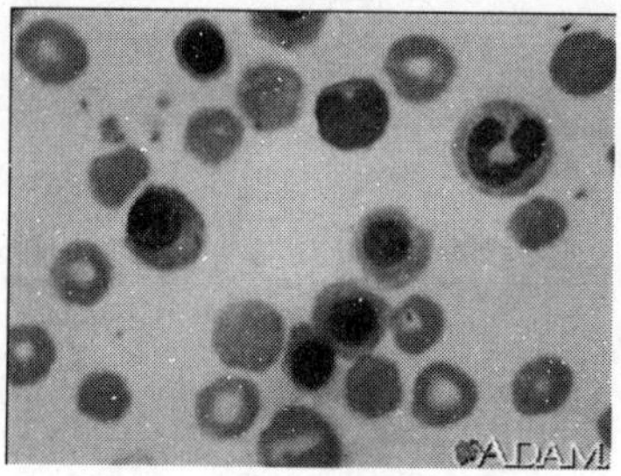

Erythroblastosis fetalis, photomicrograph

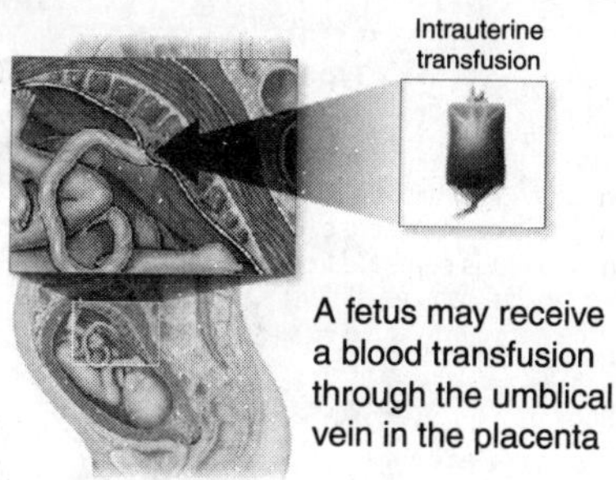

Intrauterine transfusion

Fig. 16.8

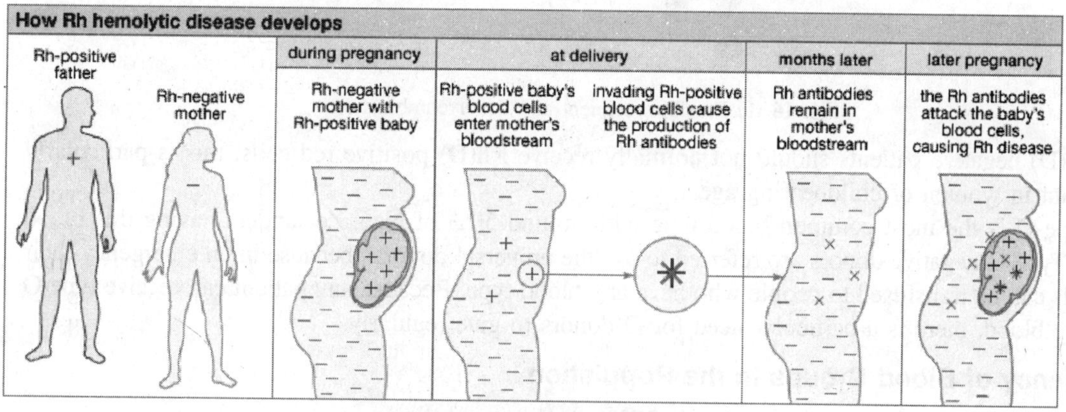

(a)

(b)

Fig. 16.9. Case of erythroblastosis (*a*) Development disease (*b*) Birth of baby with disease

		Mother's group	
		Rh^+	**Rh^-**
Father's group	**Rh^+**	Rh^+, Rh^-	Rh^+, Rh^-
	Rh^-	Rh^+, Rh^-	Rh^-

Antibodies from an Rh negative mother may enter the blood stream of her unborn Rh positive infant, damaging the red blood cells (RBCs). The infant responds by increasing RBC production and sending out immature RBCs that still have nuclei. Figure 16.8 shows normal RBCs, damaged RBCs, and immature RBCs that still contain nuclei.

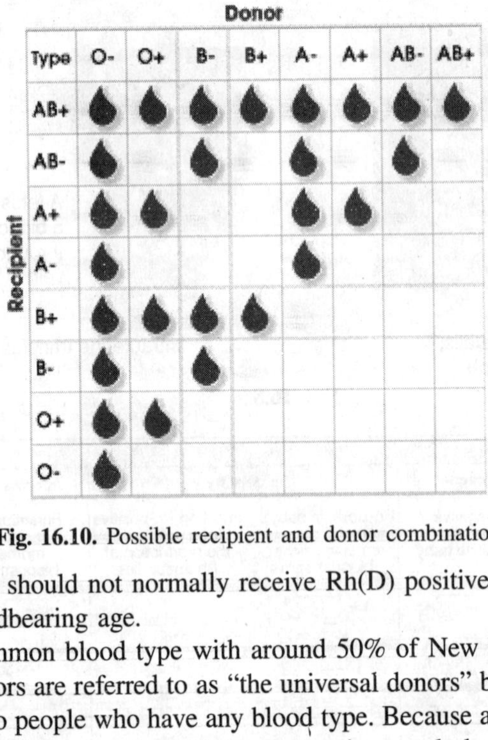

Fig. 16.10. Possible recipient and donor combination

Rh(D) negative patients should not normally receive Rh(D) positive red cells, this is particularly important in women of childbearing age.

Type O is the most common blood type with around 50% of New Zealanders having this blood group. Type O negative donors are referred to as "the universal donors" because in an emergency their red cells can be transfused to people who have any blood type. Because any patient can receive type O negative blood, there is a particular need for O donors to give regularly.

Frequency of Blood Groups in the Population

38% are O+

9% are O–

32% are A+

6% are A–

9% are B+

2% are B–

3% are AB+

1% are AB–

Fig. 16.11. The possible frequency of different blood groups

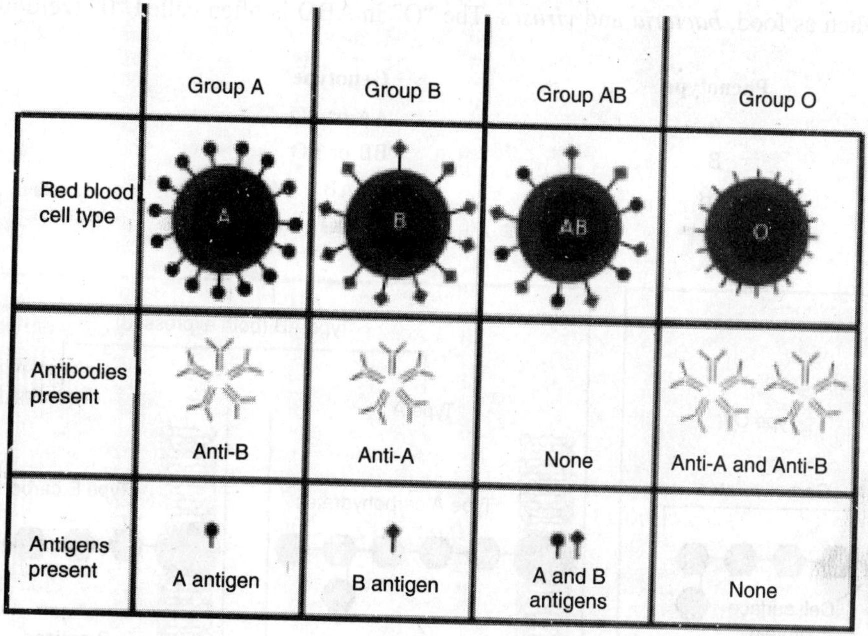

Fig. 16.12. Types of antigen, antibody present in all blood groups

Blood type (or blood group) is determined, in part, by the ABO blood group antigens present on red blood cells.

ABO Blood Group System in Greater Detail

The **ABO system** is the most important blood group system in human blood transfusion. The associated anti-A antibodies and anti-B antibodies are usually "Immunoglobulin M", abbreviated IgM, antibodies. ABO IgM antibodies are produced in the first years of life by sensitization to environmental

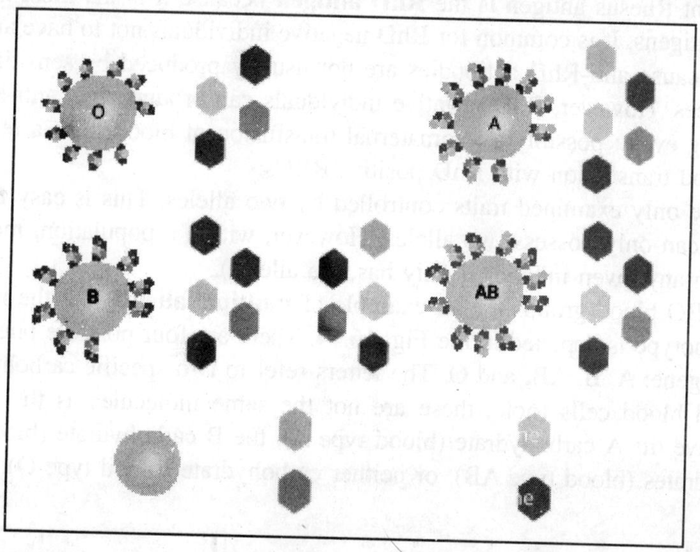

Fig. 16.13. ABO blood group system - Diagram showing the carbohydrate chains which determine the ABO blood group

substances such as food, *bacteria* and *viruses*. The "O" in ABO is often called "0" (zero/null) in other languages.

Phenotype	Genotype
A	AA or AO
B	BB or BO
AB	AB
O	OO

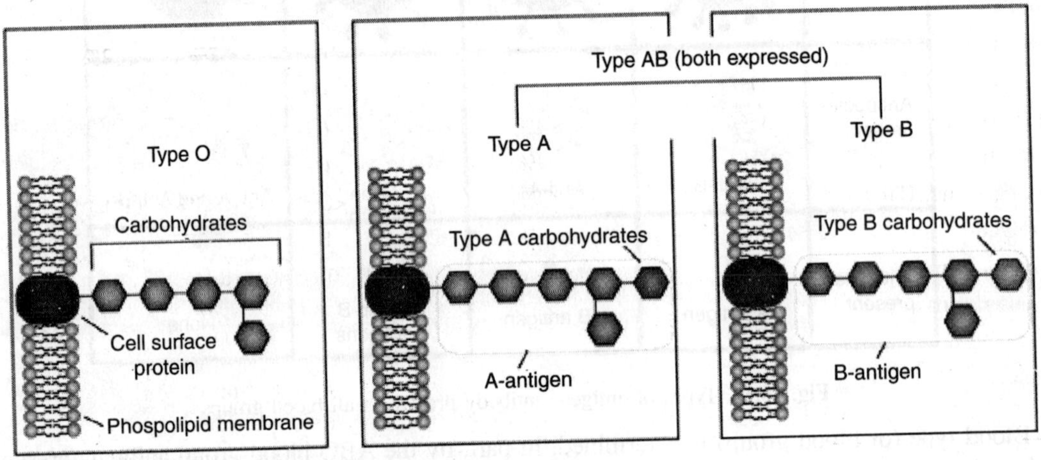

Fig. 16.14. AB antigens

16.7 RHESUS BLOOD GROUP SYSTEM

The **Rhesus system** is the second most significant blood group system in human blood transfusion. The most significant Rhesus antigen is the **RhD antigen** because it is the most immunogenic of the five main rhesus antigens. It is common for RhD negative individuals not to have any anti-RhD IgG or IgM antibodies, because anti-RhD antibodies are not usually produced by sensitization against environmental substances. However, RhD negative individuals can produce IgG anti-RhD antibodies following a sensitizing event: possibly a fetomaternal transfusion of blood from a fetus in pregnancy or occasionally a blood transfusion with RhD positive RBCs.

So far we have only examined traits controlled by two alleles. This is easy to visualize because diploid organisms can only possess two alleles. However, within a population, more than two alleles can exist (although any given individual only has two alleles).

The human ABO blood groups are an example of **multiple alleles**, and the relationship between phenotype and genotype is depicted in the Fig. 16.14. There are four possible phenotypic blood types for this particular gene: A, B, AB, and O. The letters refer to two specific carbohydrate molecules on the surface of red blood cells (note, these are not the same molecules as the MN blood groups). Individuals can have the A carbohydrate (blood type A), the B carbohydrate (blood type B), both the A and B carbohydrates (blood type AB), or neither carbohydrate (blood type O).

ABO and Rh Distribution by Country

Table 16.1: ABO and Rh Blood Type Distribution by different Nations (averages for each population)

Country	*O+*	*A+*	*B+*	*AB+*	*O−*	*A−*	*B−*	*AB−*
Austria	30%	33%	12%	6%	7%	8%	3%	1%
Australia	40%	31%	8%	2%	9%	7%	2%	1%
Belgium	38.1%	34%	8.5%	4.1%	7%	6%	1.5%	0.8%
Canada	39%	36%	7.6%	2.5%	7%	6%	1.4%	0.5%
Denmark	35%	37%	8%	4%	6%	7%	2%	1%
Finland	27%	38%	15%	7%	4%	6%	2%	1%
France	36%	37%	9%	3%	6%	7%	1%	1%
Germany	35%	37%	9%	4%	6%	6%	2%	1%
Hong Kong, China	40%	26%	27%	7%	<0.3%	<0.3%	<0.3%	<0.3%
Ireland	47%	26%	9%	2%	8%	5%	2%	1%
Korea, South	27.2%	35.1%	26.1%	11.3%	0.1%	0.1%	0.1%	0.05%
Netherlands	39.5%	35%	6.7%	2.5%	7.5%	7%	1.3%	0.5%
New Zealand	38%	32%	9%	3%	9%	6%	2%	1%
Poland	31%	32%	15%	7%	6%	6%	2%	1%
Sweden	32%	37%	10%	5%	6%	7%	2%	1%
UK	37%	35%	8%	3%	7%	7%	2%	1%
USA	37.4%	35.7%	8.5%	3.4%	6.6%	6.3%	1.5%	0.6%

16.8 CLINICAL SIGNIFICANCE

Blood Transfusion

Transfusion medicine is a specialized branch of *hematology* that is concerned with the study of blood groups, along with the work of a *blood bank* to provide a *transfusion* service for blood and other blood products. Across the world, blood products must be prescribed by a medical doctor (licensed *physician* or *surgeon*) in a similar way as medicines. In the USA, blood products are tightly regulated by the *Food and Drug Administration*.

Much of the routine work of a *blood bank* involves testing blood from both donors and recipients to ensure that every individual recipient is given blood that is compatible and is as safe as possible. If a unit of incompatible blood is *transfused* between a *donor* and recipient, a severe acute immunological reaction, *hemolysis* (RBC destruction), *renal failure* and *shock* are likely to occur, and death is a possibility. Antibodies can be highly active and can attack RBCs and bind components of the *complement system* to cause massive hemolysis of the transfused blood.

Patients should ideally receive their own blood or type-specific blood products to minimize the chance of a *transfusion reaction*. Risks can be further reduced by *cross-matching* blood, but this may be skipped when blood is required for an emergency. Cross-matching involves mixing a sample of the recipient's blood with a sample of the donor's blood and checking if the mixture *agglutinates*, or forms clumps. If agglutination is not obvious by direct vision, blood bank technicians usually check for *agglutination* with a *microscope*. If agglutination occurs, that particular donor's blood cannot be transfused to that particular recipient. In a blood bank it is vital that all blood specimens are correctly identified, so labeling has been standardized using a *barcode* system known as *ISBT 128.*

The blood group may be included on *identification tags* or on *tattoos* worn by military personnel, in case they should need an emergency blood transfusion. Frontline German *Waffen-SS* had such tattoos during *World War II*. Ironically, this was an easy form of *SS* identification.

Rare blood types can cause supply problems for *blood banks* and hospitals. For example, *Duffy*-negative blood occurs much more frequently in people of African origin, and the rarity of this blood type in the rest of the population can result in a shortage of Duffy-negative blood for patients of African ethnicity. Similarly for RhD negative people, there is a risk associated with travelling to parts of the world where supplies of RhD negative blood are rare, particularly *East Asia*, where blood services may endeavor to encourage Westerners to donate blood.

Hemolytic Disease of the Newborn (HDN)

A *pregnant* woman can make *IgG* blood group antibodies if her fetus has a blood group antigen that she does not have. This can happen if some of the fetus' blood cells pass into the mother's blood circulation (e.g. a small fetomaternal *hemorrhage* at the time of childbirth or obstetric intervention), or sometimes after a therapeutic *blood transfusion*. This can cause *Rh disease* or other forms of *hemolytic disease of the newborn* (HDN) in the current pregnancy and/or subsequent pregnancies. If a pregnant woman is known to have anti-RhD antibodies, the RhD blood type of a *fetus* can be tested by analysis of fetal DNA in maternal plasma to assess the risk to the fetus of Rh disease. Antibodies associated with some blood groups can cause severe HDN, others can only cause mild HDN and others are not known to cause HDN.

Red Blood Cell Compatibility

- **Blood group AB** individuals have both A and B antigens on the surface of their RBCs, and their *blood serum* does not contain any antibodies against either A or B antigen. Therefore, an individual with type AB blood can receive blood from any group (with AB being preferable), but can donate blood only to another group AB individual.

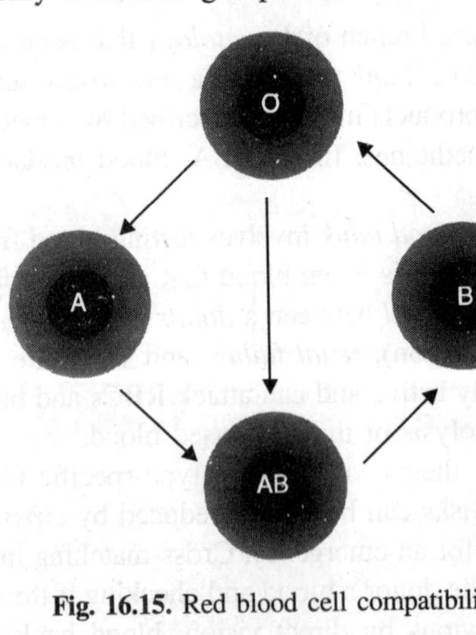

Fig. 16.15. Red blood cell compatibility

- **Blood group A** individuals have the A antigen on the surface of their RBCs, and blood serum containing *IgM* antibodies against the B antigen. Therefore, a group A individual can receive blood only from individuals of groups A or O (with A being preferable), and can donate blood to individuals with type A or AB.

- **Blood group B** individuals have the B antigen on the surface of their RBCs, and blood serum containing IgM antibodies against the A antigen. Therefore, a group B individual can receive blood only from individuals of groups B or O (with B being preferable), and can donate blood to individuals with type B or AB.
- **Blood group O** (or blood group zero in some countries) individuals do not have either A or B antigens on the surface of their RBCs, but their blood serum contains IgM anti-A antibodies and anti-B antibodies against the A and B blood group antigens. Therefore, a group O individual can receive blood only from a group O individual, but can donate blood to individuals of any ABO blood group (ie A, B, O or AB). If anyone needs a blood transfusion in a dire emergency, and if the time taken to process the recipient's blood would cause a detrimental delay, O Negative blood can be issued.

RBC Compatibility Chart

In addition to donating to the same blood group; type O blood donors can give to A, B and AB; blood donors of types A and B can give to AB.

Table 16.2: Red Blood Cell Compatibility Table

Recipient[1]	*Donor*[1]							
	O−	*O+*	*A−*	*A+*	*B−*	*B+*	*AB−*	*AB+*
O−	✓							
O+	✓	✓						
A−	✓		✓					
A+	✓	✓	✓	✓				
B−	✓				✓			
B+	✓	✓			✓	✓		
AB−	✓		✓		✓		✓	
AB+	✓	✓	✓	✓	✓	✓	✓	✓

1. Assumes absence of atypical antibodies that would cause an incompatibility between donor and recipient blood, as is usual for blood selected by cross matching.

A RhD negative patient who does not have any anti-RhD antibodies (never being previously sensitized to RhD positive RBCs) can receive a transfusion of RhD positive blood once, but this would cause sensitization to the RhD antigen, and a female patient would become at risk for *hemolytic disease of the newborn*. If a RhD negative patient has developed anti-RhD antibodies, a subsequent exposure to RhD positive blood would lead to a potentially dangerous transfusion reaction. RhD positive blood should never be given to RhD negative women of childbearing age or to patients with RhD antibodies, so blood banks must conserve Rhesus negative blood for these patients. In extreme circumstances, such as for a major bleed when stocks of RhD negative blood units are very low at the blood bank, RhD positive blood might be given to RhD negative females above childbearing age or to Rh negative males, providing that they did not have anti-RhD antibodies, to conserve RhD negative blood stock in the blood bank. The converse is not true; RhD positive patients do not react to RhD negative blood.

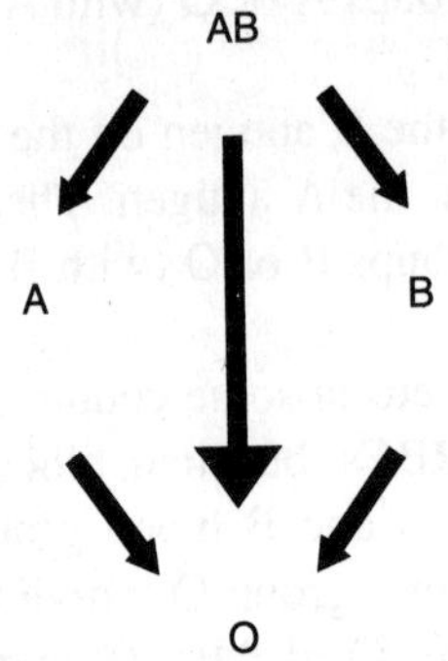

Fig. 16.16. Plasma compatibility chart

Plasma Compatibility Chart

In addition to donating to the same blood group; plasma from type AB can be given to A, B and O; plasma from types A and B can be given to O.

Recipients can receive plasma of the same blood group, but otherwise the donor-recipient compatibility for *blood plasma* is the converse of that of RBCs: plasma extracted from type AB blood can be transfused to individuals of any blood group; individuals of blood group O can receive plasma from any blood group; and type O plasma can be used only by type O recipients.

Rhesus D antibodies are uncommon, so generally neither RhD negative nor RhD positive blood contain anti-RhD antibodies. If a potential donor is found to have anti-RhD antibodies or any strong atypical blood group antibody by antibody screening in the blood bank, they would not be accepted as a donor (or in some blood banks the blood would be drawn the product would be appropriately labeled); therefore, donor blood plasma issued by a blood bank can be selected to be free of RhD antibodies and free of other atypical antibodies, and such donor plasma issued from a blood bank would be suitable for a recipient who may be RhD positive or RhD negative, as long as blood plasma and the recipient are ABO compatible.

Table 16.3: Plasma Compatibility Table

Recipient	*Donor*[1]			
	O	*A*	*B*	*AB*
O	Y	Y	Y	Y
A		Y		Y
B			Y	Y
AB				Y

1. Assumes absence of strong atypical antibodies in donor plasma.

Universal Donors and Universal Recipients

With regard to transfusions of whole blood or packed red blood cells, individuals with type O negative blood are often called **universal donors**, and those with type AB positive blood are called **universal recipients** (Strictly speaking this is not true and individuals with *Hh antigen system* (also known as the Bombay blood group) are the universal donors). Although blood donors with particularly strong anti-A, anti-B or any atypical blood group antibody are excluded from blood donation, the terms *universal donor* and *universal recipient* are an over-simplification, because they only consider possible reactions of the recipient's anti-A and anti-B antibodies to transfused red blood cells, and also possible sensitisation to RhD antigens. The possible reactions of anti-A and anti-B antibodies present in the transfused blood to the recipients RBCs are not considered, because a relatively small volume of plasma containing antibodies is transfused.

By way of example; considering the transfusion of O RhD negative blood (universal donor blood) into a recipient of blood group A RhD positive, an immune reaction between the recipient's anti-B antibodies and the transfused RBCs is not anticipated. However, the relatively small amount of plasma in the transfused blood contains anti-A antibodies, which could react with the A antigens on the surface of the recipients RBCs, but a significant reaction is unlikely because of the dilution factors. Rhesus D sensitisization is not anticipated.

Additionally, red blood cell surface antigens other than A, B and Rh D, might cause adverse reactions and sensitization, if they can bind to the corresponding antibodies to generate an immune response. Transfusions are further complicated because *platelets* and *white blood cells* (WBCs) have their own systems of surface antigens, and sensitization to platelet or WBC antigens can occur as a result of transfusion.

With regard to transfusions of *plasma*, this situation is reversed. Type O plasma can be given only to O recipients, while AB plasma (which does not contain anti-A or anti-B antibodies) can be given to patients of any ABO blood group.

All human blood may look alike but when it is tested using special agents, differences become apparent. The main red blood cell groups are A, B, AB, and O. The letters stand for two antigens (chemical substances that can be targeted by one's immune system) labeled A and B.

- Group A blood has only the A antigen
- Group B has the only the B antigen
- Group AB has both
- Group O has neither

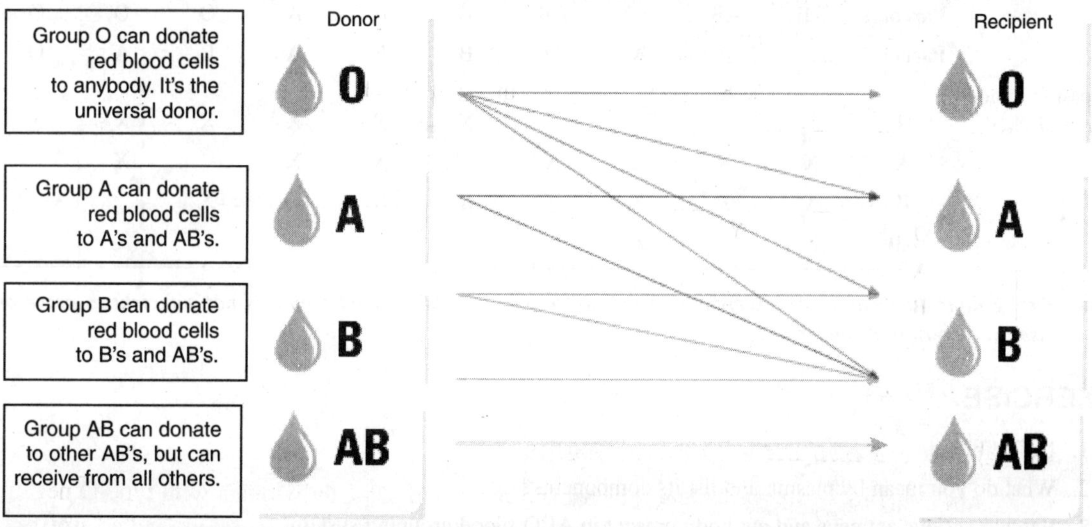

You can't donate red blood cells to just anyone and you can't receive blood from just anyone — blood groups need to be matched.

Blood groups further classified as Rh-positive or Rh-negative are called blood types. Blood is Rh-positive if the blood antigen labeled "D" is present and it is Rh-negative if the "D" antigen is absent.

O+ is the most common blood type. Not all racial and ethnic groups have the same mix of these blood types. Hispanic people, for example, have a relatively high number of O's, while Asian people have a relatively high number of B's. The mix of the different blood types in the U.S. population is:

	White	African American	Hispanic	Asian
O +	37%	47%	53%	39%
O –	8%	4%	4%	1%
A +	33%	24%	29%	27%
A –	7%	2%	2%	0.5%
B +	9%	18%	9%	25%
B –	2%	1%	1%	0.4%
AB +	3%	4%	2%	7%
AB –	1%	0.3%	0.2%	0.1%

Some patients require a closer blood match than that provided by the ABO +/- blood typing. For example, sometimes if the donor and recipient are from the same racial or ethnic background, the chance of a reaction can be reduced. That's why an African-American blood donation may be the best hope for the needs of patients with sickle cell disease, 98% of whom are of African-American descent.

Blood Type Inheritance

It's inherited—Like eye color, blood type is passed genetically from parents to children. Whether your blood type is A, B, AB or O, is based on the blood types of your mother and father.

	Parent 1	AB	AB	AB	AB	B	A	A	O	O	O
	Parent 2	AB	B	A	O	B	B	A	B	A	O
Possible blood type of child	**O**					X	X	X	X	X	X
	A	X	X	X	X		X	X		X	
	B	X	X	X	X	X	X		X		
	AB	X	X	X			X				

Note: *Testing difficulties can cause exceptions to the above patterns. ABO blood typing is not sufficient to prove or disprove paternity or maternity.*

EXERCISE

1. Define blood.
2. What do you mean by plasma and list its components?
3. Explain various antigens and antibody present in ABO blood grouping system.
4. Write a note on geno O type and phenotype of ABO blood group system.

5. Describe role played by Rh factor in blood grouping.
6. Elucidate in detail erythroblastosis foetalis in humans and their significance in pregnancy.
7. Give an account of blood transfusion system followed for considering ABO blood group and Rh factor.
8. Differentiate between antigen A and antigen B chemically.
9. Write a note on plasma incompatibility in blood transfusion.
10. List out advantages of detecting blood grouping with Rh factor.

Chapter

17

Antigen-Antibody Reactions

17.1 INTRODUCTION

For over 40 years, immunoassays have been used in hospitals, laboratory medicine, and research to improve the health and well-being of humans and animals. Information gained by clinical immunoassay testing has shortened hospital stays and decreased the severity of illness by identifying and assessing the progression of disease, thereby leading to improved therapeutic choices. In life science research, immunoassays are used in the study of biological systems by tracking different proteins, hormones, and antibodies. In industry, immunoassays are used to detect contaminants in food and water, and in quality control to monitor specific molecules used during product processing.

Antibodies (depicted as Y-shaped structures) form a heterogeneous population of molecules with different specificities. A cross-reaction of an antibody population (an anti-serum) with a foreign antigen (in the middle) occurs only, if the homologous and the foreign antigen are at least partially equipped with the same determinants. Every antibody has two identical binding sites for antigen determinants.

Antigen-Antibody Reactions

The antigen-antibody binding include hydrogen bonds, ionic bonds, hydrophobic bonds and van der Waals interactions. This interaction occurs over small distance of 1×10^{-7} mm (1 Å or less). The strength of the noncovalent interactions between antigen and antigen binding site of antibody is the **affinity** of the antibody. The strength of multiple interactions between a multivalent antibody and antigen is called the **acidity**. These antigen and antibody reactions are highly specific and a high degree of complementavity exists between them. But some times cross reactivity is possible when two different antigens are chemically similar. This can be noticed in ABO-blood group system, wherein antigens of RBC's are chemically similar to microbial antigens present as common intestinal bacteria. Hence the person whose blood group is 'A', he will be having 'B' antibodies in his serum because 'B' type of antigens are exposed by bacteria to immune system of the person and induced his 'B' - cells to produce B-antibodies.

Alloantibodies: Antibodies directed against antigen belonging to genetically different individuals of the same species.

Autoantibodies: Antibodies directed against an individuals antigenic determinants.

Detection of RBC Antigen-Antibody Reactions

Agglutination

Clumping of particles (red cells) that have antigens on their surface by antibody molecules forming bridges between antigens on different particles (Fig. 17.1 and 17.2).

- May be seen directly - IgM molecules or
- May be seen using "anti-human globulin" techniques - **Coombs test.**

17.2 STAGES OF AGGLUTINATION REACTIONS

1. Sensitization: Attachment of antibody to antigen. Sensitization is affected by
   - temperature
     * IgM: 4 - 30 °C - "cold antibodies"
     * IgG: 30-37 °C - "warm antibodies"
   - pH: neutral
   - Incubation time
     * 25% of antibody bound in first 15 minutes
     * 75% of antibody bound in 1 hour

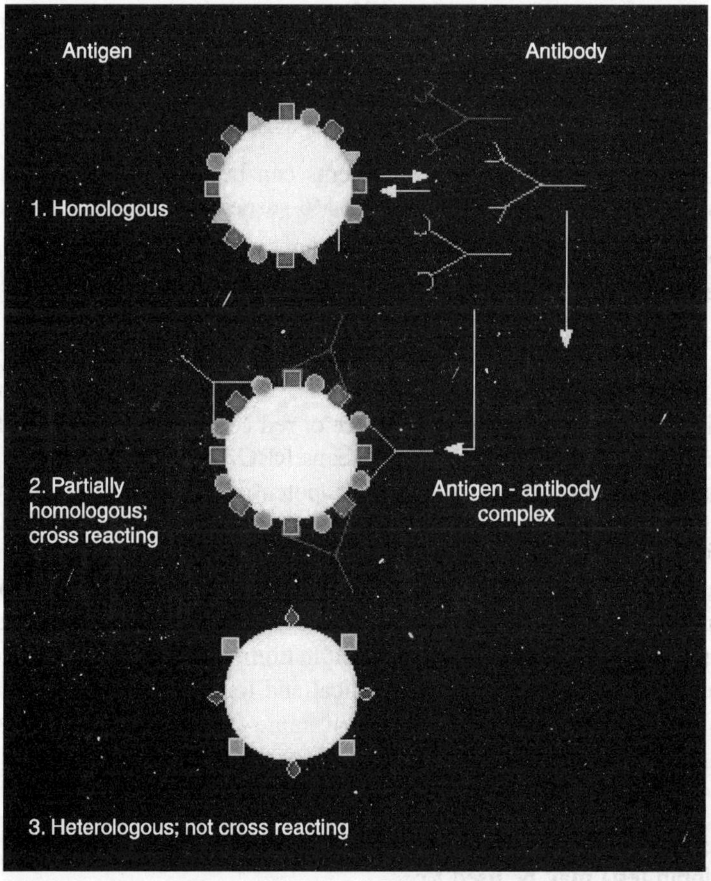

Fig. 17.1. Antigen-antibody reaction.

- Ratio of antigen/antibody

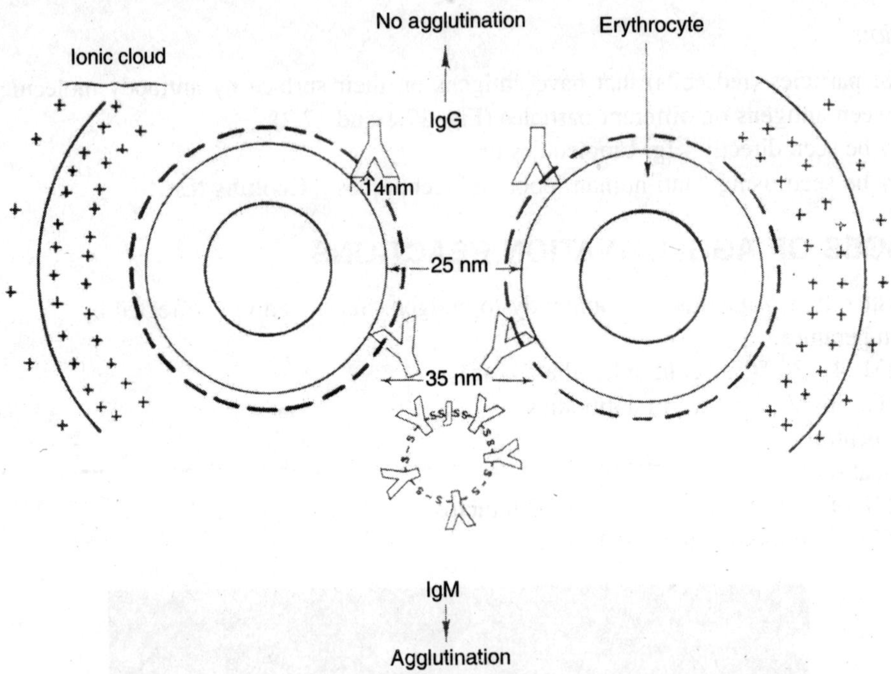

Fig. 17.2. Role of IgG and IgM in agglutination reaction.

In blood banking, prozone and postzone problems can be avoided by always using the correct proportions of cells to serum (plasma) (1 drop of 2-5% suspension with 1-2 drops serum).

Prozone: Term used to refer to the presence of excess antibody resulting in a false-negative test (rare).

Postzone: Term used to refer to the presence of excess antigen resulting in a false-negative test.

- Dosage effects: Example heterozygous Fy will type weakly.

2. Second stage
   - Zeta potential – net negative charge on surface of red cell due to ionization of N-acetylmuramic acid – a component of the cell membrane keeps RBC about 25 nm apart.
   - Use potentiators (LISS or PEG) to lower zeta potential.

Precipitation Reactions

The reaction between an antibody and a soluble antigen forming a visible precipitation in aqueous solution is known as precipitation. These antigens are referred as precipitins. This reaction takes a day or two to reach visible format (Fig. 17.3). Now precipitation reactions can be performed in liquids or gels. These tests are used as diagnostic tests in medical and forensic laboratories for detecting various antigens.

17.3 COOMBS TEST

Detection of sensitization of red blood cells with antibodies is done by using the Coombs test. The **Coombs test (antiglobin test) may be used to:**

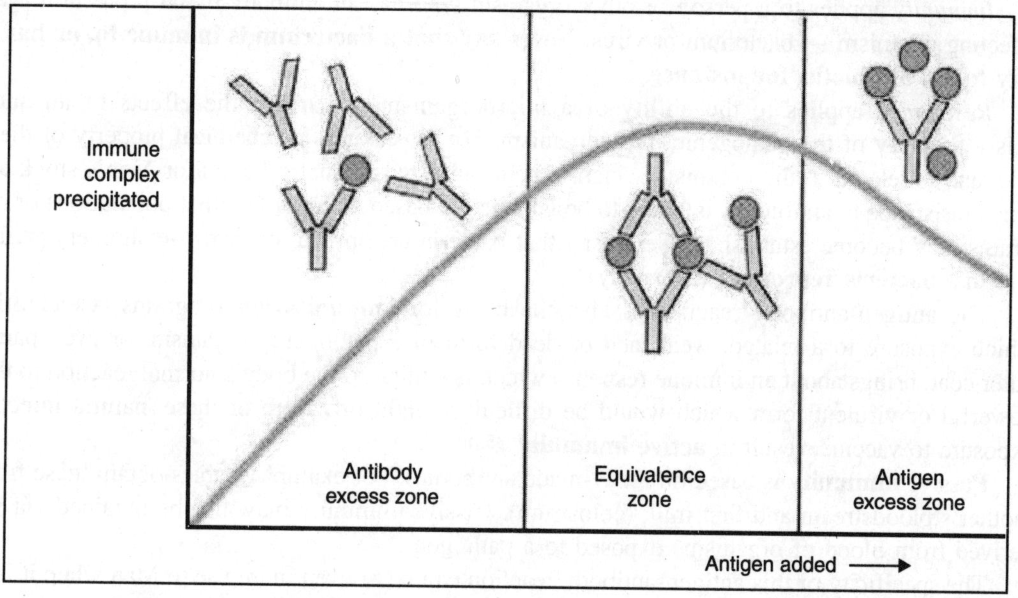

Fig. 17.3. Precipitation reaction.

- detect *in vivo sensitization* (Direct antiglobulin test-DAT) – transfusion reactions; immune hemolytic anemias.
- detect *in vitro sensitization* (Indirect antiglobulin test - IAT) – compatibility testing; screening for unexpected antibodies.

Following basic premises are made:

- antibodies are protein
- animals injected with human globulins produce anti-human antibodies. These anti-human antibodies include antibodies to complement components and IgG.
- the anti-human antibodies bind to the Fc portion of sensitizing antibodies and form bridges between the antibody-coated red cells, resulting in visible agglutination.
- anti-human antibodies will bind to human antibodies that are both bound to red cells (sensitizing) and unbound. Anti-human antibodies will react first with unbound human antibodies.

17.4 HEMOLYSIS

Degradation of blood cells is referred as hemolysis.

- Complement necessary for hemolysis.
- Complement activation is a "3rd - stage reaction".
- Cascade sequence - series of sequential assemblages of units and subunits.
  * some assemblages for active enzymes; designation by by C4b2a.
  * decay products produced are indicated by i (C4i).
- Symbol for complement is C; precursors are numbered 1 to 9 with subcomponents receiving letters a to e when cleaved.
- IgG and IgM are only immunoglobulins that can activate complement.
- Ultimately, leads to cell membrane attack and lysis.

Immunity applies to a person or other organism who has an antibody-based protection against an infecting organism — bacterium or virus. **Never say that a bacterium is immune to, or has immunity to, an antibiotic, for instance.**

Resistance applies to the ability of a microorganism to survive the effects of an antibiotic. It is a property of the pathogenic microorganism, not a weakness or chemical property of the antibiotic, and unrelated to the organism which is being infected—usually human or Man's stock animals. Since resistance to antibiotics is likely to be genetically based, there is fear that populations of resistant strains may become established. Never say that bacteria are spread, or grow—examiners prefer to be told that **bacteria reproduce** (asexually).

The antigen-antibody reaction is also the basis for **immunisation** programs (**vaccination**), in which exposure to a related, weakened or dead form of a pathogenic organism, or even parts of its outer coat, brings about an immune response which is similar to the body's normal reaction to the more powerful or virulent form which would be difficult to fight off. Both of these, natural infection and exposure to vaccine, result in **active immunity**.

Passive immunity is based on ready-made antibodies. For example, babies obtain these from their mother's bloodstream and first milk (colostrum). Passive immunity may also be obtained via *serum*—derived from blood of organisms exposed to a pathogen.

The specificity of this antigen-antibody reaction can be of even more use to Man when it is further refined and developed.

17.5 IMMUNO ASSAY

Introduction

In life science research, immuno assays are used in the study of biological systems by tracking different proteins, hormones, and antibodies. In industry, immunoassays are used to detect contaminants in food and water, and in quality control to monitor specific molecules used during product processing. For over 40 years, immunoassays have been used in hospitals, laboratory medicine, and research to improve the health and well-being of humans and animals. Information gained by clinical immuno assay testing has shortened hospital stays and decreased the severity of illness by identifying and assessing the progression of disease, thereby leading to improved therapeutic choices.

Immuno assays are quick and accurate tests that can be used on-site and in the laboratory to detect specific molecules. Immuno assays rely on the inherent ability of an antibody to bind to the specific structure of a molecule. Antibodies are proteins generated by animals in response to the invasion of a foreign molecule (antigen) into the body. Antibodies are found in blood and tissue fluids and will bind to the antigen whenever it is encountered. Because antibodies are developed based on the specific three-dimensional structure of an antigen, or analyte, they are highly specific and will bind only to that structure. Once purified from the blood, monoclonal and polyclonal antibodies are ideal assay reagents to detect and monitor specific target molecules with limited interferences from other substances.

History of Immuno Assays

Evolution of diagnostic tests began in the 1940s with colorimetric measurements of the enzymes and metabolites found in biological fluids using classical chemistry methods and agglutination reactions. In the 1950s, the radio immuno assay (RIA) was developed by Rosalyn Yalow and Solomon Berson (this group was later awarded the Nobel prize in 1977 for developing an RIA to detect and measure blood glucose levels in diabetic patients). In the 1960s, immuno assay technology was enhanced by replacing radio-isotopes with enzymes for color generation. The use of enzymes eliminates the use of radioactive materials and laboratory regulation by the Nuclear Regulatory Commission (NRC). EIAs also have faster reaction times, higher specificity to the target molecule, and longer shelf-lives compared to RIAs.

Although immuno assay techniques were first described in the 1950s, they were not readily applied outside of clinical laboratories until the advent of economical automated plate-reading systems and personal computers to analyze data. Within the past 10 years, EIAs have become increasingly popular. Scientists in fields outside of medicine are now finding EIAs to be a convenient tool that can be used on a daily basis to detect and monitor specific molecules during the processing of materials. Immuno assays make it possible for scientists to run tests in-house with a minimal investment of time and expense.

Although unskilled technicians can run immuno assays, the development of these tests requires knowledge in many areas of immunology (such as antibody specificity) and protein chemistry (such as binding interactions). As the typical scientist is a specialist in a very narrow field of study, he or she typically may not have the ability to make his or her own rugged tests. Because Immunochemistry Technologies develops assays with the end-user in mind, we try to eliminate as many dilution, mixing, and measuring steps as possible. Most of our reagents come ready-to-use and may have a shelf life of 1 year.

Radio Immuno Assay (RIA)

In the RIA, IgG subclasses are quantified as immune complexes after binding of radioactively labelled specific antibody. In case of a 'direct' technique, a radioactively labelled anti-IgG subclass-specific antibody is used. In an 'indirect' technique, an anti-IgG subclass-specific antibody directed against the first antibody. Since working with radioactively labelled reagents requires special precautions and is relatively costly,radio immuno assays have been largely replaced by enzyme-linked immuno assays.

A physicist with a special interest in radioisotopes, Rosalyn Yalow turned to nuclear medicine and from 1972 was Senior Medical Investigator for the Veterans Administration. Working with S A Berson (1918–72) in a New York hospital, she developed from the 1950s the method of **radio immuno assay** to detect and measure peptide hormones (such as insulin) in the blood. The method has proved of great value both in locating the origin of hormones in the body, and in clinical diagnosis and treatment of a variety of diseases and male and female infertility. Extension of the method in the UK has led to better control of digoxin therapy in heart disease and diagnosis of neural crest disease (e.g. spina bifida) in the fetus. The method can be used to measure very small amounts (10 –12 g) of any substance for which an antibody can be made. Its value in diagnosis and in the control of medication is immense.

Rosalyn Sussman had been a forthright child and her attempts, as a female and Jewish New Yorker, to enter her chosen career before the Second World War did not soften her. Experience as a physicist in the Veterans Hospital, dominated by medical men and service officers, made her abrasive. After Berson's death she saw that he had been assumed to be the creative member of their 22-year, highly effective professional partnership. Stung by this, she increased her research output in the 20 years before her retirement and became even more confrontational, unsoftened by her Nobel Prize in 1977.

Radio immuno assay (RIA) is a term employed to describe any of several methods for determining small concentrations of substances in biological fluids based on the utilization of radioactively labeled substances which form immunochemical complexes with antibodies to that substance. Various RIA techniques are known for measuring concentrations of both antibodies and substances for which there exist antibodies. The RIA of a substance for which there exists antibodies specific to that substance is based on the observation that a known amount of that substance which has been radioactively labeled will compete equally with an unknown amount of that substance which is unlabeled for a limited number of complexing sites on antibodies specific to the substance.

RIA is especially suited for the detection of extremely minute amounts of substances in biological fluids and would, therefore, be extremely valuable in the detection of drugs or their metabolites in the biological fluids of the body. With conventional analytical methods, it is extremely difficult to measure the minute amounts of such metabolites which occur in human body fluids due to the intake of the drugs.

As noted above, RIA procedures are based upon the characteristics that an antibody binds equally to labeled or unlabeled antigen. The concentration of the non-labeled form in the solution determines the relative amount of labeled-non-labeled antigen which will bind to antibody. By maintaining the concentration of antibody and labeled antigen constant and conducting the RIA procedure using a series of known amounts of non-labeled antigen, a standard curve can be constructed. Subsequently, when an unknown amount of antigen in a serum or other biological sample is reacted in the same way, its concentration can be determined by relating the value obtained to the standard curve.

The technique may be represented by the following scheme: ##STR1## where, Ag is the antigen to be assayed

Ag* is the labeled antigen and
Ab is the anti body

The amount of Ab is arranged to be insufficient to react with all of Ag+Ag*. The amount of Ag* added and Ag present in the standards or the unknown sample will compete for the limited binding sites on Ab. If Ag* is separated from Ag*Ab in the above reaction and the level of activity of each separated part measured, then their individual values or their ratios would be related to the amount of Ag present in the standards or the sample. By the use of standard preparations of Ag, a calibration curve is generated.

A radio immuno assay (RIA) method according to the present invention is capable of detecting as little as 5-10 ng of benzoylecgonine, a major cocaine metabolite in human urine, thereby rendering it the most efficient and sensitive method for screening biological fluids for cocaine metabolites perfected to date. The technique enables the monitoring of cocaine intake by man in a relatively simple and highly efficient manner.

It is preferred to employ sheep gamma globulin, bovine or porcine thyroglobulin as the protein carrier for the ecgonine in the preparation of the immunogen. However, it is to be understood that any suitable protein, peptide or polysaccharide (amino sugar containing polysaccharides) may be employed as the carrier.

Specific examples include: bovine, rabbit, pork, sheep or human serum albumin; bovine, rabbit, pork, sheep or human gamma globulin; poly-L-lysine, hemocyanin, etc.

The ecgonine or acetate thereof may be bound to the protein, peptide or polysaccharide carrier by reaction in a suitable reaction medium in the presence of a suitable catalyst such as a water soluble carbodiimide; e.g., 1-ethyl-3-(3-dimethylaminopropyl) carbodiimide HCl, dicyclohexylcarbodiimide, etc.

The ecgonine or acetate forms an amide linkage through the 2-carboxyl group on the ecgonine molecule with the amine groups of the proteins, peptides or amino sugar containing polysaccharides where sheep gamma globulin is employed as the carrier and a carbodiimide as the catalyst, the reaction may be represented: ##STR2##

The anti-serum may be prepared by immunization of an animal with the antigenic conjugate followed by bleeding the animal and separating the anti-serum therefrom. Preferably, the antigenic conjugate is dissolved in a saline solution to give a concentration ranging from 1.0 to 4.0 mg/ml. Animals, e.g., horses, rabbits, sheep, goats, etc., may be immunized by injection of the solution such that the animal receives a dose of from about 2 to about 20 mg of the conjugate. Subsequent boosters of smaller amounts of the conjugate may be administered according to any predetermined schedule. As noted above, it could not be predicted that the anti-serum would be specific to benzoylecgonine as opposed to ecgonine from which the conjugate was prepared.

The anti-serum may be utilized in a radio immuno assay method for the detection of cocaine metabolites in organ homogenates and in biological fluids, particularly human urine. Standard calibration curves may be prepared employing known amounts of anti-serum, radiolabeled benzoylecgonine and non-labeled benzoylecgonine. When utilizing the radio immuno assay method for analyzing an unknown sample, the values obtained are compared with the calibration curves in order to ascertain the amount of cocaine metabolite in the unknown sample.

Fig. 17.4. Munich, April 1976. From *left* to *right*, Dr. Eva Engvall (Sweden), Dr. Anton Schuurs (The Netherlands), Dr. Peter Perlmann (Sweden), Dr. Bauke van Weemen (The Netherlands), and Prof. Johannes Büttner (Germany), President of the German Society of Clinical Chemistry.

It is to be understood that the invention includes compositions comprising the anti-serum and varying amounts of the radiolabeled antigen or the non-labeled antigen for preparing the standard calibration curves for utilization in the radio immuno assay.

17.6 ENZYME-LINKED IMMUNOSORBENT ASSAY (ELISA)

Definition

Enzyme-linked immunosorbent assay (ELISA): ELISA stands for "enzyme-linked immunosorbent assay." This is a rapid immunochemical test that involves an enzyme (a protein that catalyzes a biochemical reaction). It also involves an antibody or antigen (immunologic molecules).

ELISA tests are utilized to detect substances that have antigenic properties, primarily proteins (as opposed to small molecules and ions such as glucose and potassium). Some of these include hormones, bacterial antigens and antibodies.

Enzyme immuno assay (EIA) and **enzyme-linked immunosorbent assay** (ELISA) have become household names for medical laboratories, manufacturers of *in vitro* diagnostic products, regulatory bodies, and external quality assessment and proficiency-testing organizations. This brief historical note spotlights the development of enzyme labels in immuno assay from the invention of this method in the 1960s through its development and early use during the 1970s and 1980s.

The first published EIA and ELISA systems differed in **assay** design, but both techniques are based on the principle of immuno assay with an enzyme rather than radioactivity as the reporter label. Two scientific research groups independently and simultaneously developed this idea and executed the necessary experiments to demonstrate its feasibility. The ELISA technique was conceptualized and developed by Peter Perlmann, principal investigator, and Eva Engvall at Stockholm University, Sweden, and the EIA technique by Anton Schuurs, principal investigator, and Bauke van Weemen at the Research Laboratories of NV Organon, Oss, The Netherlands.

RIA was first described in 1960 for measurement of endogenous plasma insulin by Solomon Berson and Rosalyn Yalow of the Veterans Administration Hospital in New York. Yalow would later be awarded the 1977 Nobel Prize for Medicine for "the development of the RIA for peptide hormones", but because of his untimely death in 1972, Berson could not share the award. Also in 1960, Dr. Roger Ekins of Middlesex Hospital in London published his findings on "saturation analysis" used to estimate thyroxine in human plasma.

The immuno assay technique with a radioactive label immediately caught the imagination of many researchers and clinicians, and in the ensuing decade RIAs for new analytes were published at a rapid pace and variants of the method were rapidly developed. In 1968, Miles and Hales published their first results of an "immunoradiometric" technique with radioactive labeled antibodies rather than labeled antigen for measuring insulin in human plasma.

In many laboratories around the world, special facilities were built in which investigators could safely work with the amounts of radioactivity required for the labeling of antigens or antibodies, but concern persisted with regard to the safety of laboratory personnel, the radioactive waste problem, the requirements of building special laboratory facilities, and the procurement of expensive counting equipment. It should be recalled that in the original studies iodine-131 (ß and γ radiation) was used for the labeling because no alternatives were available at that time. The potential health problems related to the use of radioactive materials were greatly diminished when manufacturers such as Amersham and NEN began marketing iodine-125 (weak γ radiation) preparations of sufficiently high specific activity and purity.

There are variations of this test, but the most basic consists of an antibody attached to a solid surface. This antibody has affinity for (will latch onto) the substance of interest, for example, human chorionic gonadotropin (HGC), the commonly measured protein which indicates pregnancy. A mixture of purified HCG linked (coupled) to an enzyme and the test sample (blood, urine, etc) are added to the test system. If no HCG is present in the test sample, then only HCG with linked enzyme will bind. The more HCG which is present in the test sample, the less enzyme linked HCG will bind. The substance the enzyme acts on is then added, and the amount of product measured in some way, such as a change in color of the solution.

ELISA tests are generally highly sensitive and specific and compare favorably with radio immune assay (RIA) tests. They have the added advantages of not needing radioisotopes or a radiation-counting apparatus.In this technique, enzyme-labelled antibodies are used for detection. In the 'sandwich'-type of enzyme immuno assay, the IgG subclass of interest is captured by anti-human IgG subclass-specific antibody and quantified by an enzyme-labelled anti-IgG antibody. Because of its high sensitivity and specificity, this assay allows accurate measurement of very low levels of IgG subclasses. In Fig. 17.5, a schematic outline of the ELISA technique is shown. The sensitive ELISA comprises many incubation- and washing steps. Because of the need for high dilutions when measuring IgG sublasses in sera,ELISA assays may be less reproducible in comparison with RID and nephelometry. For measurement of IgG and its subclasses in large numbers of samples,ELISA is increasingly being replaced by nephelometry. ELISA may be advocated for measuring IgG subclasses in other body fluids than serum/plasma, e.g. saliva, cerebrospinal fluid and broncho-alveolar lavage fluid.

Brief Outline of the Method

An ELISA is generally performed in wells of microplates.

- Wells of the plates are coated with unlabelled monoclonal antihuman IgG subclass-specific antibody and washed (Fig. 17.5A);
- Test samples,standard-and control sera are introduced in the respective wells and incubated; The IgG subclass to be determined will bind to the solid phase and non-bound IgG is removed by washing (Fig. 17.5B);

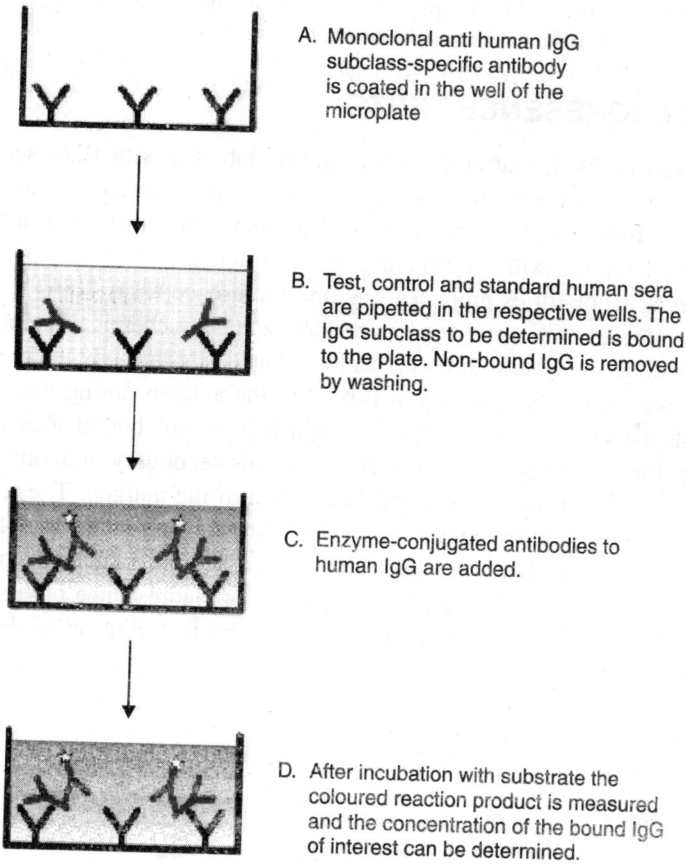

Fig. 17.5. Subclass ELISA.

- Enzyme-labelled anti-human IgG antibodies are added to each well and non-bound conjugate is removed by washing (Fig. 17.5C);
- Plates are incubated with substrate solution;

- After incubation,the coloured reaction product is measured photometrically (Fig. 17.5D);
- The concentration of the IgG subclasses in the test samples is calculated relative to the values of the calibration curve.

The ELISA typically works by binding antigens to a solid surface and then adding tissue samples, secondary antibody, and a reporter label, with wash steps in between. Antibodies in the tissue will bind to both the antigen and the antibody (or another antigen if it's a sandwich assay). The secondary antibody will bind to the label, which fluoresces. After washing, the only fluorescence will be due to antibodies, since they will anchor the secondary antibody and labels to the antigen, which is bound to the solid surface, which is also known as the ***solid phase***.

Though the ELISA can be used to demonstrate pharmacokinetic or hormonal effects, its chief use is in diagnostics, in which capacity it detects antibodies to antigens expressed by ***infectious agents***, including viruses, bacteria, fungi, and protozoa, as well as parasites. The most common use of the ELISA in humans is the detection of HIV-1 in serum samples. The ELISA is also used for many different animal species. For this writeup, I will focus on the ELISA's capacity to detect infectious agents in sera.

17.7 IMMUNOFLUORESENCE TESTS

Albert Coons showed in 1944, that antibodies could be labelled with fluorescent molecules. If antibody molecules are tagged with fluorescent dye (fluorochrome), detection of antigens immunofluorescence. Many fluorochromes such as fluorescein, rhodamine and phycoerythsin can be conjugated to the Fc regions of an antibody without affecting the specificity.

Many commercial laboratories produces immunofluoresence assay (IFA) test plates and slides by immobilising cells infected with virus or parasitic tachyzoites by chemical fixation. Samples are added to the test plates or slides at several dilutions and incubated at core body temperature; if antibodies to the antigen are present in the sample they will bind to the antigen during this incubation (Fig. 17.6). The tests are washed to remove any antibodies which have not bound then a suitable anti-species antibody conjugated to a fluorescent marker is added. this secondary antibody will bind to any antibodies which were present in the sample and had bound to the antigen. The tests are incubated then washed again before being read by microscopy using ultra violet illumination. The fluorescent marker glows with a bright apple green fluorescence under the UV light indicating the presence of specific antibodies bound to the antigen (Fig. 17.7). The pattern of fluorescence observed often confirms the specificity of the reaction. The sample antibody titre is the last dilution of the sample to show this specific fluorescence.

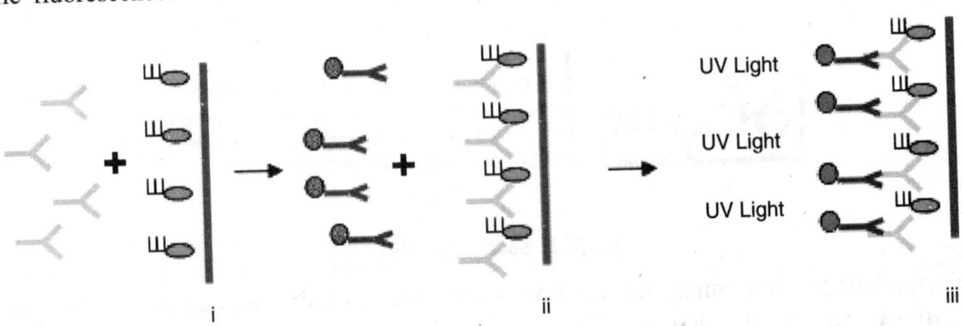

Fig. 17.6. Detection of antibodies by immunofluorescence.

(*i*) Test serum is incubated with antigen immobilised on a 96-well plate or microscope slide.
(*ii*) Secondary antibodies labelled with a fluorescence are added.
(*iii*) After washing, any bound secondary antibodies can be detected by shining UV light on the slide. The label used may be a fluorochrome (IFA test) or be an enzyme which induces a colour change when the substrate is added (ELISA).

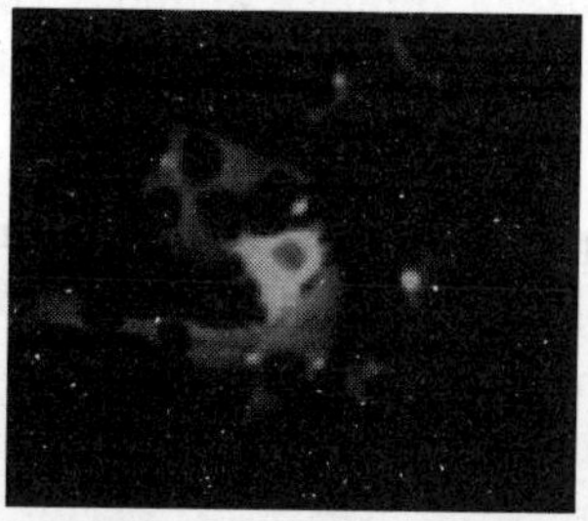

Fig. 17.7. FCoV Serology by IFA.

An FCoV infected culture of cells is used to show the presence of antibodies to the virus in cat serum. Infected cells fluoresce against a background of uninfected cells. In cats without antibodies.

Table 17.1: Currently Available Immunodiagnostic Tests for the SARS-associated Coronavirus

Detection Method	*Clinical Material/Specimen*	*Technical Details*	*Diagnostic Significance*
Virus detection			
Virus isolation on cell culture	Respiratory tract samples: sputum, BAL	Suitable cell lines: Vero; biosafety level 3 facility required	Indicates presence of infectious virus; negative result does not preclude SARS!
Polymerase chain reaction (PCR)	Respiratory tract samples: sputum, BAL, throat swab, throat washing, stool	Different primer sequences and protocols available from the WHO website*	Indicates presence of viral genome, not necessarily of infectious virus; negative result does not preclude SARS!*
Antibody detection			
Immunfluorescence assay (IFA)	Serum	For detection of specific IgG or IgM antibodies or both	IgM IFA usually positive from day 10 after the onset of symptoms
Enzyme-linked immunosorbent assay (ELISA)	Serum	May be designed to detect specific IgG or IgM antibodies or both	Usually positive from day 21 after the onset of symptoms
Neutralization test (NT)	Serum	Requires BSL-3 facility ("live" virus)	Under investigation; study use only

Immunofluorescence is a technique allowing the visualization of a specific protein or antigen in cells or tissue sections by binding a specific antibody chemically conjugated with a fluorescent dye such as fluorescein isothiocyanate (FITC). There are two major types of immunofluorescence staining methods: (1) direct immunofluorescence staining in which the primary antibody is labeled with fluorescence dye, and (2) indirect immunofluorescence staining in which a secondary antibody labeled with fluorochrome is used to recognize a primary antibody. Immunofluorescence staining can be performed on cells fixed on slides and tissue sections. Immunofluorescence stained samples are examined under a fluorescence microscope or confocal microscope.

Fig. 17.8. Immunofluorescence assay (IFA): SARS-CoV-infected Vero cells incubated with patient serum (1:50 dilution) obtained 11 days after the onset of symptoms, showing cytoplasmatic fluorescence.

Fig. 17.9. Immunofluorescence assay (IFA): SARS-CoV-infected Vero cells incubated with negative control serum.

The sensitivities of various immunoassays are given in the Table 17.2 and it shows clearly ELISA and RIA are better immuno assays that detects antigen.

Table 17.2: Immuno Assays and Their Sensitivity

Immuno Assay	*Sensitivity (μg) Antibody/ml*
Precipitation (liquid)	20-200
Precipitation (gel)	2-200
Agglutination	0.006-0.3
Radio immuno assay (RIA)	0.0006-0.006
Enzyme-linked immunocorbent assay (ELISA)	< 0.0001-0.01
Immunofluorescence	1.0

EXERCISE

1. What do you mean by antigen-antibody reaction?
2. Differentiate between alloantibodies and autoantibodies.
3. Explain affinity and acidity of antibodies.
4. Define cross reactivity with suitable example.
5. Compare agglutination and precipitation.
6. Give an account of Coomb's test and its advantages.
7. Describe homolysis and its steps.
8. What are immuno assays? Mention their usage in medicine and forensic fields.
9. Write a note on history of immuno assays.
10. Elucidate radio immuno assay and its usage in diagnostic area.
11. List out advantages of RIA.
12. With principles explain enzyme-linked immunosorbent assay and its application.
13. What is meant by immunofluorescence? Write a note on its benefits.
14. Give a note on sensitivities of different immuno assays used in immunology.

Part C
Plant Biotechnology

Chapter

18

Plant Tissue Culture

18.1 INTRODUCTION

Since time immemorial man involved in the improvement of plants for his basic needs. Generally, crop improvement by conventional breeding methods are tedious and long time process. Nowadays plant tissue culture procedures are used for fulfilling the objectives, which is not possible or impossible through the conventional methods. The purelines or inbreds production require six to seven generations of selfing. But two generations are sufficient for production of haploids through distant crosses or using pollen, anther or ovary culture, followed by chromosome doubling. This results in saving of four to six years. The production of plant derived valuable drugs such as vincristine, vinblastine, digoxin, taxol are produced at the concentration required for the growing market demand. The regular plant growth and their extraction from the plants not only takes long time but also is tedious one. The transfer of insecticidal gene, *cry* gene (crystal toxin protein) from *Bacillus thuringiensis,* into a plant cell and regeneration of whole plants containing and expressing this gene (transgenic plants) is method of improving the crop plant. Only by combination of tissue culture and genetic engineering it can be done. In conventional breeding methods it is impossible to produce such a recombinant plant having genes from prokaryotic microorganisms.

18.2 TERMS AND DEFINITIONS, USE OF GROWTH REGULATORS

The **tissue culture** is defined as *in vitro* culture of plant cells, tissues as well as organs on suitable medium. *In vitro* means in glass i.e., inside the glass vessel and *in vivo* means in life or living organisms i.e., inside the living organisms. The excised plant portion (bits of leaf, stem, root or any part of plant) (Fig. 18.1) used to start tissue culture is known as **explant.** Explant selection, the timing of excision and pretreatment are important factors which determines the success of culture. The process of dissection and removal to culture these tissues or organs is called explantation. The *in virto* culture of single or relatively small groups of plant cells is known as **cell culture** e.g., suspension cultures. Generally any plant cells or tissues cultured on suitable medium with proper environmental condition divides into group of unorganised mass of simple parenchymatous cells. This undifferentiated and unorganised mass of tissue is known as **callus.** Sometimes the term cell culture is also often used for callus cultures. The plant structures such as root tips, shoot tips, embryos etc. are cultured *in vitro* to obtain their development as organised structures, it is called **organ cultures.**

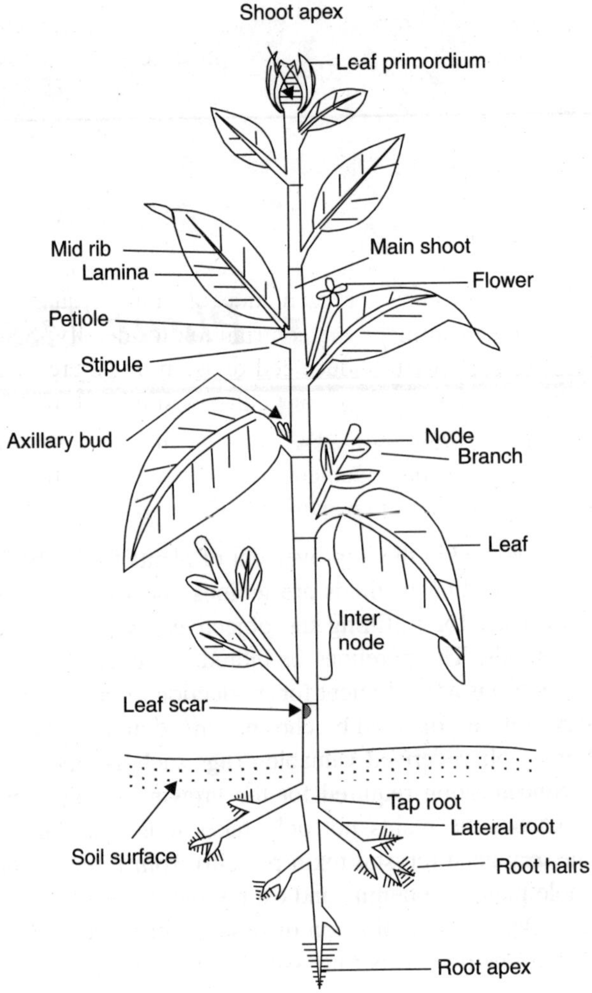

Fig. 18.1. Diagrammatic representation of the plant body showing different parts.

18.3 HISTORICAL DEVELOPMENT OF TISSUE CULTURE

The idea of growing isolated plant cells or tissue was started by Vochting (1878) and Rechinger (1893). **Totipotency** is the potential (inherent capacity) of a plant cell to develop into a complete plant. All cells of an organism having the ability to show the totipotency character. **Gottlieb Haberlandt** a German botanist in 1902, tried to culture isolated mesophyll cells from leaves of *Lamium purpureum* in Knop's salt solution enriched with asparagine, peptone and sucrose. The cells survived up to one month and increased in size, accumulated starch but failed to divide. He wanted to show the totipotency of the cells and not got the success, but able to predict the requirements for cell divisions *in vitro.* Thus, Haberlandt is considered **father of tissue culture.**

Robbins and Kotte in 1922 independently started working on root-tip culture and showed that meristematic tissues can be used for the success of culture system. Later White (1934) with tomato roots first showed indefinite culture of isolated roots. Efforts made to prove totipotency led to the development of techniques for cultivation of plant cells under defined conditions. This was made possible by the brilliant contributions from R.J. Gautheret in France and Phillip R. White in U.S.A. during the third and the fourth decades of 20th century. Most of the modern tissue culture media have been derived from the work of Skoog and coworkers during 1950s and 1960s.

Snow (1935) and Went and Thimann (1937) discovered hormonal characteristics of **auxin,** indole acetic acid which gave greater boost to maintain callus culture of plants indefinitely. First crude embryo culture was done by Hanning in 1940 and he cultured nearly mature embryos of certain Crucifers and grew them to maturity. The same method was utilised by Laibach (1925) to recover hybrid progeny from an interspecific cross in *Linum.* Steward's group at Cornell University (1948) made number of studies related to nutrition, technique of culture, quantitative analyses of culture growth and morphogenesis. They also made the contribution to identify the **coconut milk** as a major natural product which have the capacity to induce cell division in culture. Muir in 1953 clearly shown that plant cells can be grown in liquid (suspension) cultures similar to microorganisms.

Steward (1958) and Reinert (1959) discovered for the first time **somatic embryos production** (production of embryo-like structure from somatic cells) in carrot cultures independently. In 1960s Botany school at University of Delhi started working on *in vitro* culture of ovary, ovule and other reproductive organs of flowering plants. Kanta, Rangaswamy and Maheshwari (1962) first in India worked on *in vitro* pollination and fertilization of angiosperms.

First in 1964, Maheshwari and Guha by culturing anthers of *Datura* produced haploid plants from pollen grains (male gamete in plants). This made the beginning of **anther culture** or **pollen culture** for the production of haploid plants. This technique has been further developed by many workers including Nitch, J.P., Nitch, C. and coworkers. These workers showed that isolated microspores of tobacco produce complete plants.

Plant **protoplasts** are naked plant cells from which cell wall has been removed. In 1960, Cocking produced large quantities of protoplasts by using cell wall degrading enzymes. Protoplast production techniques have been considerably refined afterwards. It is now possible to regenerate whole plants from protoplasts and also to fuse protoplasts of different plant species. In 1972, Carlson *et al.,* produced for the first somatic hybrid plant by fusing the protoplasts of *Nicotiana gauca* and *N. infasoedii.* Since then many divergent somatic hybrids have been produced.

A successful establishment of callus cultures depended on the discovery during mid-thirties of IAA (indole-3-acetic acid), the endogenous auxin, and of the role of B vitamins in plant growth and in root cultures. First Skoog in 1944 reported continuously growing callus cultures were established from tobacco pith tissues cultured *in vitro.* This made possible by identifying the **kinetin** as major factor induced cell division. Later in 1957, Skoog and Miller discovered root and shoot differentiation in culture system was regulated by auxin and cytokinin ratio. The first plant from a mature plant cell was regenerated by Brqun in 1959. Vasil and Hilderbrandt in 1965 clearly demonstrated a single isolated cell can divide and ultimately give rise to a whole plant (known as **regeneration**).

Ball in 1946 showed the greater potential of shoot apex for regeneration of plants. Later Morel and Wetmore regenerated whole plants from shoot apex measuring 100-250 mm length having two leaf primordia. This was successfully used for orchid propagation.

Within a brief period of time, the tissue culture techniques showed a tremendous progress. From the prime objective of proving the totipotency of differentiated plant cells, the technique now finds application in both basic and applied researches in a number of fields of enquiry.

18.4 TISSUE CULTURE METHODOLOGY (*In Vitro* Methods)

During culture of plant cells or organs mainly two basic problems have to be solved. First, to keep the plant cells and organs free from contaminating microbes (bacteria and fungi) and second, to ensure the desired development in the cells and organs by providing suitable concentration of nutrient and other environmental conditions. Contamination can be eliminated by using modern equipment and careful handling during various cultural operations. The second problem remains the area of active research and is likely to be so for quite sometime in the future. It relies mainly on the manipulation of culture medium, especially growth regulators, and, to a much lesser extent, of other factors, including environmental conditions. The general technique of tissue, cell and organ culture is briefly outlined below in a rather simplified manner.

Laboratory Space

The tissue culture laboratory construction depends mainly on the nature and the type of activity. In general space for the following is very much needed: (a) washing, drying and storage of vessels, (b) preparation, sterilization and storage of media, (c) aseptic handling of explants and cultures, (d) maintenance of cultures and (e) observation of cultures. Now-a-days, provisions for washing and preparation can be made in a single room (media room), while thc remaining work can be done in another room (culture room).

Tissue Culture Room

Tissue culture room should have the following facilities: (i) controlled temperature (usually at 25° ± 2° C with the help of air-conditioners and room heaters; higher or lower temperatures may be desirable in some cases), (ii) culture racks fitted with light (generally, 1000 lux or lower light regenerated by fluorescent tubes; automatic timers can be used to regulate photoperiod), and (iii) a shaker for agitation of liquid cultures. It is desirable to have a generator set for providing power to the culture room in the event of electricity cuts/failures which may often ruin the experiments.

Preparation of Culture Vessels

Glass culture vessels are used regularly as they are cheaper, reusable, clear and autoclavable. It is always better to use only borosilicate or pyrex glasswares as the ordinary soda glass may be toxic to some tissue types. Culture vessels made of clear plastic are available for a variety of purposes; these vessels are generally presterilized and disposable but certain types are autoclabale and, therefore, reusable. In general, plastic vessels are much costlier than glass vessels in the long run.

Plant tissues are generally cultured in culture tubes (rimless tubes of 25 × 150 mm or larger), flasks (long neck or even ordinary concial flasks) and Petri plates; but many specially designed dishes are also used. Wide mouth bottles are often employed especially for micropropagation work. Suspension cultures, of necessity, are maintained in long neck culture flasks. Culture tubes and flasks are usually stoppered with cotton plugs which are often wrapped in cheese cloth, but preparing such plugs on a large scale may be time consuming and inconvenient. Many workers use caps made of either polypropylene or a metal (aluminium/stainless steel) for this purpose; these caps have proved quite effective but their design may considerably affect their performance.

Culture vessels and other labwares are generally soaked in a suitable nontoxic detergent solution overnight, washed with a suitable brush, thoroughly rinsed clean with tap water followed by a rinse with distilled water and dried in a hot air oven (75-80° C). Culture vessels having contamination or some dried agar sticking to the wall are first autoclaved. Washed culture vessels should be stored preferable in a dust-proof cupboard or cabinet.

18.5 CULTURE MEDIUM

Cultured plant cells are similar to those required by whole plant. The nutrient mixture used in the medium depends on the type of material used for culture *i.e.*, cells, tissues, organs and protoplast. Most of the nutrient medium consist of inorganic nutrients (macronutrients and micronutrients), carbon source(s), vitamins, plant growth regulators and other organic compounds (Table 18.1). If medium composed of chemically defined compounds, such medium is referred as a synthetic medium. If any

Table 18.1: Nutritional Composition of Some Plant Tissue Culture Media commonly Used in Laboratory

Nutrients	*White's*	*MS*	*B5*	*Nitsch's*	*N6*	*E1*
				Amount (mg l^{-1})		
Macronutrients						
$MgSO_4 \cdot 7H_2O$	750	370	250	185	185	400
KH_2PO_4	–	170	–	68	400	250
$NaH_2PO_4 \cdot H_2O$	19	–	150	–	–	–
KNO_3	80	1900	2500	950	2830	2100
NH_4NO_3	–	1650	–	720	–	600
$CaCl_2 \cdot 2H_2O$	–	440	150	–	166	450
$(NH_4)_2 \cdot SO_4$	–	–	134	–	463	–
Micronutrients						
H_3BO_3	1.5	6.2	3	–	1.6	3
$MnSO_4 \cdot 4H_2O$	5	22.3	–	25	4.4	–
$MnSO_4 \cdot H_2O$	–	–	10	–	3.3	10
$ZnSO_4 \cdot 7H_2O$	3	8.6	2	10	1.5	2
$Na_2MoO_4 \cdot 2H_2$	–	0.25	0.25	0.25	–	0.25
$CuSO_4 \cdot 5H_2O$	0.01	0.025	0.025	0.025	–	0.025
$CoCl_2 \cdot 6H_2O$	–	0.025	0.025	0.025	–	0.025
KI	0.75	0.83	0.75	–	0.8	0.8
$FeSO_4 \cdot 7H_2O$	–	27.8	–	27.8	27.8	–
$Na_2EDTA \cdot 2H_2O$	–	37.3	–	37.3	37.3	–
EDTA Na ferric salt	–	–	43	–	–	43
Sucrose (g)	20	30	20	20	50	25
Organic supplements						
Vitamins						
Thiamine HCl	0.01	0.5	10	0.5	1	10
Pyridoxine HCl	0.01	0.5	1	0.5	0.5	1
Nicotinic acid	0.05	0.5	1	0.5	0.5	1
Myoinositol	–	100	100	100	–	250
Others						
Glycine	3	2	–	2	–	–
Folic acid	–	–	–	0.5	–	–
Biotin	–	–	–	0.05	–	–
pH	5.8	5.8	5.5	5.8	5.8	5.5

Agar 8 grams per liter

Growth regulators depending upon required concentration.

Large number of media composition are specified for a particular type of tissues and organs of plant. White's medium was designed for root culture. Murashige and Skoog (MS) medium and Linsmaier and Skoog (LS) medium was generally used for induction of organogenesis and regeneration of cultured plant tissues. Cell suspension and Callus cultures was carried out by B5 medium. Modified B5 medium known to be very useful for the protoplast culture and protoplast regeneration. The Nitsch's medium and N6 medium are much used for anther culture and other tissues. The MS medium is widely used in tissue culture work because of its success with large number of plant species and explant (Table 18.2).

Table 18.2: Method for Stock Solution Preparation of Murashige and Skoog (MS) Medium in Laboratory

Nutrient	*Concentration in MS Medium (mg/1)*	*Concentration in the Stock Solution (mg/1)*	*Volume to be taken/litre of Final Medium (ml)*
Macronutrients (10 ×)			
NH_4NO_3	1650	16500	100
KNO_3	1900	19000	
$MgSO_4 \cdot 7H_2O$	370	3700	
$CaCl_2 \cdot 2H_2O$	440	4400	
KH_2PO_4	170	1700	
Micronutrients (100 ×)			
H_3BO_3	6.2	620	10
$MnSO_4 \cdot 4H_2O$	22.3	2230	
$ZnSO_4 \cdot 7H_2O$	8.6	860	
KI	0.83	83	
$Na_2MoO_4 \cdot 2H_2O$	0.25	25	
$CuSO_4 \cdot 5H_2O$	0.025	2.5	
$CoCl_2 \cdot 6H_2O$	0.025	2.5	
Iron source			
Fe EDTA-Na salt	40	Added fresh	
Vitamins			
Nicotinic acid	0.5	50 mg/100 ml	1
Thiamine HCl	0.1	50 mg/100 ml	0.2
Pyridoxine HCl	0.5	50 mg/100 ml	1
Others			
Myoinositol	100	Added fresh	
Sucrose	30,000	Added fresh	
Agar	8000	Added fresh	
pH 5.8			

18.6 PLANT GROWTH REGULATORS

Natural organic compounds produced by plants which influence (in low concentration) growth and development is known as **phytohormones.** Now synthetic compounds are also known to have such effects similar to natural ones. These are collectively called **plant growth regulators** (Table 18.3). There are four classes of growth regulators namely, auxin, cytokinins, gibberellins and abscisic acid. These compounds affect growth, differentiation and organogenesis of plant tissues in culture condition.

Auxins

The major function of auxin is induction of cell division and formation of callus in culture condition. It induces cell division, cell elongation and swelling of tissues and formation of adventitious roots in nature and *in vitro*. At low auxin concentrations, adventitious root formation occur, but at high concentrations root formation will not occur and callus formation takes place. The NAA (naphthalene acetic acid), IAA (Indole acetic acid), IBA (Indole butyric acid), 2,4,5-T (2,4,5-trichlorophenoxy acetic acid), 2-4-D (2,4-dichlorophenoxy acetic acid), *p*CPA (*p*-chlorophenoxy acetic acid), pictoram (4-amino-3,5,6-trichloro picolinic acid) are commonly utilized, but 2,4-D is most frequently used and highly effective.

Table 18.3: Characteristics of Commonly Used Growth Regulators in Tissue Culture

Growth regulator	*Chemical Formula*	*Molecular Weight*	*Solubility Media*
p-Chorophenoxy acetic acid (*p*CPA)	$C_8H_7O_3Cl$	186.6	Alcohol
2,4-Dichlorophenoxy acetic acid (2,4-D)	$C_8H_4O_3Cl_2$	221.0	Alcohol
Indole-3 acetic acid (IAA)	$C_{10}H_9NO_2$	175.2	1 N NaOH
Indole-3 butyric acid (IBA)	$C_{12}H_{13}NO_2$	203.2	1 N NaOH
α-Naphthalene acetic acid (NAA)	$C_{12}H_{10}O_2$	186.2	1 N NaOH
β-Naphthoxy acetic acid (NOA)	$C_{12}H_{10}O_3$	202.3	1 N NaOH
Adenine	$C_5H_5N_5 \cdot 3H_2O$	189.1	H_2O
Adenine sulphate (AS)	$(C_5H_5N_5) \cdot H_2SO_4 \cdot 2H_2O$	404.4	H_2O
Benzyl adenine (BA)	$C_{12}H_{11}N_5$	225.2	1 N NaOH
N-isopentenylamino purine (1PP)	$C_{10}H_{13}N_5$	203.3	1 N NaOH
Kinetin	$C_{10}H_9N_5O$	215.2	1 N NaOH
Zeatin	$C_{10}H_{13}N_5O$	219.2	1 N NaOH
Gibberellic acid (GA)	$C_{19}H_{22}O_6$	346.4	Alcohol
Abscisic acid (ABA)	$C_{15}H_{20}O_4$	264.3	1 N NaOH
Colchicine	$C_{22}H_{25}NO_6$	399.4	H_2O

Cytokinins

Cytokinin are adenine derivatives and helps in cell division, shoot induction and somatic embryo formation. The BAP (6-benzyl aminopurine), BA (6-benzyladenine), Kinetin (N-2-λ[furfurylamino]-1-H-purine-6-amine), Zeatin (6-[4-hydroxy-3-methyl-trans 2-butanyl amine and 2-iP (isopentyl adenine) are used frequently. Cytokinins with auxins promote cell division. The cytokinins at high concentration (1 to 10 mg/l) adventitious shoot formation is induced but root formation is inhibited. It also promotes axillary shoot formation by decreasing apical dominance. Cytokinins activates RNA synthesis and stimulates protein and enzyme activity in tissues.

Gibberellins and Abscisic Acid

Gibberellins induce shoot elongation, enhance callus growth and growth of meristems or buds *in vitro*. GA_3 usually inhibit adventitious root and shoot formation.

Abscisic acid (ABA) either stimulates or inhibits callus growth depending upon species. It is also used in embryogenesis.

Sterilization Methods

All the materials, e.g., vessels, instruments, medium, plant material etc., used in culture work must be freed from contaminating microbes. This is achieved by one of the following approaches; (i) dry heat, (ii) incineration, (iii) autoclaving, (iv) filter sterilization, (v) wiping with 70% ethanol, and (vi) surface sterilization.

Dry Heat: The glasswares and Teflon plastic wares (empty vessels), and instruments may be sterilized by dry heat in an oven at 160-180° C for 3 hr. It is generally preferred to autoclave glass and plastic wares etc. and flame sterilize instruments like forceps etc. Recently glass bead sterilizers (300° C) are being employed for the sterilization of forceps, scalpels etc.; these devices use dry heat.

Incineration: Instruments like forceps, scalpels, needles etc. are ordinarily flame sterilized by dipping them in 95% alcohol followed by flaming. This process of killing by burning is called incineration. It is customary to flame the mouths of culture vessels prior to inoculation/subculture.

Autoclaving (Moist heat): Culture vessels alone or with medium, small materials and equipment are generally sterilized by heating in an autoclave or a pressure cooker to 121°C at 15 p.s.i. (pounds per square inch; 1.06 kg/cm^2) for 15 (20-50 ml medium) to 40 (2 litre medium) minutes, the time being longer for larger medium volumes. Sterilization during autoclaving depends mainly on temperature. Certain types of plasticwares and some instruments, e.g., micropipettes etc. are also autoclavable. Care should be taken to properly stopper all the vessels and to open the autoclave only when its pressure gauge indicates zero pressure.

Sterilization by filtration: All heat liable compounds such as growth regulators, e.g., GA_3, zeatin, ABA (absicisic acid), urea, certain vitamins, and enzymes should not be subjected to higher temperatures. Such compounds are filter sterilized by passing their solution through a membrane filter of 0.45 µm or lower pore size. The membrane filter is held in a suitable assembly; the assembly together with the filter is sterilized by autoclaving before use. Filter sterilized heat labile compounds are added to autoclaved media after cooling. In case of agar medium they are added when the medium has cooled to about 40° C.

Laminar air-flow chambers are used to maintain aseptic condition in working space by blowing filter sterilized air through an enclosed (on all sides except one side) space. Air is first filtered through a coarse prefilter to remove large particles and then passed through a HEPA (high efficiency particulate air) filter which filters out all particles larger than 0.3 mm. This sterilized air blows through the cabinet at 1.8 km/hr which is sufficient to keep the enclosed working are in aspectic condition.

Wiping with 70% Ethanol: The surfaces that are not possible to be sterilized by any other methods e.g., platform of the laminar air-flow chamber, hands of the worker etc., are sterilized by wiping them thoroughly with 70% alcohol and then alcohol is allowed to evaporate.

Surface Sterilisation: All explants are treated with an appropriate sterilizing agent to kill contaminating microbes present on their surface; this is called surface sterilization. An explant is the excised piece of tissue or organ of plant used for culture. Surface sterilization procedure will depend mainly on the source and the type of tissue of the explant which will determine the contamination load and tolerance to the sterilizing agent.

The common sterilizing agents used for surface sterilization are calcium hypochlorite (9-10%), Sodium hypochlorite (2%), mercuric chloride (0.1-1%), silver nitrate (1%), bomine water (1-2%), H_2O_2 (10-12%) and antibiotics (4-5 mg/l) (Table 18.4). Among these calcium or sodium hypochlorite and $HgCl_2$ are the most commonly used with satisfactory results. The period of treatment varies from 5-30 min. As these agents are also toxic to plant tissues, the time and concentration used should be such that it causes minimum tissue damage. The rinsing of surface sterilized explant with pure sterile water after treatment should remove them as completely as possible. Otherwise they will not show any response in the culture condition.

Table 18.4: Disinfectant Chemicals Used for Surface Sterilization of Explants

Disinfectant	*Concentration*	*Exposure Period*
Benzalkonium chloride	0.01-0.1%	5-20 minutes
Bromine water	1-2%	2-10 minutes
Calcium hypochlorite	9-10%	3-30 minutes
Ethyl alcohol	75-95%	30-60 seconds
Hydrogen peroxide	3-12%	5-15 minutes
Mercuric chloride	0.1-1.0%	2-10 minutes
Silver nitrate	1%	5-30 minutes
Sodium hypochlorite	0.5-5%	5-30 minutes

In general procedure for disinfection of stem bit may be as follows: stem bits explants are selected for culture. Explants obtained from field should be first thoroughly washed under running tap water to remove the dust and epiphytotic microbes (microbes present on the plant surfaces) for 30 min. and may be first treated with 1% Cetavlon solution for upto 10-15 min to reduce bacterial contamination. These bits are treated with 0.2% $HgCl_2$ (acidified with a few drops of 1N HC1) for 10-15 min. This is followed by a rinse with 70% ethanol, and finally 4-6 rinses with sterilized distilled water (Fig. 18.2). The entire protocol is carried out in an aseptic area in a laminar air-flow cabinet. Most explant are large enough to be processed directly. But many explants, e.g., ovules, embryos, pollen grains etc., are rather small for efficient handling. In such cases, the organs containing them, e.g., ovaries for ovules, ovaries or developing seeds for embryos etc., are surface sterilized following which the desired explants are dissected out aseptically by using sterile materials.

18.7 TISSUE CULTURE RESEARCHES IN INDIA

The development of tissue culture methods throughout the world has initiated the interest in the Indian scientists. Professor P. Maheshwari of University of Delhi in 1950s has realized quite early the importance of *in vitro* culture technique and its application in the field of crop improvement. This made the beginning of tissue culture research in India. He considered complex tissues of plants such as ovary, ovule and control of fertilization for initiation of research in the field of tissue culture. This initial exploratory work on these organs formed the strong foundation in plant cell, tissue and organ culture in India.

This created lot of interest among the scientists in University of Delhi and other laboratories. Efforts were made to initiate the culture of pollinated ovaries of *Aerva tomentosa, Allium cepa, Althaea rosea, Anethum graveolens, Foeniculum vulgare, Hyoscyamus niger, Iberis amara, Linaria maroccana* and *Tropaeolum majus* between 1956 and 1966. Grate success with the culture of pollinated ovaries made it possible to culture ovaries and ovules at even earlier stage of growth. In case of *Zephyranthes* cultured ovaries from pollinated flowers formed fully developed seeds with completely differentiated embryos. Excised ovules containing zygote and primary endosperm nucleus also matured into normal-looking seeds but were without fully-differentiated embryos. Kappor (1959) noticed that these seeds were always smaller than the *in vivo* seeds. Whereas the *in vitro* formed fruits of *Anethum* were found to be larger than the naturally produced fruits. The occurrence of polyembryony due to cleavage and/or budding of zygotic embryos, on a culture medium was another important feature observed. This polembryony is important in the multiplication of plants which are known to form a limited number of seeds. Polembryony is of seeds. Polembryony is of most significant in umbellifers group because of low seed-set and germination.

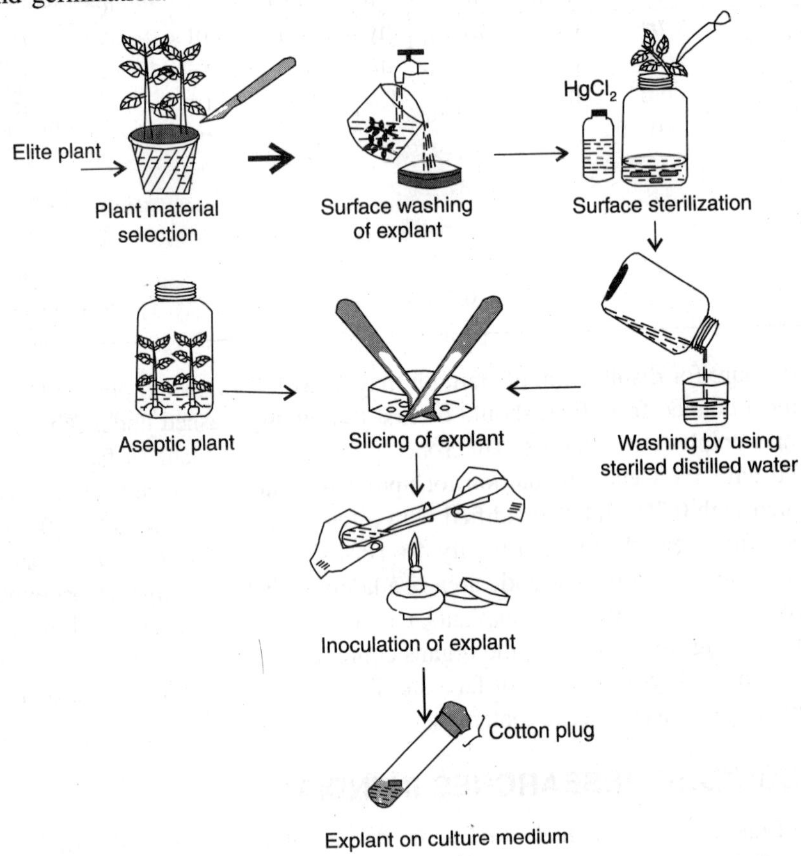

Fig. 18.2. Diagrammatic presentation of the steps followed for surface sterilization of plant material and inoculation of explant for culture initiation.

For the production of hybrid seeds, ovary and ovule culture is most suitable method. Joshi (1962) got success in growing *Gossypium hirsutum* ovules with 12-celled proembryo and 500 endosperm nuclei to maturity. The hybrid ovules of a cross between *G. arboreum* and *G. hirsutum* could be grown to maturity and viable seeds were obtained.

The adventive embryos from nucellus which is genetically uniform and reproduce the characters of maternal parent is very much important. Rangaswamy (1961) excised nucellus from micropylar-half of pollinated ovules of *Citrus microcarpa* (polyembryonic species) proliferated on a culture medium and the callus differentiated pseudobulbils which, eventually, developed into plantlets. Mitra and Chaturvedi (1972) reported embryogenesis from nucelli of unfertilized ovules of *Citrus sinensis* and *C. aurantifolia*. The embryos arose from nuclear cells or the callus. It was demonstrated that the pollina tion/fertilization stimulus is not essential.

Johri and Bhojwani (1965) shown organogenic potentiality of endosperm in *Exocarpus cupressiformis.* In *Scurrula* and *Taxillus,* shoot buds differentiated on a cytokinin medium, whereas auxin is also essential for *Dendrophthoe* and *Tolypanthus.* Continuously-growing tissue and organogenesis have also been observed in the mature endosperm of autotrophic taxa such as *Croton honplandianum, Jatropha panduraefolia, Putranjiva roxburghii, Codiaeum variegatum, Santalum album, Oryza sativa, Nigella damascena, Euphorbia geniculata* and *Emblica officinalis.*

The methods for *in vitro* growth of immature hybrid embryos of *Carica papaya* × *C. cauliflora* and development of F_1 plants shown by Khuspe *et al.* (1980). These seedlings obtained can be finally transferred to pots and field condition. Growth of young floral primordia in culture condition has provided a potential tool for understanding the morphogenic sequences involved in flowering. Such studies in floral morphogenesis have been conducted in *Ranunculus scleratus, Pholx drumondii, Kalanchoe pinnata* and *Browallia demissa.* Cultured floral buds proliferated rapidly in *Ranunculus* spp. and they differentiated roots, shoot buds and embryoids. The buds developed into shoots which finally ends in a flower bud. The callus directly produced flowers in culture. Rarely, mature seeds were also formed. The embryoids passed through the stages comparable to *in vivo* embryogenesis. These plants also flowered *in vitro.*

Haploids production provided a novel experimental system for speeding up of homozygosity, detection of mutations and analyses of genetic combinations. Guha and Maheshwari (1967) noticed numerous embryoids emergence on Nitsh-medium in cultured anthers of *Datura innoxia* (with uni- or bi-nucleate pollen) which ultimately formed plantlets. George and Rao (1979) got success induction of triploids through anther cultures of *Physalsis* spp. Sharma and Bhojwani (1985) found direct microspore-derived embryos in anther culture of two cultivars of *Brassica juncea.* They later developed into plantlets on B5 medium with sucrose. *B. nigra* also produced similar results. *In vitro* flowering of these plants in 2-months-old cultures considered most significant result.

Embryoids or plantlets production by anther culture was noticed in *Petunia axillaris* by Doreswamy and Chacko (1973), *Solanum melongena* by Raina and Iyer (1973), *Petunia obconica* by Bajaj (1981), and *Physalis ixocarpa, Datura innoxia* and *Petunia hybrida* by Chandra and coworkers (Chandra and Kothari, 1987). The anther culture technique leads to homozygous recombinants of *Hyoscyamus niger* with higher alkaloid content. Similar methods has great potential in the crop improvement. Haploids and used for one-step transfer of genotypes of cereals inbred lines and for the induction and characterization of mutants. For the isolation of protoplasts pollen grains are extensively used in number of plants.

18.8 OVARY CULTURE

Ovary is a female part of the flower which consists of either single or many ovules (egg containing part). La Rue (1942) made the first successful attempt to raise aseptic cultures of excised flowers and upto certain extent the limited growth of ovaries. Later Nistch in 1951 cultured pollinated ovaries of tomato on synthetic medium into mature fruits with viable seeds (Fig. 18.3). He also reported ovary culture of *Cucumis anguria, Nicotiana tabacum* and *Phaseolus vulgaris* with reduced fruit size than fruits produced in nature. In India, similar efforts are made to culture ovaries of *Aerva tomentosa,*

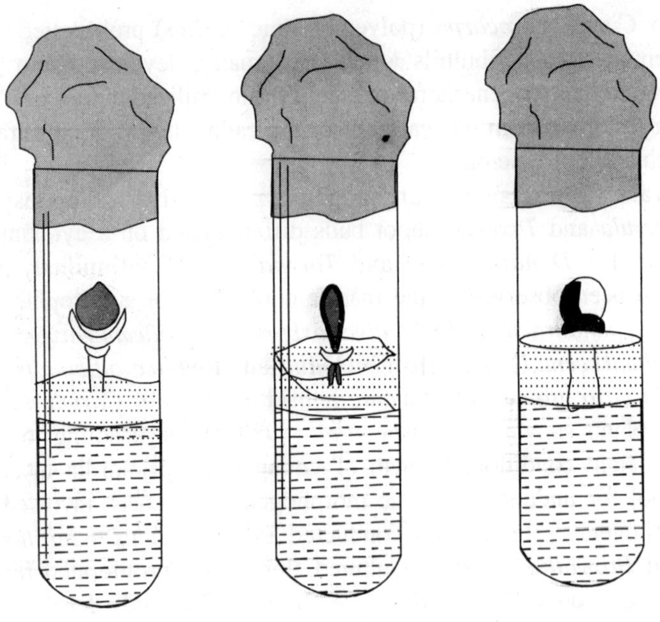

Fig. 18.3. Diagrammatic representation of development of ovary on synthetic medium a, b and c—Three different stages of ovary culture; c—Small fruit developed *in vitro* condition.

Allium cepa, Althaea rosea, Anethum graveolens, Foeniculum vulgare, Hyoscyamus niger, Iberis amara, Linaria maroccana and *Tropaeolum majus,* especially in Delhi University by using Nitsch's or White's medium supplemented with vitamins, hormones, amino acids and other growth inducing substances. Johri and Sehagal in 1966 by using Nitsch's medium amended with coconut milk produced larger size fruits in *in vitro* than naturally produced fruits.

From the studies it is discovered that auxins generally acts as pollination stimulus. Because production of seedless fruits is possible from the unpollinated ovaries in case of tomato (Nitsch, 1963). Such studies also indicated involvement of perianth or calyx (protective covering of flower) along with their protective role they also play a major role in the fruit development. Production of viable seeds from cultured ovaries is a big problem. In most of the ovary cultures, non-viable seeds are formed inside the fruits. However, in 1962 Sachar and Guha produced viable seeds in ovary culture of *Ranunculus scleratus.*

Ovary culture is quite important method for horticulturists to obtain complete knowledge of fruit physiology for improving the quality of fruits. It is successfully used for the production of haploids and rare hybrids. Mess and Wagner (1976) produced parthenogenetic haploids of *Mimitus lutens* by pollinating their exposed ovules in ovary culture. It is important because haploid in this plant could not be produced by any other method. To study chemical effects on fruit development ovary culture can be used.

18.9 OVULE CULTURE

Ovules are zygote containing part of ovary, which later develops into seeds. White in 1932 first started ovule culture by using snapdragon (*Antirrhinum majus*). Later in 1958 Prof. Maheshwari made perfect efforts to culture ovules. Chopra and Sabharwal (1963) by using *Gynandropsis* and ***Impatiens*** and

Guha and Johri (1966) by using *Allium cepa* successfully cultured ovules under *in vitro* condition. In 1961 Maheshwari and group produced viable seeds from excised ovules in culture for *Papaver somnifera* by using Nistch medium with growth substances. By using the ovary culture methods in orchids, Indian scientist Arnoldi (1960) studied from pollen tube entry into the ovule and up to the development of embryo. She used only 10% of sucrose solution to trace the details.

This type of studies also indicated some of the fruit juices such as cucumber, water melon, parent plant or ovules extract known to have beneficial effects on the ovules development (Joshi and Johri, 1972). Recently, ovule culture is utilized for the interspecific hybrids that fail to grow in nature. Punir (1967) successfully produced hybrids by using cross between *Gossypium arboreum* × *G. hirsutum*.

18.10 *IN VITRO* POLLINATION AND FERTILIZATION

Pollination means transfer and deposition of pollen grains (male gamete) on the stigma of the flower. Pollination occurs by means of various agencies such as air (anemophily), water (hydrophily), insects (entomophily), birds (ornithophily), bat (cheiropterophily) and snail and slug (zoophily). The pollen after deposition on the stigma grows and produces the pollen tube that contain male gametes. Pollen tube grows down the stigma through style, ovule and last fuse with the female gamete. This fusion of male and female gametes is known as **fertilization**. These processes occurs in natural condition between the compatible species. Thus *in vitro* pollination and fertilization offer an opportunity for production of hybrid embryos among plants that cannot cross by conventional plant breeding methods or between incompatible species. This is due to barriers which blocks the growth of the pollen tube on the

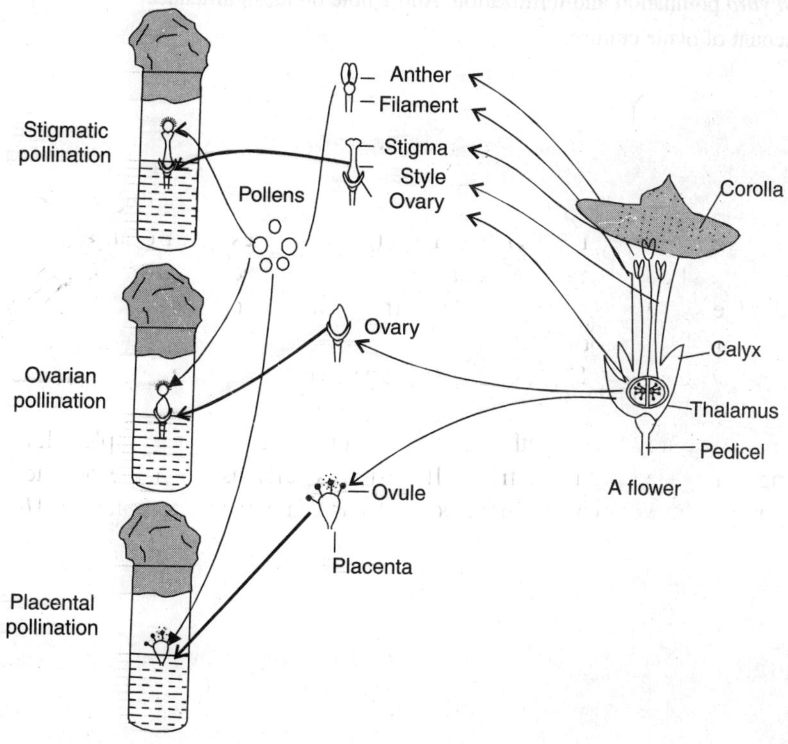

Fig. 18.4. Diagrammatic representation of various types of *in vitro* pollination.

stigma or style. In such cases the pollen grains are deposited on stigma (stigmatic pollination) (Fig. 18.4), or applied on excised ovaries which are not having stigma or style (ovarian pollination), or applied on ovaries which are attached with placenta (placental pollination). These techniques are developed at University of Delhi to produce hybrids among species of the Solanaceae and Papaveraceae by Maheshwari and coworkers. Now *in vitro* pollination has been done in *Nicotiana tabacum, N. alata, Petunia hybrida* (Solanaceae), *Papaver somnifera, Eschscholtzia california, Argemone mexicana* (Papaveraceae) and *Melandrium album, M. rubrum, Agrostemma githago, Dianthus caryophyllus* (Caryophyllaceae).

All the materials such as pollen grains, ovary, ovule or entire pistil (ovary containing stigma and style) are surface sterilized and used with the suitable medium in the aseptic condition. Before collection of all these materials proper care has to be taken in selection of suitable materials for the *in vitro* culture techniques.

EXERCISE

1. Define callus.
2. What is meant by tissue culture?
3. Give the meaning of totipotency and protoplast.
4. How tissue culture work started in India?
5. Explain ovary culture.
6. Write a short note on maintenance of aseptic condition. (BU April 2001)
7. Describe *in vitro* pollination and fertilization. Add a note on its significance. (BU April 2001)
8. Give an account of ovule culture. (BU April 2001)

Chapter 19

Callus Cultures, Suspension Cultures and Micropropagation

19.1 INTRODUCTION

The cultured explants on medium with suitable growth regulators combination undergo division and forms many cells. In culture condition certain mature and differentiated cells (parenchyma and often colenchyma) undergo many physiological changes to form meristematic (ability to divide) cells and is known as **dedifferentiation.** During dedifferentiation renewed and enhanced RNA and protein synthesis along with many chemical changes leading to the formation of new cellular components needed for meristematic activity. In the beginning cell divisions are confined to the cut ends, but later it covers the entire explant. The resulting cell mass is ordinarily unorganized, but it often consists of several cell types including fibers, and vascular elements. The tissues and cells cultured on an agar medium form an undifferentiated and unorganized mass of cells called **callus** (Fig. 19.1). For every 3-5 weeks callus cultures are subcultured because of cell growth, nutrient deficiency and medium drying. Therefore, calluses are easy to maintain and are the most widely used.

The tissues and cells cultured in a liquid medium produce a suspension of single cells and cells clumps of a few cells these are known as **suspension cultures.** To increase aeration and separation of cell clumps into individual cells, the liquid cultures are continuously agitated by gyratory shaker at 100-250 revolution per minute (rpm). Suspension cultures show faster growth than the callus cultures. Thus subculture should be done for very week.

19.2 CALLUS CULTURE

The callus tissue is basically a more or less organised mass of cells that usually arises on wounds of differentiated tissues and organs of the plant. Hence, in principle, **callus** can be defined as an unorganised and less differentiated parechymatous tissue. Generally callus culture is not homogeneous mass of cells, because it is made up of both differentiated (matured) and non-differentiated (immatured) tissues. The differentiated tissues of plant (roots, stem, leaves, flowers, anthers etc.) can be used as a starting

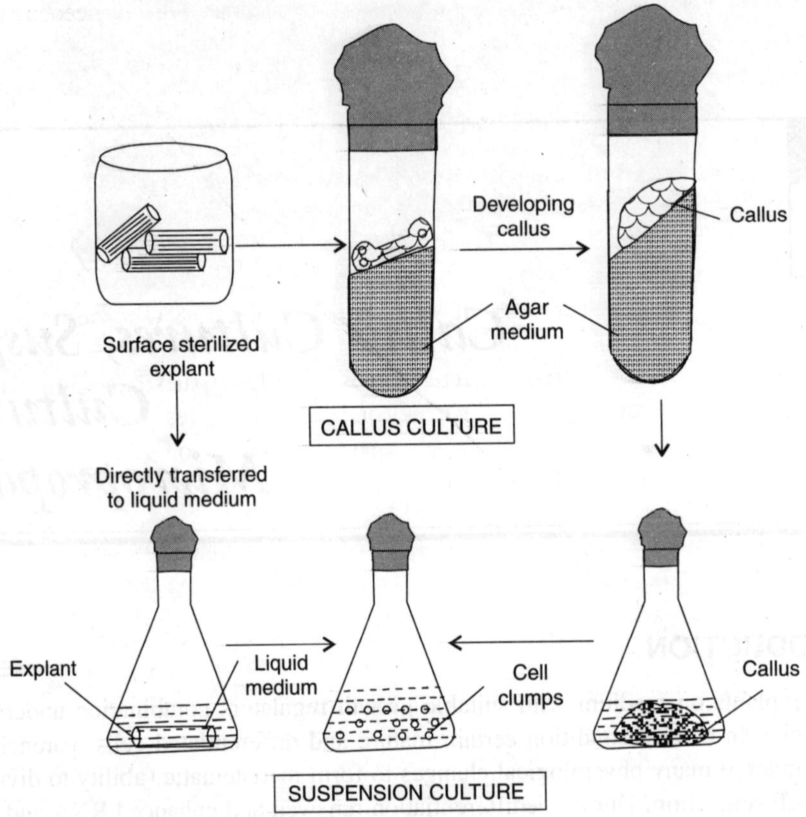

Fig. 19.1. Diagrammatic representation of callus and suspension cultures initiation.

material (explant) for callus induction. These explant tissues generally show distinct planes of mitotic cell division, cell proliferation and organisation into specialised structures such as vascular systems. If only differentiated cells are present in an isolated explant, then **dedifferentiation** (*i.e.*, cells undergo changes to become meristematic) must take place before cell division can occur. The mature parenchyma cells (and also collenchyma cells) present in explant usually undergo this process of dedifferentiation. If the explant already contains meristematic tissue when isolated, then this can divide immediately without dedifferentiation process. During dedifferentiation process, mature adult cells are temporarily able to revert from adult to the juvenile state. The rejuvenated cells have a greater growth and division potentiality. These cells under special circumstances are able to regenerate organs and/or embryos.

Exogenously supplied growth regulators present in the nutrient medium influence the callus formation from the explant. The type of growth regulator requirement and its concentration in the medium depends strongly on the genotype and endogenous hormone content of an explant. These requirements can be put into three categories:

1. Auxin alone (especially in monocotyledons)
2. Cytokinin alone
3. Both auxin and cytokinin

In cases where the callus is difficult to induce, or if young callus is needed, then immature embryos or seedlings or parts of these are used. It should be taken into account that the type of starting material (juvenile or adult) and the original position of the explant in the plant reflects the endogenous hormone level which have an important influence on processes such as cell division, organ and embryo formation.

The genotype, composition of the nutrient medium, physical growth factors (light, temperature, etc.) are also important for callus formation. To induce callus the Murashige and Skoog (1962) mineral medium, or modifications of this are used. Sucrose or glucose (2-4%) is usually employed as the sugar source. The effect of light on callus formation is dependent on the plant species; light may be required in some cases and darkness in other cases. The temperature of 22-28°C is normally employed for callus formation.

The different plant species may form different types of callus tissues that shows varied structure and growth habit; white or coloured, soft (watery) or hard, friable (easy to separate into cells) or compact. The callus growth within a plant species may also depend on factors such as the original position of the explant within the plant, age of the plant, physiology of the plant and the growth conditions.

Once the callus tissue formed from the explant, the callus is further transferred on to a new medium. This process of transfer is referred to as **subculturing**. When subcultured regularly on a agar medium, callus cultures will exhibit a S-shaped or sigmoid pattern of growth during each passage. There are five phases of callus growth:

1. Lag phase, where cells prepare to divide.
2. Exponential phase, where the rate of cell division is highest.
3. Linear phase, where cell division slows down but the rate of cells expansion continues to increase.
4. Deceleration phase, where the rate of cell division and elongation continues to decrease.
5. Stationary phase, where the number and size of cells remain constant.

Applications of Callus Cultures

Callus cultures provides an information on the regeneration potentiality of the plant.

It acts as a good source material for the protoplast isolation.

Callus can be used as inoculum for the initiation of the suspension cultures.

The callus can serve as a better source of cells for the genetic manipulation of the plant genomes.

Callus cultures can be maintained indefinitely in the *in vitro* condition for longer period of time.

19.3 SUSPENSION CULTURE

The individual cell or cell aggregates dispersed and growing in moving liquid media is known as **suspension culture**. In regular practice suspension culture is initiated by transferring pieces of undifferentiated and friable calli to a liquid medium. Which is continuously agitated by using rotary shaker. Suspension cultures have also been started from sterile seedlings or imbibed embryos or leaves by mechanical method. Leaves or the other tissue can be gently grinded or soft tissues can be broken up in a hand operated glass homogenizer. This homogenate, containing intact living cells, dead cells and cell debris is cleared by filtration and centrifugation, which is then transferred to liquid (broth) medium. In an enzymatic method for isolation of single cells pectinases which digest the pectin cell wall was employed. Generally the single cell cultures used in basic and applied research is isolated from **cultured tissues**.

At first stage cultures are initiated by placing freshly cut sections of plant organs (root, stem, leaves) on a solidified nutrient medium containing suitable growth regulators. In this condition the explant are transferred to a fresh medium to build up reasonable amount of callus tissue. This callus is transferred to a liquid medium and agitated to raise fine suspension of cells. Successful establishment of the suspension culture depends upon the initial callus. A wide variety of explants and media compositions have been used with success for callus induction viz. MS, B_5, LS (Linsmaier and Skoog) and Blaydes medium etc. For these media vitamins, inositol, sucrose and growth regulators, especially auxin (~1–5 μM) are incorporated for induction of cell division. The inclusion of 1μM kinetin is also found beneficial for the induction. The growth rate and friability of callus produced can vary widely due to explants. The culture medium for suspension culture should be such that it maintains good growth of callus for the longer period.

In the suspension culture system orbital shakers are widely employed for the initiation and serial propagation of plant cells. They should have a variable speed control (30-150 rpm) and the stroke should be in the range of 4-8 cm orbital motions. The shaking of medium on a shaker serves two purposes. First it exerts a mild pressure on cell aggregates, breaking them into smaller clumps and single cells. Secondly, agitation maintains uniform distribution of cell and cell clumps in the medium. Movement of the medium also provides good gaseous exchange between the culture medium and air. The shaker should be kept in the air-conditioned room with good thermostat. The Erlenmeyer flasks with wide-mouth are commonly used as culture vessels. The volume of the culture medium should be appropriate to the size of the culture vessel e.g., 20 ml per 100 ml flask and 70 ml per 250 ml flask. The flasks are normally sealed with aluminium foil. Flask closure must maintain sterility, allow gas exchange and reduce evaporation. Cotton wool plugs may be used for sealing flasks during autoclaving but not for culturing cells. They are a common source of contamination in flasks that are sitting on a shaker for several weeks.

Suspension cultures are used for production of secondary metabolites from different plant cell culture (Table 19.1).

Table 19.1: Different Secondary Metabolites Produced by Plant Species used for Increased Production Through Tissue Cuiture Techniques with their Bioactivity.

Compound	*Plant Species*	*Bioactivity*
Pyrethrins	*Chrysanthemum cincerariefolium*	Insecticidal
	Tagetus erecta	Insecticidal
Nicotine	*Nicotiana tabacum*	Insecticidal
	Nicotiana rustica	Insecticidal
Rotenoids	*Derris elliptica*	Insecticidal
	Lonchocarpus utilis	Insecticidal
	Tephrosia vogaeli	Insecticidal
	Tephrosia purpurea	Insecticidal
Azadirachtin	*Azadirachita indica*	Insecticidal
Phytoecdysones	*Trianthema portulacastrum*	Insecticidal
Baccharine	*Baccharis megapotamica*	Antineoplastic

(Contd.)

Bruceantine	*Brucea antidysenterica*	Antineoplastic
Cesaline	*Caesalpinia gillisessi*	Antineoplastic
3-deoxycolchicine	*Colchicum speciosum*	Antineoplastic
Ellipticine, 9-methoxyellipticine	*Ochrosia moorei*	Antineoplastic
Fararonine	*Fagara zanthoxyloides*	Antineoplastic
Harringtonine	*Cephalotaxus harringtonia*	Antineoplastic
Indicine *N*-oxide	*Heiotropium indicum*	Antineoplastic
Maytansine	*Maylenus bucchananii*	Antineoplastic
Podophyllotoxin	*Podophyllum peltatum*	Antineoplastic
Taxol	*Taxus brevifolia*	Antineoplastic
Thalicarpine	*Thalictrum dasycarpum*	Antineoplastic
Tripdiolide, triotolide	*Tripterygium wilfordii*	Antineoplastic
Vinblastine, ajmalicine, vincristine	*Catharanthus roseus*	Antineoplastic
Quinine	*Cinchona officinalis*	Antimalarial
Digoxin, reserpine	*Digitalis lanata*	Cardiac tonic
Diosgenin	*Dioscorea deltoidea*	Antifertility
Morphine	*Papaver somniferum*	Analgesic
Thebaine	*Papaver bracteatum*	Source of codeine
Scopolamine	*Datura stramonium*	Antihypertension
Atropine	*Atropa belladonna*	Muscle relaxant
Codeine	*Papaver* spp.	Analgesic
Shikonine	*Lithospermum erythrorhizon*	Dye, pharmaceutical
Anthraquinones	*Morinda citrifolia*	Dye, laxative
Rosamarinic acid	*Coleus blumei*	Spice, antioxidant
Jasmine	*Jasmium* spp.	Perfume
Stevioside	*Stevia rebaudiana*	Sweetner
Crocin, picrocrocin	*Crocus sativus*	Saffron dye
Capsaicin	*Capsicum frutesces*	Chilli dye
Vanillin	*Vanilla* spp.	Vanilla flavour

TYPES OF SUSPENSION CULTURE

Batch Culture

The cell suspension culture grown in a fixed volume of nutrient culture medium is known as **batch culture**. This is also known as **closed system** because they are incubated in a closed culture vessel within a single batch of medium. During incubation cell suspension increases in biomass by cell division and cell growth, this continues until a factor in the culture environment (nutrient or oxygen availability) becomes limiting and then growth ceases. The cells in culture exhibit the following five phases of a growth cycle and are represented in the Fig. 19.2.

(*i*) Lag phase, where cells prepare to divide by synthesising various materials required for cell division.
(*ii*) Exponential (log) phase, where the rate of cell division is maximum and increases in log scale.
(*iii*) Linear phase, where cell division slows but the rate of cells expansion increases.
(*iv*) Deceleration phase, where the rates of cell division and cell elongation starts decreasing.
(*v*) Stationary phase, where the number and size of cells remain constant by stopping cell growth.

During the subculturing of cells into fresh medium there is a lag phase in the cells. It is the initial period of a batch culture when no cell division is apparent. It may also be used with reference to the synthesis of a specific metabolite or the rate of a physiological activity. Then follows a period of cell division (exponential phase). It is a finite period of time early in a batch culture during which the rate of increase of biomass per unit of biomass concentration (specific growth rate) is constant and measurable. Biomass is usually referred in terms of the number of cells per ml of culture. After 3 to 4 cell generations the growth declines. Finally, the cell population reaches a stationary phase during which cell dry weight is also lowered. It is the terminal phase of batch culture growth cycle where no net synthesis of biomass or increase in cell number is apparent. Cultures can be maintained continuously in exponential phase by frequent (every 2-3 days) subculture of the suspensions. Cells may remain viable for many days by utilising intracellular reserves or recycling metabolites released by lysed cells. The stability of cells in stationary phase depends upon the species and the nature of the growth limiting factor (cells brought to stationary phase by nitrogen limitation retain viability longer than when carbohydrate starved).

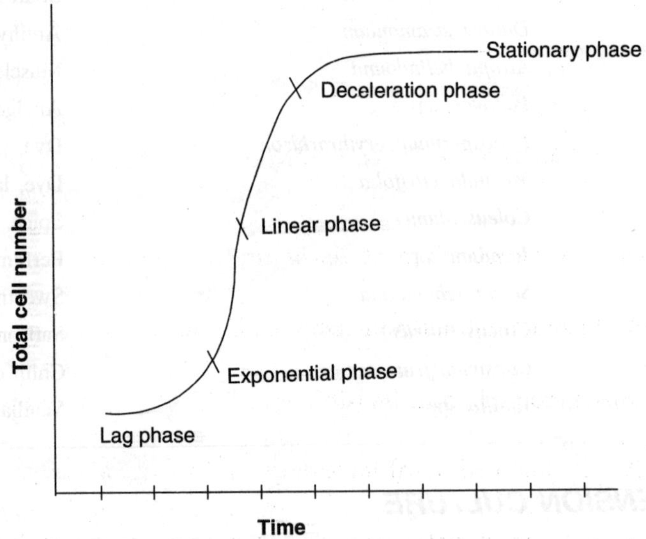

Fig. 19.2. Total number of cells per unit volume of culture to time in a batch suspension culture.

When such a stationary phase cell suspension is subcultured, the cells in succession pass through lag phase, a short period of exponential growth, a period of declining relative growth rate and then again enter stationary phase. Prior to the last phase, the cells expand in size. Traditionally such cultures at each subculture initiate from a relatively high cell density and the cells therefore accomplish only a limited number of divisions (cell number doublings) before entering stationary phase again. For example, a suspension culture of *Acer pseudoplatanus* initiated at 2×10^5 cells/ml will reach a final cell density of ca. 3×10^6 cells/ml corresponding to four successive doublings of the initial population. The cell generation time (doubling time) in suspension cultures varies from 24 to 48 hours in well established cell cultures. **Doubling time** (t_d) is the time required for the concentration of biomass of a population of cells to reach double.

For general purposes, the objective with cell suspension cultures is to achieve rapid growth rates, and uniform cells, with all cells being viable. Frequent subculturing in a suitable medium ensures these qualities. Cells should be subcultured at weekly intervals or less if they are to be used for experimental purposes. The size of the inoculum (dilution rate) will determine the lag period. Optimally suspension cultures passage length is 1-2 weeks, but the exact time and the dilution required must be determined for each cell line. Dilutions of 1:4 after 1 week or 1:10 after 2 weeks are commonly used. There is generally a minimum inoculum size, below that a culture will not recover.

Continuous Culture

In this method culture is continuously supplied with nutrients by the addition of fresh medium but the culture volume normally remains constant. Here cells are continuously in the exponential growth phase and are also known as **open system** of culture. A **continuous culture** is one in which inflow of fresh medium is balanced by outflow of corresponding volumes of culture including harvest of cells. Cells are constantly washed out with the out-flowing liquid. The rate of inflow of medium and culture harvest are adjusted so that the cultures are maintained indefinitely at a constant sub-maximal growth rate. In a steady state, the rate of cells washout equals the rate of formation of new cells in the system. A situation of balanced growth is achieved, i.e., majority of cells in the culture are in a similar metabolic state. The growth rate and cell density are held constant by a fixed rate of input of growth limiting nutrients and removal of cells and spent medium.

The continuous culture system may be of two types viz., chemostat and turbidostat.

Chemostat: In this system growth rate and cell density are held constant by a fixed rate of input of a growth limiting nutrient medium (nitrogen, phosphorus or glucose). In such a medium, all the constituents other than growth limiting nutrients are present at concentration higher than that required to maintain the desired rate of cell growth. The growth limiting substance is so adjusted that it increase or decrease in the growth rate of cells.

Turbidostat: In this fresh medium flows in response to increase in the turbidity so as to maintain the culture at a fixed optical density of suspension. A preselected biomass density is maintained by the washout of cells.

Applications of Suspension Cultures

- Suspension cultures are ideal for the production of secondary metabolites productions which are of therapeutic value.
- It also serves as a major source for obtaining individual isolated cells used for protoplast production.
- It also gives isolated cells which can be used for hybrids or cybrids production.
- Suspension cultures provide cells that are used for the immobilisation of plant cells.
- Suspension cultures are also used for the induction of mutation and genetic manipulation of plant cells.
- Cells obtained from suspension cultures are easily preserved by using cryopreservation technique than other tissues or organs of the plant.

19.4 MICROPROPAGATION (Clonal Multiplication of Elite Species)

Generally plants are propagated by sexual and asexual methods. In sexual methods seeds are produced by fertilization of egg by pollen grains and shows higher degree of heterogenicity. In asexual methods (vegetative propagation) new plants are obtained by mitotic division of earlier plant parts and they are genetical identical to parent. Multiplication of genetically identical copies of plant is known as **clonal propagation.** Individual plants obtained by this asexual methods is known as **clone**. Usually vegetative propagation is done by cuttings, budding and grafting. This is known as *in vivo* clonal propagation of plants and is often difficult, expensive and even unsuccessful in majority of time. Tissue culture is a better option for this purpose. Clonal propagation of plants through tissue culture is called **micropropagation.** This can be done in a short time and space. George Morel in 1960 used micropropagation methods for orchid plants production at commercial level.

In micropropagation shoot-tip, meristem and auxillary buds are the general explants used for the initiation of the culture. Shoot-tip and meristem consists of a group of actively dividing meristematic cells at the shoot apical portion (Fig. 19.3). Cutter in 1965 clearly defined shoot-tip and meristem. **Meristem** or **apical meristem** is the terminal portion of shoot contain active dividing meristematic cell mass and measures about 100-250 μm in length.

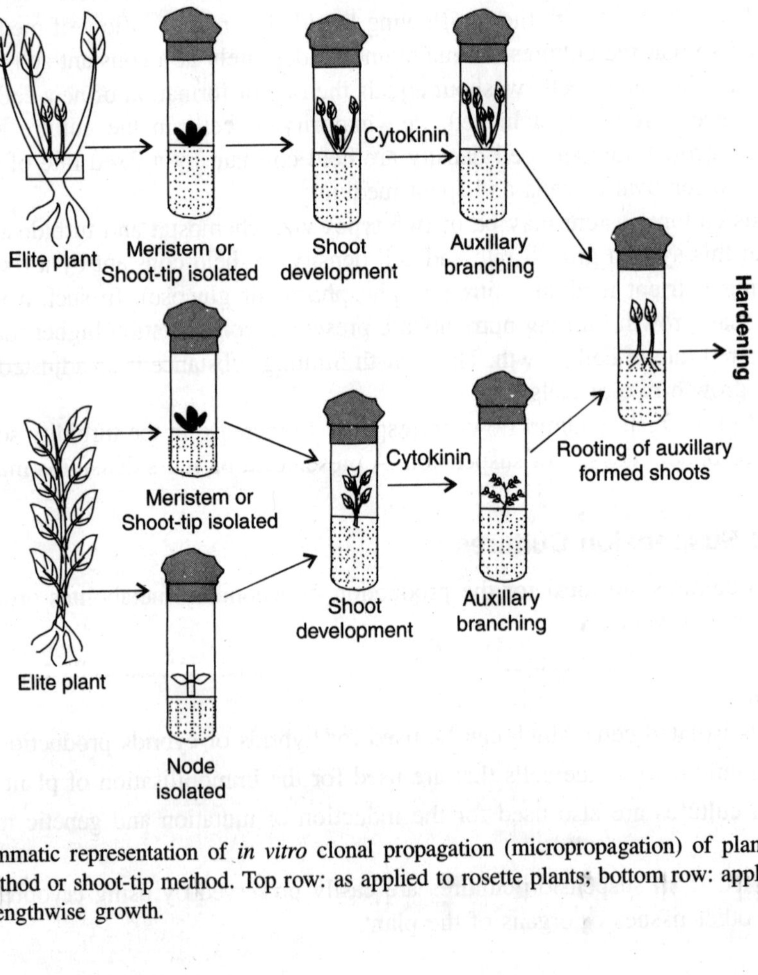

Fig. 19.3. Diagrammatic representation of *in vitro* clonal propagation (micropropagation) of plants by the auxillary bud method or shoot-tip method. Top row: as applied to rosette plants; bottom row: applied to plants which show lengthwise growth.

Shoot-tip or **shoot apex** is the terminal meristematic portion along with the one to three leaf primordia and measures about 100-500 μm in length. For *in vitro* clonal propagation shoot apex is enough as explant, whereas in case of meristem culture for the production of disease free plants clear excision of only meristem without surrounding tissue is very much necessary. **Auxillary buds** are the meristematic cells present in the axile portion of the node. These are either in dormant or active state depending upon the physiological state of the plant. Auxillary bud method was first used for carnation, but now this method is rarely used for carnation. This method is widely employed for the large scale micropropagation of strawberry and gurbera.

Micropropagation or *in vitro* clonal propagation is a highly complicated process consisting of different steps or stages. Murasighe in 1978 recognized four stages for overall micropropagation of plants. Wherein stage I to III are followed under *in vitro* condition and stage IV is carried out in the green house condition. In 1981 Debergh and Maene introduced stage 0 for micropropagation systems.

Stage 0: Selection of Elite/Stock Plants

This is the first step of micropropagation, in this elite/stock plants (plants having a desired characters for the large scale multiplication) are selected and maintained in controlled environmental conditions for at least three months to initiate the culture. The elite plants are grown under low humidity, irrigation and without systematic microbial infections.

Stage I: Establishment of Aseptic Culture

In this stage selected explants are prepared for the inoculation and established on a suitable culture medium. The selected shoot-tip, meristem and auxillary bud explants are surface-sterilized by using the chemicals such as 0.1% $HgCl_2$ or 5% sodium hypochlorite or 70% alcohol either in combination or alone depending upon the surface contamination. These surface-sterilized explants are inoculated on to MS medium (or other suitable basal medium) supplemented with vitamins, sucrose and growth regulators. The cytokinins (1-3 mg/1 BAP) with less concentration of auxin (IAA) seems to be good for micropropagation. The auxin 2, 4D is good because it stimulates callus formation and suppresses organogenesis. These culture are incubated at 3,000-10,000 lux light intensity with 16 hours photoperiod.

Stage II: Multiplication of Explants on Defined Medium

Majority of time is taken for multiplication through the stimulation of shoots regeneration from the explants. Sometimes single shoots develops from the apical shoots and in such cases such shoots are used for excision of number of nodal explants. These nodal explants are further inoculated on cytokinin rich medium to proliferate the multiple shoots. During this process multiple shoots can also be obtained by organogenesis or somatic embryogenesis directly on explants. In most plant species each explant is known to produce five to six shoots in four to five weeks. This would give up to 5^{10}-6^{12} plants in one year from the single explant if all the plants survives.

Stage III: Rooting of Regenarated Shoots or Germination of Somatic Embryo

Multiple shoots produced during the stage II are induced to produce rooting in fresh medium. Individually separated shoots are inoculated to rooting medium (with auxins) or in some species rooting is induced directly in the soil in high moisture condition. The rooted plants are highly fragile and sensitive to moisture. In case of somatic embryos, they are allowed to germinate to form the plantlets and then transferred to soil. These plantlets are slowly transferred to soil by hardening process of the plantlets. The medium for hardening should be either peat/vermiculite/pearlite which generally hold more moisture and maintained high humidity condition until it transferred to the soil.

Stage IV: Hardening

These plantlets are first prepared for soil condition from its *in vitro* conditions. This makes plants to become resistance to moisture stress and disease, making plantlets completely autotrophic from the heterotrophic nature in cultural conditions. The plantlets must be protected from the direct sunlight and the relative humidity should be gradually decreased over a period of time. During this acclimatization period plantlets develops well developed roots and also forms cuticular wax in the aerial tissues. Thus becomes suitable for complete transfer into field.

Some species or shoots formed *in vitro* condition appear glassy, brittle and water soaked and this is known as **vitrification**. Vitrification is due to poorly developed vascular bundles, abnormal functioning of stomata and abnormal wax quality and thus results in loss of plantlets. This can be overcome by addition of high concentration of agar (1%), use of growth retardants, bottom cooling of culture tubes and addition of hydrolysate compounds.

Meristem culture

Cultivation of shoot apical meristem *in vitro* is known as **meristem culture.** Meristem culture was developed by Morel and Martin in 1952 for elimination of rivers from *Dahlia.* Morel in 1965 employed meristem culture for micropropagation of orchid *Cymbidium.* Meristem culture involves the growth of an already existing shoot meristem and subsequent regeneration of adventitious roots from these shoots (Fig. 19.4).

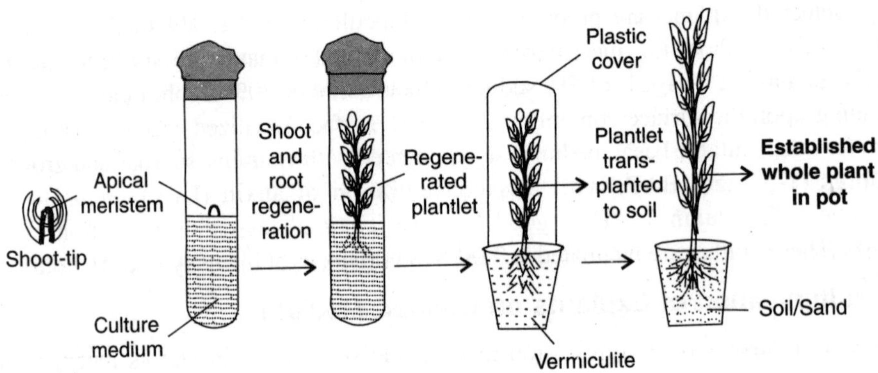

Fig. 19.4. Schematic representation showing steps in meristem-tip culture of plants.

Meristem lies in the 'shoot-tip' beyond the youngest leaf or first leaf primordium. It measures up to about 100 mm in diameter and 250 mm in length. Hence, a shoot-tip of 100-500 mm would contain one to three leaf primordia in addition to the apical meristem. Normally shoot-tips of up to 10 mm are used when the objective is virus elimination to obtain disease free plant. The shoot-tip culture is followed for rapid clonal propagation for which much larger (5-10 mm) explants are used. Thus, most of meristem culture are essentially shoot-tip cultures. Nodal explants of various sizes are also employed for rapid clonal propagation.

If micropropagation is the main objective, the size of shoot-tip used for culture is not important. But when the objective is to make stock free from a virus (or other pathogenic microorganisms) contamination, it is essential that the apical meristem should be excised with a minimum of the surrounding tissue. The shoot-tip may be cut into fine pieces to obtain more than one plantlet from each shoot-tip. In case of some species such as cauliflower pieces of curd (the inflorescence) are used, while in plants having underground stems shoot-tips or tissue pieces bearing buds of such stems may be used. Usually explants taken from actively growing plants at the beginning of growing season are the most suitable.

Meristem explant are then placed on MS medium which is known to be effective in majority of the species. For some species lower salt concentration is suitable. During the growth several fungicides (bavistin) or antibiotics (streptomycin/choramphenicol) can be added to the medium to completely remove the endophytic contamination. Similar to micropropagation, in meristem culture also follows similar steps namely (*i*) culture initiation, (*ii*) shoot multiplication, (*iii*) rooting of the developed shoots, (iv) transfer of plantlets to soil/pots.

Applications of Micropropagation and Tissue Culture

- A small amount of plant tissue is sufficient for production of millions of clones in a year. Whereas it would take more years to get same number of plants.
- Micropropagation offers an alternative method of vegetative propagation for mass propagation. The large number of plants can be obtained in short period. Any particular varieties can be obtained in large quantity and time period to develop new varieties was reduced to 50%.
- Large number of plants can be maintained in less space and timely need can be fulfilled by this method. These plants are maintained in good health.
- This helps in germplasm storage and saving the endangered species.
- Diseased free plants are produced from this method.
- Plants with increased yield and vigour in floriculture species are produced.
- This method provides speedy international exchange of plant material without problem of plant quarantine.

19.5 ORGANOGENESIS AND EMBRYOGENESIS

Introduction to Organogenesis and their Practical Applications

Induction of adventitious roots, shoots or other organs in cultured tissues is very important in the plant tissue culture methodology. The formation of shoots, roots, flowers can be initiated from the cultured plant cells or tissues. The induction of different adventitious organs of the plant from cultured tissues *in vitro* condition is known as **organogenesis** (Fig. 19.5). Generally in tissue culture two types of organogenesis was noticed or induced. They are caulogenesis and rhizogenesis. Induction of adventitious shoot buds from the cultured tissues is known as **caulogenesis.** This was usually made after the formation of callus in the culture condition or directly from the explant which was placed on the medium. Whereas induction of adventitious roots from the cultured tissues is known as **rhizogenesis.** Rhizogenesis was induced only after the caulogenesis in the plant tissue culture methods. After both the type of organogenesis (shoot first and root next) the separate individual plant was transferred to the soil condition through the process of hardening.

Organogenesis has been obtained from many species of plants and from varieties of explants, namely, cotyledons, hypocotyl, stem, leaf, shoot apex, root, youn inflorescence, flower petals, petioles, embryos etc. which are cultured *in vitro* conditions. Explants from both immature and mature parts are made to induce to form callus. Then callus was regenerated to make complete plant.

Sometimes anomalous structures develops during the organogenesis and these structures are called organoids. The group of cells which give rise to the shoots and roots is known as meristemoids or

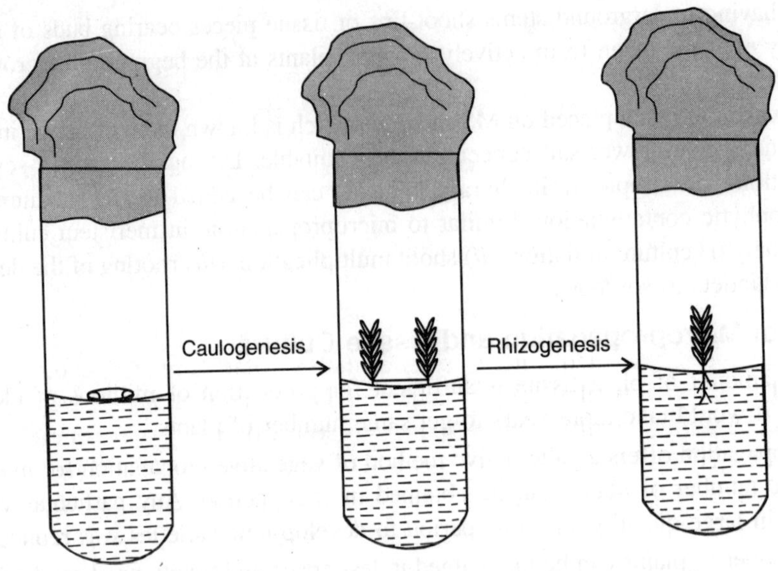

Fig. 19.5. Diagrammatic representation of organogenesis in plants.

nodule. Meristemoids are always developed in areas where the cells accumulates the high concentration of starch. These meristemoid generally forms roots from the inner cells thus they are endogenous in origin, while shoots formed from outer cells and are called exogenous in origin.

Organogenesis is due to dedifferentiation of the mature cells and followed by redifferentiation of cells. Dedifferentiation favours unorganised cell growth and the resultant developed callus has meristems randomly divided. Most of these meristems if provided proper *in vitro* conditions would redifferentiate shoot buds and roots. The whole plant regeneration from the cultured cells may occur either through shoot bud differentiation or somatic embryogenesis.

Different types of tissue culture media are used for organogenesis such as MS, B5, White's and SH (Schenk and Hilderbrant, 1972) medium. The MS medium should contain high concentration of nitrogen as ammonium when compared to other media. It is observed difficulty in transfer of cells from B5, White's or SH medium to MS medium. The season of the year, donor plant conditions, age and physiological state of the parent plant have great impact on organogenetic potential of explant.

The organogenesis in case of plants is poorly understood and this involves various factors such as plant growth status, source of explant, culture medium, addition of growth regulators and environmental conditions. Skoog and Miller in 1957 confirmatively showed that growth regulators control the organogenesis in case of tobacco pith. They showed balance between auxin and cytokinin ratio helps in induction of different types of organogenesis. The high concentration of cytokinin known to induce shoot formation. Whereas high concentration of auxin induce the root formation. The inhibitory effect of one molecule of auxin on caulogenesis could be counteracted by two molecules of cytokinin. This is not always true with all the plant species.

In addition to auxin and cytokinin, other growth regulators is also known to have effect on organogenesis. Gebberellins commonly have inhibitory effect on shoot or root organogenesis in culture condition. Ethylene is shown inhibitory effect on shoot initiation, but promotes the bud growth once it is initiated. The physical condition of the medium, genotype of the plant materials, light, concentration of the sucrose and inorganic phosphate and abscissic acids are having profound influence on the organogenesis in the varieties of the plants.

Applications of Organogenesis

- By organogenesis any plants are made to produce plantlets in very short period.
- It also provides a method for the production of large number of plantlets in case of tree species or any hard species where somatic embryogenesis is not possible.
- This helps in understanding the interplay of various chemical factors involved in morphogenetic change in plants.
- This acts as a simple bioreactor system where organ specific chemicals are produced.

19.6 SOMATIC EMBRYOGENESIS

Somatic embryogenesis means induction of somatic embryo other than the zygotic embryo from the culture tissues. The embryo developing from zygote is known as zygotic embryo or simply embryo. The somatic embryos are commonly called **embryoids** or **nonzygotic embryos**. Generally embryogenesis is not considered under organogenesis because the embryo will not be having vascular connection with parent plant tissues and have independent existence. But other organs such as shoot, root or flower always having the vascular connection with parent plant tissues. Steward *et al.* (1958) and Reinert (1959) first observed first somatic embryogenesis in carrot suspension and agar cultures respectively. Sharp and others in 1980 concluded the somatic embryogenesis may be initiated directly from the explant or indirectly from the callus tissues.

Auxin is known to induce the somatic embryo formation. This always occurs in two phases viz., first induction phase and second maturation phase. Induction phase requires high auxin (0.5-40/60 mg/lt 2, 4 D) and reduced nitrogen (NH_4^+) in the induction medium and low auxin or growth regulator free medium (with high sucrose-6-40%) for maturation phase.

Initially somatic embryos derived from the single cells which later divide to form meristematic cell mass known as meristemoid. These cell mass contain dense cytoplasm, large starch grains, large nucleus and nucleolus with high dehydrogenase activity. They later become cutinized and losts cytoplasmic connection with the surrounding cells. This cell mass are called proembryo and develop in a definite sequence with characteristic morphology. First it forms globular structure and named as globular embryoid. This by further division forms heart-shaped embryoids and torpedo embryoid. Finally it forms cotyledonary somatic embryos. The cotyledonary embryoid is having a definite polarity i.e., a lower radicular end forms roots and plumular end forms shoot (Fig. 19.6). Thus they have full potential to develop into complete young plantlets.

During the development sometimes abnormalities may occur and forms embryogenic clumps or multicotyledons or irregular shaped embryos. This is due to high concentration of auxins and can be controlled by using abscissic acid with mannitol in the medium. In some species first formed somatic embryos (primary somatic embryo) further regenerate to produce second group of somatic embryos on them and these are called secondary somatic embryos, the process is known as **secondary embryogenesis** or **recurrent embryogenesis** (Fig. 19.7).

Applications of Somatic Embryogenesis

- Somatic embryogenesis have tremendous advantage for mass production of plants in large numbers when compared to both regular clonal propagations (e.g., grafting, cuttings) and other tissue culture methods (e.g., micro propagation).
- By somatic embryogenesis unlimited numbers of embryos can be produced from a single explant and it can be maintained for longer time.

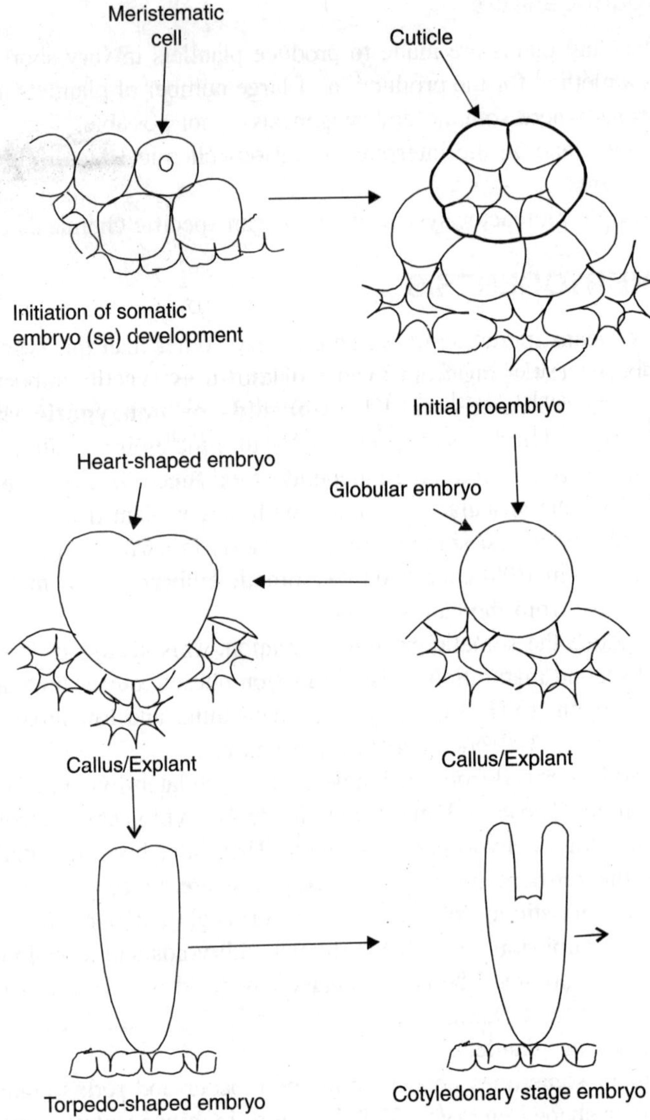

Fig. 19.6. Scheme showing development of a somatic embryo from a single superficial cell of the explant material.

- Somatic embryos are true to their parents and shows no or less tolerance to mutations.
- Like organogenesis, there is no need of shoot productions separately. Because they have complete potential of forming embryonic shoots, roots and leaves.
- This helps in scale-up of propagules with the automated culture systems because the somatic embryo can be produced in the liquid medium.
- The bioreactors can be used with somatic embryo production which helps in design of closed system for separation, sizing and distribution of propagules into culture vessel and even helps in automation of transplanting plantlets to soil mix in greenhouse tray at a rate of 8000 per hour. This cuts the cost, reduces contamination, saves space, time and labour.

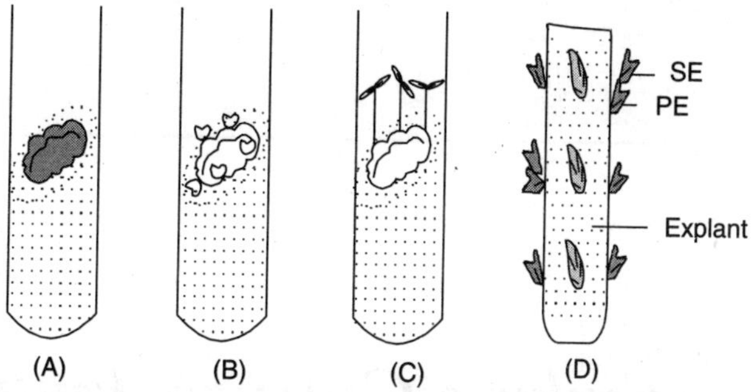

Fig. 19.7. Figures showing development of embryoid in tissue culture. (A)—callus, (B)—numerous embryoids in six week old cultures, (C)—plantlets developed from embryoids and (D)—an enlarged portion showing secondary embryoids (SE) on primary embryoids (PE) produced on explant.

- Somatic embryos serve as better source for protoplasts. Because cells of embryo have good potentiality of regeneration (in case of grass species, horticultural and forestry species protoplasts have less regenerative capacity.
- It also provide cells genetic transformations where callus induction is very difficult.

EXERCISE

1. What is meant by micropropagation?
2. Differentiate between shoot-tip and meristem.
3. Explain different stages of micropropagation.
4. Describe meristem culture and add a note on its application. (BU April 2001)
5. Give the meaning of organogenesis and embryogenesis.
6. What is somatic embryogenesis? Give an account of its induction and importance. (BU Oct. 2001)
7. How somatic embryos are produced *in vitro*?

Chapter

20

Embryo Culture, Embryo Rescue and Their Application

20.1 INTRODUCTION

Embryo is the zygotic tissue obtained after fertilization of ovum by the male gamete. This embryo develops properly when nourishing tissue, the endosperm was present in the seed during the development. But in case of the cross between the two distant species result in the degeneration of endosperm tissue. Thus it hampers the embryo development and viable plant will not develop. To recover such hybrid embryo, culturing of embryo was followed. This culturing of embryo (hybrid) *in vitro* condition is known as **embryo rescue** and widely used for crop improvement. This method is also used to culture where the embryo is incompletely developed (orchids), to skip the seed dormancy for shortening the breeding cycle, determination of seed viability and microcloning of the source material.

20.2 CULTURING OF EMBRYO

Hanning in 1904 first cultured isolated mature embryo of cross between *Cochleria* × *Raphanus* and successfully obtained the plants. Hybrid seeds from *Triticum aestivum* × *Hordeum vulgare* are not obtained naturally. But hybrid seedlings have been obtained by embryo culture.

The embryo was dissected out carefully from the surface sterilized seeds (5-10% chlorox or 0.45% sodium hypochlorite for 5-10 minutes). Both immature and mature embryo can be used. Sometimes before placing embryo directly on to the medium they can be once again surface sterilized by using 70% ethanol for 30-60 seconds. Placing the embryo on endosperm tissue present on the medium was known to show good growth when compared to placing of embryo directly on the nutritive medium (embryo-nurse endosperm technique). Excision of the embryo along with suspensor (is an ephemeral structure found at radicular end of the proembryo) is always beneficial for the embryo development *in vitro.* The MS, White's and B5 medium with little modification is widely used for the embryo culture. The medium rich in sucrose/glucose, ammonium nitrate, vitamins, casein hydrolysate, natural plant extracts with growth regulators at pH range of 5 to 7.5 known to show good responses in the culture. The cultures are incubated at 25 ± 2°C with light exposure of 4000 lux for 4 hours at the beginning. The callus obtained from the embryo tissue is generally having high regenerative capacity compared to those derived from the leaf, stem and root. The callus tissue is made to produce different organs (organogenesis) or embryoids (somatic embryos) and ultimately complete plants are produced.

20.3 APPLICATIONS OF EMBRYO CULTURE AND EMBRYO RESCUE

Viable Embryo Production from Incompatible Crosses

Interspecific and intergeneric crosses results in the non-viable seeds. Isolation and culturing of such hybrid embryo results in the production of rare viable plants.

Dormancy Breaking

Seeds which are undergoing long periods of dormancy delay the breeding work in case of horticultural and crop plants. By embryo rescue this can be solved and breeding cycle can be shortened.

Embryo Production from Sterile Seeds

Embryo will be immature stage in case of early ripening fruits as a result seeds will not germinate. In such cases by embryo culture methods seedlings can be produced from sterile seeds, e.g., plum, apricot and peach.

Haploid Production

Interspecific crosses between *Hordeum vulgare* × *Hordeum bulbosum* results in the normal diploid after normal fertilization. But later during development chromosomes of *H. bulbosum* lost and result in the haploid embryo.

Micropropagation

As an embryo forms potential (young) organ for regeneration and thus can be used in micropropagation of hard species. This method was followed for the development of plantlets in case of *Pinus palustris.*

EXERCISE

1. List the applications of embryo rescue.
2. Define embryo.
3. What do you mean by embryo culture?
4. Why is endosperm necessary for development of embryo?
5. How embryo culture is helpful for hybrid production?
6. What is meant by embryo rescue?
7. Name the plants where successful embryos are produced from hybrids.
8. Briefly explain production of seedling from sterile seeds.

Chapter

21

Haploids and Their Application

21.1 INTRODUCTION

Haploid plants (haploids) are the plants bearing only one complete set of chromosomes (n) whereas diploids contain two complete set of chromosomes (2n). Haploid plants was first discovered by Bergner, A.D. during 1921 in *Datura stramonium.* In natural condition haploids arises from parthenogenesis and they rarely produce character of male parent. The production of haploid individual from male cells/gametes is known as **androgenesis.** The haploid plants can be obtained *in vitro* from pollen grains by placing anthers or separated pollen grains on a suitable medium, this is known as **anther culture** or **pollen culture** respectively.

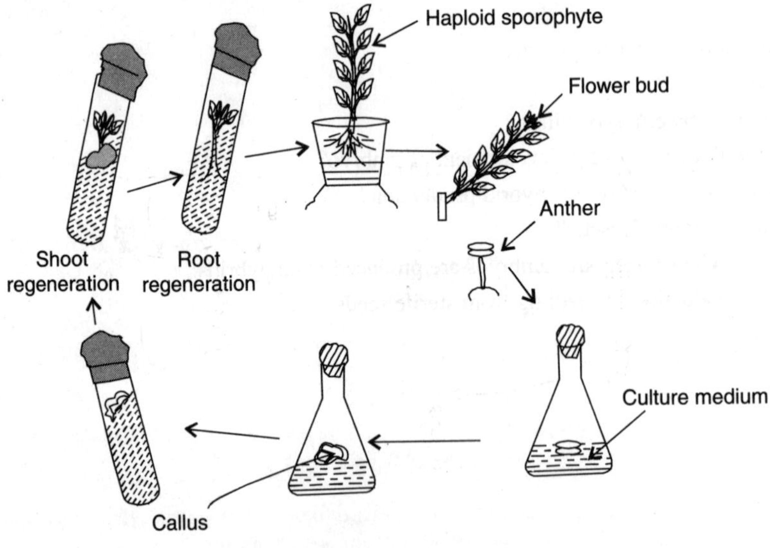

Fig. 21.1. Technique of anther culture.

21.2 PRODUCTION OF HAPLOIDS

Tulecke in 1953 first observed formation of haploid callus on medium in *Gingko biloba,* a gymnosperm.

Later Guha and Maheshwari (1966) observed haploids in varieties of angiosperms. Anthers may be taken from the plants grown in open field or in pots which are preferably grown under controlled temperature, humidity and light.

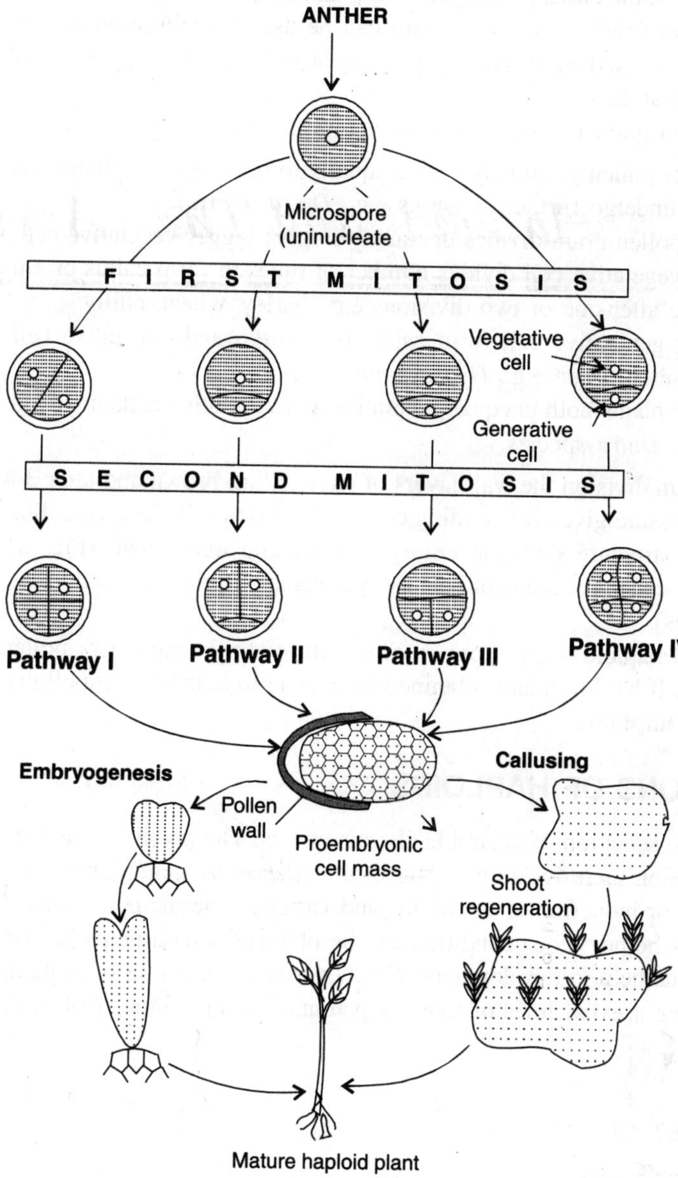

Fig. 21.2. Diagrammatic representation of different patterns of early divisions in pollen grains during androgenesis and the subsequent development of proembryogenic cell masses so obtained.

The particular stage of anthers at the time of culture is the very important factor which decides the success. Flower buds which are having premitotic microspores are collected, surface sterilized (Fig. 21.1). Then anthers are excised and placed horizontally on the culture medium. Adequate care should be taken to avoid any injury to anthers, otherwise it induces callus formation from the diploid, sporophytic anther wall cells. Sometimes pretreatment of anther such as exposure of excised flower buds to a low temperature (3-5°C for two days) or exposure to high temperature (35°C for 24 hours), irradiation with gamma rays or X-rays, centrifugation are known to have increased haploids production in culture condition. In some cases pollen grains are taken out from the anther and directly cultured on a culture medium. Both solid or liquid medium can be used to culture anthers or pollen grains. The anther cultures are maintained in alternating periods of light (12-18 hr; 5000-10000 lux M^2) at 28°C and darkness (12-6 hr) at 22°C.

The cultured pollen grain follow mainly four (Fig. 21.2) pathways:

(*i*) The uninucleate pollen grain may divide equally to form two daughter cells of equal size. Later both the cells undergo further divisions e.g., *Dutra innoxia.*

(*ii*) In few plants pollen grain divides unequally to give bigger vegetative cell and smaller generative cell. The vegetative cell divides number of times to form callus or embryo and generative cell degenerate after one or two divisions e.g., barley, wheat, chilly.

(*iii*) In some other generative cell form callus or embryo and vegetative cell does not divide or forms only limited tissue e.g., *Hyoscyamus.*

(*iv*) Lastly in some plants both unequal generative and vegetative cells involved in callus or embryo formation e.g., *Dutra* species.

After the maximum division the wall layers of anthers turn brown and after 3-8 weeks anther burst open due to excess pressure given by dividing cells. This forms fully formed callus or differentiate into globular embryo-like structure (somatic embryo) depending upon plant (Fig. 21.2). Later callus or somatic embryo is induced to form haploid plants by using appropriate concentration of growth regulators (auxins and cytokinins).

The production of haploid plants from *in vitro* culture of microspores or pollen grains is known as **androgenic haploids**. If haploid plants obtained from *in vitro* culture of unpollinated ovule or ovary is known as **gynogenic haploids.**

21.3 APPLICATIONS OF HAPLOIDS

Haploid plants help in selection of desirable characteristics. The phenotype is the expression of single copy genetic information there by no masking of a character by gene dominance. The haploid plants act as good material for induction of mutations and various gene manipulations. Through doubling of haploid chromosomes homozygous condition can be obtained which helps in plant breeding and crop improvement. Haploids help in development of new varieties, selection of mutants which are resistant to diseases, developing asexual lines of trees or perennial plants, transfer of desirable genes.

EXERCISE

1. What are haploids?
2. Define anther culture. (BU Oct. 2002)
3. Elucidate the different pathways followed by cultured pollen grain.
4. Give the applications of haploids. (BU Oct. 2001)
5. Who discovered haploids?
6. Name a few plants in which haploids are produced.
7. What are androgenic haploids? Discuss the technique of their production and application. (BU Oct. 2001)
8. Describe *in vitro* process of haploid production. Add a note on its application. (BU Oct. 2002)

Chapter

22

Somaclonal Variations

22.1 INTRODUCTION

Somaclonal variations are defined as the genetic variability present among cultured somatic cells. Plants derived from such cells or progeny of such plant is called **somaclones**. The term somaclonal variation is first used by **Larkin** and **Scowcroft** in 1981. The term somaclonal variation is used for genetic variability present among all kinds of cells or plants obtained from cells cultured *in vitro*. Plants regenerated from tissue and cell cultures show heritable variation for both qualitative and quantitative traits. Somaclonal variation has been described in apple, sugarcane, potato, tomato etc. Some variants are obtained in homozygous conditions in the plants regenerated from cells cultured *in vitro* (R_0 regeneration) but most variants are recovered by progeny of tissue culture regenerated plants (R_1 generation). Chaleff in 1981 considered variants from tissue culture as R or R_0 plants and selfed progeny of R_0 plants as R_1 plants. Somaclonal variation considered a novel method of genetic variation and having much interest as a basic genetic process, as it contradicts the theory of clonal uniformity. But it is disadvantageous when the objective is micropropagation of considered genotypes.

22.2 PROCEDURE FOR ISOLATION OF SOMACLONAL VARIANTS

Mutants or variants for several traits can be far more easily isolated from cell cultures than from whole plant populations in the field condition. This is because a large number of cells, as high as 10^6-10^9, can be easily and effectively screened for mutant traits. Screening of as many plants would be very difficult to perform, ordinarily impossible. The mutants can be effectively selected for disease resistance, improvement of nutritional quality, adaptation of plants to stress conditions (e.g., Saline soils, low temperature), toxic metals (e.g., aluminium), resistance to herbicides and to increase the biosynthesis of plant products used for medicinal or industrial purposes (Fig. 22.1). The various approaches to the isolation of somaclonal variants can be grouped into two broad categories. They are screening and cell selection (Fig. 22.1) methods.

Screening

It involves the observation of a large number of cells or regenerated plants from tissue culture for the detection of variant individuals. This approach is only feasible technique for the isolation of mutants for yield and yield traits, in general R_1 progeny (progeny of regenerated, R_0 plants) are scored for the identification of variant plants, and then R_2 progeny lines are evaluated for confirmation. This has been employed in number of crop plants (Table 22.1).

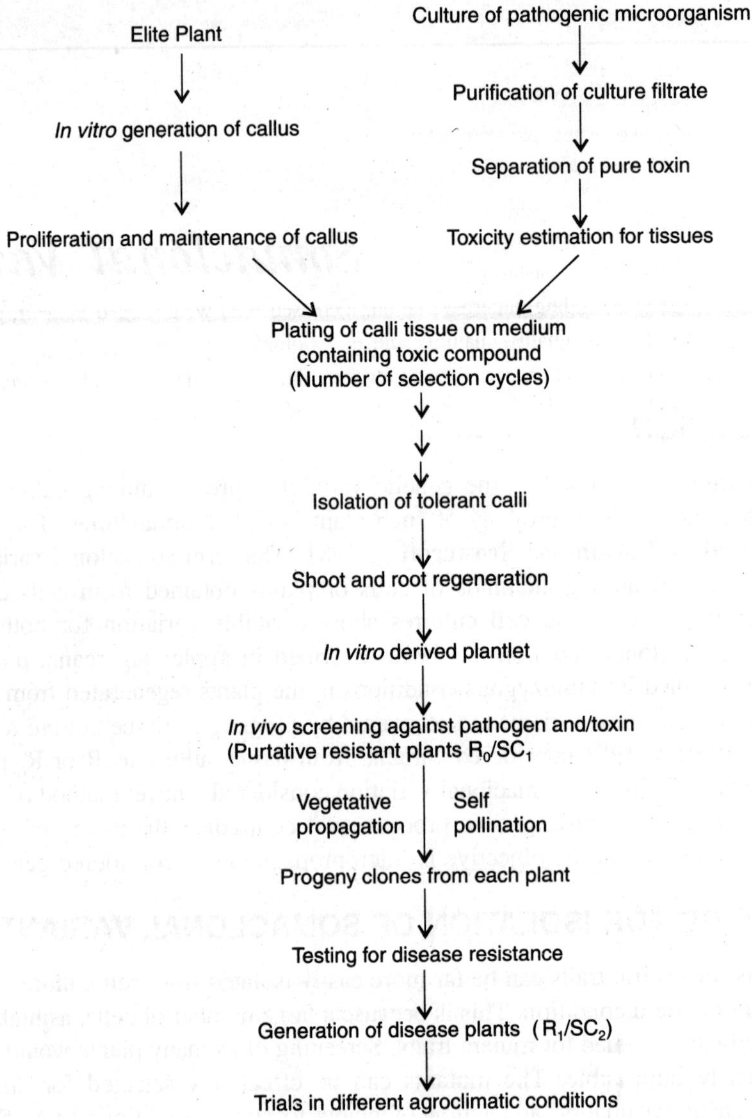

Fig. 22.1. Schematic representation of procedure followed for the isolation of disease resistant plants from culture.

The screening method has been profitably and widely employed for the isolation of cell clones that produce higher quantities of certain important biochemical agents; computer based automated cell sorting devices have also been used to screen as many as 1000-2000 cells/second from which desirable variant cells were automatically separated. The variant cells was further regenerated to produce complete plantlets.

Table 22.1: List of Plant Species for Which Somaclonal Variations Obtained for Different Beneficial Characters.

Plant	*Characters*
Barley	Grain yield, test weight, leaf shape, 1000 grain weight, height, lodging resistance, heading date, β-glucan content, ash
Brassica napus	Growth habit, glucosinolate content
Carrot	Higher carotene content
Datura	Solfonylurea, imidazolinone and triazolopyrimidine herbicide resistance
Fennel	Chlorotic foliage, altered stem height, reduced fruit production, varied oil composition
Maize	Glyphosate resistance
Oats	Avenin banding pattern variation, increased seed weight, seed number, grain yield
Potato	Yield, vine quality, maturity and morphology
Rice	Tiller number, particle size, flowering date, plant height, heading date, culm length, leaf shape, leaf colour, ear shape, sterility mutants
Sorghum	Improved acid soil stress tolerance
Soyabean	Varied height, lodging, maturity, seed protein and oil content
Sugarcane	Yield, sugar content, disease resistance
Tomato	Violet cotyledon colour, dwarf habit, modification of first floral node, early flowering, orange fruit colour
Wheat	Grain colour, height, tiller number, seed storage protein

Cell Selection

In the cell selection method an appropriate selection pressure is applied which permits the preferential survival/growth of variant cells only during culture. Some examples of cell selection are, selection of cells resistant to various pathogen derived toxins, herbicides, acidity high salt concentration etc., when the selection pressure allows only the mutant cells to survive or divide, it is called **positive selection**.

On the other hand, in the case of **negative selection** method, the wild normal type cells divide normally and therefore are killed by the counter selection agent used in the medium, e.g., arsenate; the mutant cells are unable to divide as a result of which they are unaffected by the counter selection agent. These cells are subsequently rescued by removal of the counter selection agent from the medium. Negative selection approach is generally employed for the isolation of auxotrophic mutants.

The positive selection approach may be further sub-divided into four categories, namely, direct selection, rescued method, step-wise selection and double selection.

Direct Selection

The cells resistant to the selection pressure survive and divide to form mature colonies; the selection agent used shows lethal effect on the wild type cells. The mutant cells are not affected by the selection agent and shows growth. This is the most common selection method employed for the isolation of cells resistant to toxins (produced by pathogens), herbicides, elevated salt concentration, antibiotics, amino acid analogues etc. in varieties of crop plants.

Rescue Method

The selection agent used in the medium kills the wild type cells, whereas the variant cells remain alive but, usually, do not divide due to the unfavorable environment created by the selection agent. This selection agent is then removed to recover the variant cells and hence the method is known as Rescue Method of selection. This approach has been widely used to recover low temperature and aluminium resistant variant cells in number of plants.

Stepwise Selection

The selection pressure, selection agent e.g., salt concentration may be gradually increased from a relatively low level to the cytotoxic level in the medium; the resistant clones isolated at each stage are further subjected to the higher concentration of selection pressure. Such a selection approach is called stepwise selection. It may often favour gene amplification (which is unstable) in the cells or mutations in the DNA of organelles.

Double Selection

In certain cases, it may be feasible to select for survival and/or growth on one hand and some other feature reflecting resistance to the selection pressure on the other hand, this is called Double Selection. An example, of such double selection is provided by the selection for resistance to the antibiotic streptomycin, which inhibits chlorophyll development in cultured cells. The selection was based on cell survival and colony formation in the presence of streptomycin (first selection trait) as well as for the development of chlorophyll in these colonies (second selection trait; only green colonies should be selected). This approach has been used for the selection of cells resistant to the herbicide amitrole, 2-4 D, and aluminium.

During *in vitro* selection procedure of cell variants by using selection agent was done at concentration that shows lethality up to 70-90% to wild type of cells. The concentrations that are too low should not be used, because cells fail to grow due to physiological reasons are wrongly considered as true variants. Whereas selection agent at too high concentrations also not advised because it may result in potential variant cells death. Cells should be grown in presence of selection agent for sufficient duration to ensure expression and growth of the resistant somaclones.

22.3 CHARACTERISATION OF SOMACLONAL VARIANTS

Majority of somaclonal variants, isolated through cell selections are often unstable, the frequency of stale variables may range from 8–62% perhaps depending on the species and selection agent employed in the test.

(a) Many selected somaclones are unable to exhibit their resistance during further screening or selection, obviously these clones are susceptible and are wrongly classified as **resistant somaclones (escapes)**.

(b) Several somaclones lose their resistance to the selection agent after a period of growth in the absence of selection pressure. Such clones are called **unstable variants** and may result from changes in gene expression and from gene amplification.

(c) Some variant phenotypes are quite stable during the cell culture phase, but not true, when plants are regenerated from the variant cultures, or when the regenerated plants reproduce sexually in case they are expressed in regenerated plants. Such changes are known as **epigenetic changes.**

Basis for somaclonal variation in plants

Number of factors are considered as responsible for somaclonal variation in the culture condition. They include:

- Pre-existing chromosomal ploidy in the explant cells used for culture initiation.
- Gene mutations in the nuclear and cytoplasmic genomes that may be translocations, duplication, deletions and inversions.
- Nuclear fragmentations during first divisions at callus induction stages.
- Endomitosis or endoduplication during culture initiations.
- Mitotic crossing over during culture induction step.
- Activation of transposable elements in the cultured cells.
- Changes in gene expression and gene amplication.

22.4 SOMACLONAL VARIATIONS AND INDUCED MUTATIONS

Somaclonal variation are preferable to induced mutation for several reasons:

- **Chimaerism** is a major problem in induced mutations but not in somaclonal variation obtained in culture condition.
- Induced mutations are often associated with undesirable features like sterility, unusual character etc.
- *New alleles* and even *'new' mutations* have been recovered through somaclonal mutations which are generally beneficial in nature.
- The frequency of beneficial mutation is very high in the case of somaclonal variations.
- Highly effective selection can be employed *in vitro* for several economically important characters, which is virtually impossible in the case of mutation breeding.
- As astronomically large number of individuals can be effectively screened *in vitro*.

Advantages of Somaclonal Variations

1. Somaclonal variation occurs in rather stable and high frequencies, which is a great advantage over conventional mutagenesis.
2. Some 'new' alleles or even 'novel' mutations may be isolated which were not available in the germplasm or through mutagenesis, e.g., jointless pedicel mutant in tomato.
3. This can be performed in all types of plants i.e., vegetatively or sexually or asexually propagated plants.
4. Use of somaclonal variation may reduce by two years the time required for the release of new variety as compared to mutation breeding. This is because somaclonal variations are usually free from undesirable features like sterility, while induced mutations are generally associated with such defects which necessitates one or two back crosses with parent variety.
5. A very effective selection can be practised at the cell level for several traits, e.g., disease resistant, salt tolerance and herbicide resistance.
6. This is the only approach for the isolation of biochemical mutants, especially auxotrophic mutants in plants, salt tolerance and herbicide resistance.

Disadvantages of Somaclonal Variations

1. The somaclones production is applicable only to those species cell cultures which regenerate complete plants.
2. Selected cell lines often show reduced or no regeneration potential.
3. Many selected clones show undesirable features such as reduced fertility, growth rate and even overall performance of the plant.
4. Many of the selected clones are unstable, that necessitates careful and detailed evaluation for identification of stable clones.
5. Most agronomic traits, e.g., yield quality etc., are not expressed at the cell level and hence cannot be selected for *in vitro.*
6. The variation observed is generally cultivar dependent.
7. The variation observed is not always stable and heritable.
8. The unpredictable and uncontrollable variations occur without much significance.

EXERCISE

1. What is meant by somaclonal variations? (BU Oct. 2001)
2. Give the definition for epigenetic changes.
3. Describe different selection methods of variants.
4. Give advantages of somaclonal variation.
5. Mention various limitations of somaclonal variation.
6. Explain how somaclonal variation is preferred to induced mutation.
7. What are somaclonal variations? Briefly describe their isolation, molecular basis and applications. (BU Oct. 2002)
8. Name the plant for which somaclonal variants are produced.

Chapter

23

Somatic Hybridization and Endosperm Culture

23.1 INTRODUCTION

The production of hybrid plants through fusion of two different plant protoplasts (wall-less naked cells) is known as **somatic hybridization** and such hybrids are called somatic hybrids. Carlson and others (1972) first produced somatic hybrids by using two different species of *Nicotiana* (*N. glauca* × *N. langsdorfii*). Somatic hybridization involves the following four steps: (*i*) isolation of protoplasts, (*ii*) fusion of the protoplasts of desired species/varieties, (*iii*) selection of somatic hybrid cells, and (*iv*) culture of the hybrid cells and regeneration of hybrid plants from them. A brief, elementary consideration of these steps is presented below (Fig. 23.1).

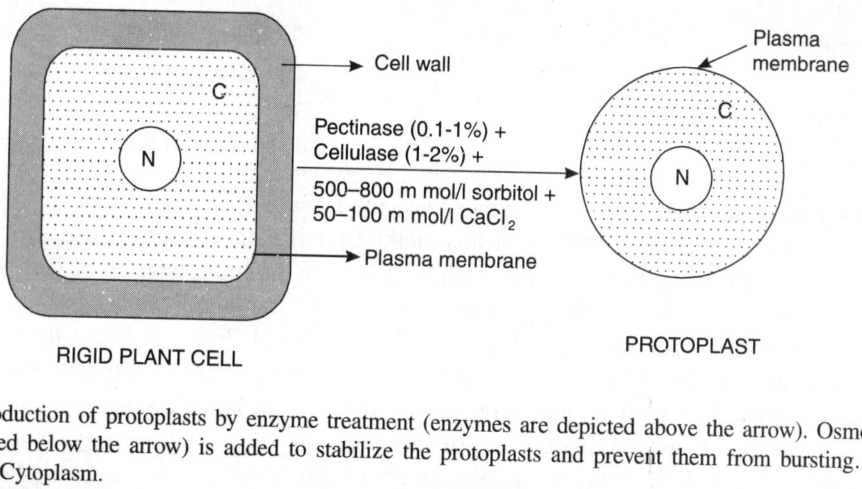

Fig. 23.1. Production of protoplasts by enzyme treatment (enzymes are depicted above the arrow). Osmoticum (indicated below the arrow) is added to stabilize the protoplasts and prevent them from bursting. N, Nucleus; C, Cytoplasm.

23.2 METHODS FOR ISOLATION OF PROTOPLAST

The word **Protoplast** was coined by Hanstein in 1880 for the living matter surrounded by the cell membrane. The isolated protoplast is highly fragile and outer plasma membrane is fully exposed. The plasma membrane is the only barrier between the interior of the living plant cell and the external environment. Isolation of protoplast can be done by three methods (i) mechanical, (ii) sequential enzymatic and (iii) mixed enzymatic methods.

Mechanical Method of protoplast isolation was first done by Klercker (1982). This was done first by cutting the tissues which are earlier plasmolysed with a sharp knife into small pieces. Then these pieces were deplasmolysed by using dilute solution to release the protoplasts. But in this method very low concentration of protoplasts were obtained.

Sequential Enzymatic Method was first used by Takebe and others in 1968, wherein they used commercially available enzymes (Table 23.1) in two steps. The macerated tissue was first incubated in pectinase (degrade pectin cell wall) and then treated with cellulase (degrade cellulosic cell wall) for liberation of protoplasts.

Table 23.1: Various Commercially Available Enzymes Commonly Used for Isolation of Protoplast

Enzyme	*Source Organism*
Cellulase onozuka R-10	*Trichoderma viride*
Cellulase onozuka RS	*Trichoderma viride*
Cellulysin	*Trichoderma viride*
Drisclease	*Irpex lacteus*
Hemicellulase	*Aspergillus niger*
Hemicellulase H-2125	*Rhizopus* species
Helicase	*Helix pomatia*
Macerase	*Rhizopus arrhizus*
Macerozyme R-10	*Rhizopus arrhizus*
Meicelase P-1	*Trichoderma viride*
Pectolyase	*Aspergillus japonicus*
Rhozyme HP 150	*Aspergillus niger*
Zymolyase	*Arthrobacter luteus*

In **mixed enzymatic method** both the enzymes were used together to reduce time and chances of contamination. Power and Cocking (1968) used this method for isolation of protoplasts from number of plant species and it gives very good yield as high as 2.5 × 10^6 protoplasts/gram of leaf tissue.

Protoplasts can be isolated by treating cells with a suitable mixture of cell wall degrading enzymes. The mixture of pectinase or macerozyme (0.1-1.9%) and cellulase (1-2%) is suitable for majority of plant parts. Additionally hemicellulase may be necessary for some tissues. Commercially available enzyme solution is adjusted to the pH value between 4.7 and 6.0 and is kept at temperature 25-30°C. The osmotic concentration of enzyme mixture and of subsequent media is increased by adding sorbitol/mannitol (500-800 m mol L^{-1}) to improve the plasma membrane stability. The enzyme mixture with cells and tissues are incubated for few to several hours (16-18). Then naked protoplasts devoid of cell wall are gradually released in the enzyme mixture (Fig. 23.1).

Protoplasts have been isolated from all plant parts, but leaf mesophyll is the most convenient tissue for this purpose. The fully expanded leaves are surface sterilized, their lower epidermis is peeled off with a pair of forceps and the peeled areas are cut into small (ca. 1 cm^2) pieces with a scalpel and suspended in the enzyme mixture. If epidermis cannot be peeled, leaf may be cut into ca. 1 mm^2 pieces and treated with the enzyme mixture vacuum infiltration may be used to increase the entry of enzymes into the tissues. After the period of incubation, protoplasts are washed with a suitable washing medium in order to remove the enzymes and the debris. The protoplasts may be cultured on a suitable medium in a variety of ways: (*i*) in agar medium (Bergmann's plating technique), (*ii*) in a thin layer of liquid

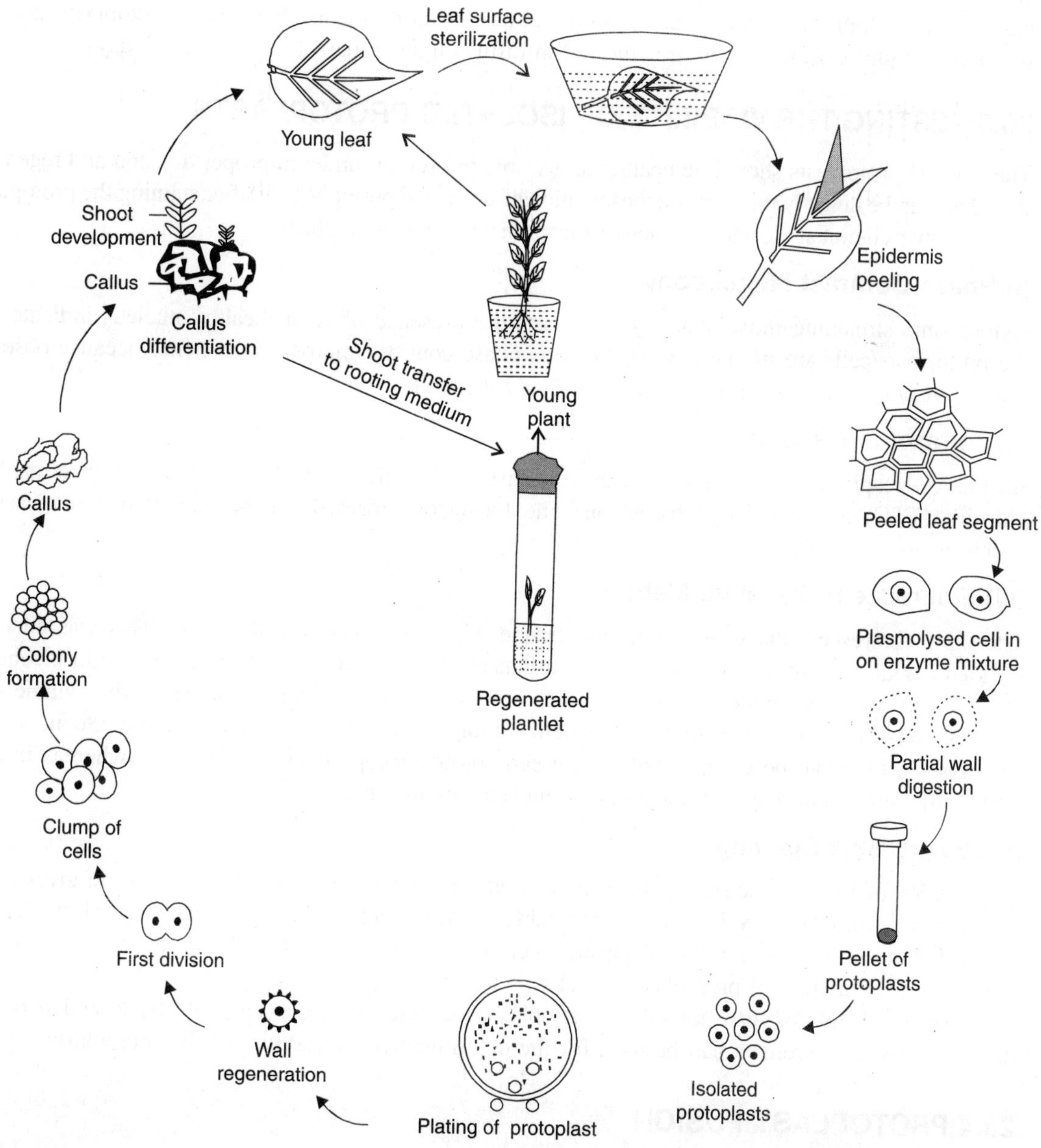

Fig. 23.2. Diagrammatic representation of the technique used for isolation, culture and regeneration of plants from leaf protoplasts.

medium or (*iii*) in small microdrops of 50-100 μl. Protoplasts readily regenerated cell wall (within 2-4 days) and undergo mitosis to form macroscopic colonies which can be induced to regenerate whole plants.

Most of the time MS and B5 media, and their modifications are used for protoplast culture. The media are supplemented with a suitable osmoticum with an auxin and a cytokinin, their types and concentrations depending mainly on the plant species. After 7-10 days of culture, protoplasts regenerate cell wall and the osmolarity of medium is gradually reduced to that of normal medium. The macroscopic colonies are transferred onto normal tissue culture media (Fig. 23.2). Protoplasts are very sensitive to light; therefore, they are cultured in diffuse light or dark for the first 4-7 days.

23.3 TESTING THE VIABILITY OF ISOLATED PROTOPLASTS

The isolated protoplasts should be healthy and viable in order to undergo proper division and regeneration. This can be done by microscopic observation of untreated protoplasts or after staining the protoplasts with suitable chemicals to indicate active metabolism in the protoplasts.

(i) Phase Contrast Microscopy

Cytoplasmic streaming movement (cyclosis) and the presence of clear, healthy nucleus indicate that the protoplasts/cells are in viable state. For this phase contrast microscope is better because observation of unstained cells under bright field is highly difficult.

(ii) Tetrazolium Reduction

In this test respiratory efficiency of cells is measured by reduction of 2, 3, 5-triphenyl tetrazolium chloride (TTC) to the red dye formazon. The formazon formed can be extracted and measured spectrophotometrically.

(iii) Fluorescein Diacetate Method

The 0.5% fluorescein diacetate (FDA) in acetone is prepared and stored at 0°C. This was added at 0.01% of final concentration to protoplasts suspension with osmatic stabilizer. After 5 min incubation the cells are observed under microscope with suitable filter. Esterase present in living viable cells cleaves non-fluorecing FDA into fluorescein (fluorescent compound) and accumulate inside the protoplast, which can be clearly visible as green colour protoplast under UV light, whereas in dead protoplast fluorescein does not accumulate and remains invisible.

(iv) Evans Blue Staining

The 0.025% of Evan's blue stain solution was used for staining the protoplasts. The stain gives colour to the dead protoplasts by becoming permeable to dead ones. Whereas viable protoplasts remains colourless due to impermeability of plasma membrane to the stain. This is the most simple method used for differentiation of protoplasts or cells.

Along with this oxygen uptake by protoplasts measured by an oxygen electrode and photosynthetic activity measurement can be used for the determination of viability of the protoplasts.

23.4 PROTOPLAST FUSION

As protoplasts are devoid of walls it makes them easy for undergoing fusions *in vitro*. In protoplasts fusion incompatibility barriers do not exit during the fusion process at interspecific, intergeneric or even interkingdom levels. Thus it forms major tool for somatic cell genetics and crop improvement programme.

Number of methods have been used to induce fusion between protoplasts of different strains/ species and successful result are obtained in many crop and commercial plants (Table 23.2). The spontaneous fusion, mechanical fusion and induced fusion are the types of protoplasts fusion. In **spontaneous fusion**, the adjacent protoplasts in enzyme mixture have tendency to fuse together to form homokaryons (having same type of nucleus). Similar type of fusion is possible with the two different protoplasts but the frequency is very low. In **mechanical fusion**, gentle tapping of protoplasts suspension in a depression slide results in protoplasts fusion. In **induced fusion** protoplasts are induced to fuse in presence of various chemical or physical agents (fusogen). The major fusogen used include sodium nitrate, high pH/Ca^{2+}, polyethlene glycol, polyvinyl alcohol, dextran sulphate, fatty acids and esters, lysozyme, concavalin, antibodies and electrofusion. Out of these the following methods have been relatively more successfully used for protoplasts fusion.

Table 23.2: Different Genetic Traits Transferred via Protoplast Fusion in Various Crops

Somatic Hybrids Obtained		*Resistance Trait Plants*
Nicotiana tabacum	+ *N. nesophila*	Tobacco mosaic virus
Solanum tuberosum	+ *S. chacense*	Potato virus X
N. tabacum	+ *N. nesophila*	Tobacco horn worm
S. tuberosum	+ *S. brevidens*	Potato leaf roll virus
		Late blight and PLRV Potato virus Y
S. circalifolium	+ *S. tuberosum*	*Phytophthora*
S. melongena	+ *S. saniteongsei*	*Pseudomonas solanacerum*
S. tuberosum	+ *S. commersonii*	Frost tolerance
B. napus	+ *B. nigra*	*Phoma lingam*
B. oleracea	+ *Sinapis alba*	*Alternaria brassiceae*
B. napus	+ *Sinapis alba*	*Alternaria brassiceae*
Citrus sinensis	+ *Poncirus trifoliata*	*Phytophthora*
Lycopersicon esculentum	+ *L. peruvianum*	TMV, spotted wilt virus, cold tolerance
S. lycopersicoides	+ *L. esculentum*	Cold tolerance
Solanum ochranthum	+ *L. esculentum*	Tomato diseases and insect pests
Brassica oleracea	+ *B. napus*	Black rot *(Xanthomonas campestris)*
B. oleracea spp. capitata	+ *B. oleracea* (Ogura CMS line)	Cold tolerance
S. tubersum	+ *S. cammersonii*	Frost tolerance
R. sativa (Japanese radish)	+ *B. oleracea var. botrytis*	Club rot disease
B. oleracea var. botrytis	+ *S. alba* + *B. carinata*	*Alternaria brassicola* and *Phoma lingam*
Citrullus lanatus	+ *Cucumis melo*	Club rot resistance

(Contd.)

Somatic Hybrids Obtained		*Resistance Trait Plants*
Hordeum vulgare	+ *Daucus carota*	Frost and salt tolerance
Solanum melongena	+ *S. sisymbrifolium*	Nematode
S. tuberosum	+ *S. bulbocastamum*	Nematode
S. tuberosum	+ *S. bulbocastamum*	Root knot nematode
Raphanus sativus	+ *Brassica napus*	Beet cyst nematode
Sinapis alba	+ *R. saticus* + *B. napus*	Beet cyst nematode
Quality characters		
N. rustica	+ *N. tabacum*	High nicotine content
B. napus	+ *Eruca sativa*	Low erucic acid
Agronomic characters (transferred through cybrid formation)		
N. tabacum	+ *N. syvestris*	Streptomycin resistance
N. tabacum spp.		Triazine resistance
S. nigrum	+ *S. tuberosum*	Triazine resistance
B. nigra	+ *B. napus*	Hygromycin resistance
B. napus spp.		Triazine resistance
N. tabacum	+ *N. sylvestris*	CMS
B. campestris	+ *B. napus*	CMS
N. tabacum	+ *N. sylvestris*	CMS
B. napus + *B. campestris*	+ *Raphanas sativa*	CMS and Triazine resistance
Oryza spp.		CMS
L. esculentum	+ *Solanum acaule*	CMS
B. napus	+ *B. tournefortii*	CMS

High pH/High Ca^{2+} Treatment

Protoplasts of desired strains/species are mixed in almost equal proportion; generally they are mixed while still suspended in the enzyme mixture. The protoplast mixture is then subjected to a high pH at 10.5 and high Ca^{2+} concentration (50 m mol 1^{-1}) at 37°C for about 30 min. This technique is quite suitable for some species, while for some others it may be toxic. This was first developed by Keller and Melchers (1973) and used for many intraspecific or interspecific somatic hybrids production.

Sodium Nitrate Treatment

The isolated protoplasts are transferred to 0.25 M sodium nitrate solution and then centrifugation results in protoplasts fusion. Power and others in 1970 developed this method.

Polyethylene Glycol (PEG) Treatment

Polyethylene glycol (PEG) induced fusion is the most commonly used as it induces frequency of fusion and low toxicity to most plant cells. The protoplast mixture is treated with 28-50% PEG (MW 1, 500-6000) for 15-30 min followed by gradual washing of the protoplasts to remove PEG; protoplast fusion occurs during the washing. The washing medium may be alkaline (pH 9-10) and contain a high Ca^{2+} ion concentration (50 m mol 1^{-1}); this approach is a combination of PEG and high pH- high Ca^{2+} treatments, and is usually more effective than either treatment alone. PEG is negatively charged and may bind to cation like Ca^{2+} which, in turn, may link to the negatively charged molecules present in plasma lemma; they can also bind to cationic molecules of plasma membrane. During the washing process, PEG molecules may pull out the plasma lemma components bound to them. This would disturb plasma lemma organisation and may lead to the fusion of protoplasts located close to each other (Fig. 23.3).

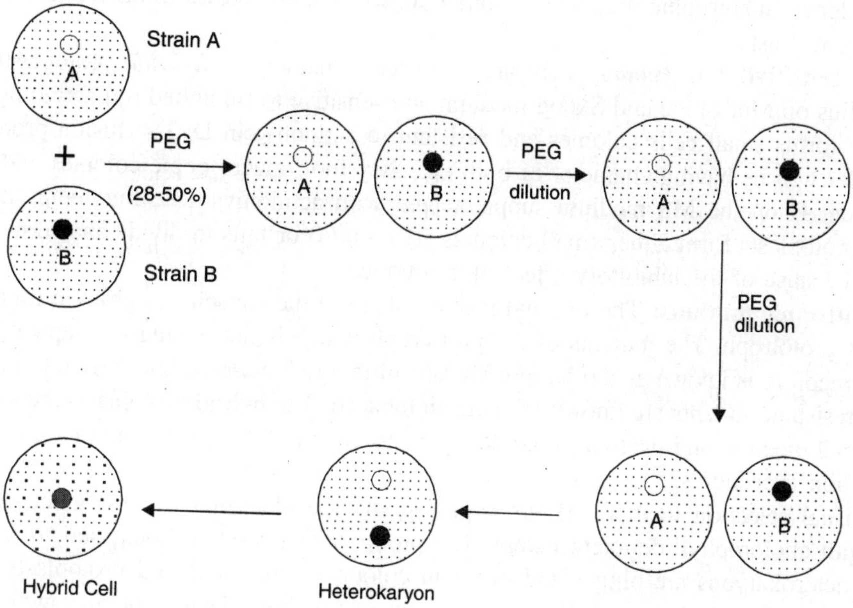

Fig. 23.3. Polyethylene glycol induced protoplast fusion. Protoplasts are first brought close together (aggregation) by PEG. Fusion occurs during PEG dilution due to disturbances created in plasma membrane.

Electrofusion Technique

All the above fusion techniques are nonselective in that they induced fusion between two or more protoplasts. A more simple, selective and less drastic method is the electrofusion technique which utilizes low voltage electric current pulses to align the protoplasts in a single row like a pearl-chain. The aligned protoplasts are pushed, with a micromanipulator, at a gentle phase through the narrow gap between the two electrodes. When two protoplasts that are to be fused are appropriately oriented opposite the electrodes, a short pule of high voltage is released which induces the protoplasts to fuse. The high voltage creates transient disturbances in the organisation of plasma lemma, which leads to the fusion of neighbouring protoplasts. The entire operation is carried out manually in a specially designed equipment, called electroporator, under a microscope. Many workers feel that this fusion technique is more desirable than the others for a number of important reasons. This method was first followed by Senda *et al.* (1979) and later modified by Zimmermann and Scheurich (1981).

The techniques for protoplast fusion are pretty refined and highly effective for almost all the systems.

23.5 SELECTION OF HYBRID CELLS

During the process of protoplasts fusion there is a possibility of formation of homokaryons, heterokaryons and unfused parental protoplasts are present in the mixture. Hence proper selection of hybrid protoplasts or cells is of utmost important for the improvement. The biochemical, visual, albino or flow cytometric methods are employed for the selection process.

In biochemical methods some of the biochemical compounds are used in the medium (selection medium) which helps in the selection of parental and hybrid cells/protoplasts based on their difference in the expression of the characters. Some workers used antibiotics (drug sensitivity) or various salts (nitrate/chlorate-auxotrophic mutants) as compounds which are known to have effect on the growth of cultured protoplasts.

Drug sensitivity: In *Petunia* there are two species namely, *P. hybrida* protoplasts forms macroscopic callus on Murashige and Skoog medium and sensitive to (inhibited by) actinomycin D, whereas *P. parodii* forms small cells colonies and resistant to actinomycin D. The fusion protoplasts of these two species will be having character of both i.e., they form macroscopic colonies and are resistant to actinomycin D on the MS medium supplemented with actinomycin D, thus helps in selection. But parental protoplasts form either small colonies (*P. parodii*) or fails to divide and form the colonies (*P. hybrida*) because of the inhibitory effect of antibiotic.

Auxotrophic mutants: The original protoplasts have the capacity to grow in minimal medium is known as prototroph. The mutants of the prototroph which is not having the capacity to grow in the minimal medium is known as auxotroph. Various nitrate reductase deficient (nitrate non-utilising) and chlorate resistant mutants are known to occur in tobacco. The hybrid protoplasts are known to grow in the minimal medium and parental protoplasts are not able to grow in the minimal medium. It helps in the selection procedures.

In visual selection method the fused protoplasts are identified by fusing the cholorophyll rich parent with cholorophyll-deficient parent. The products of fusion are identified by using microscope because heterokaryons are bigger and green in colour, whereas parental protoplasts are either small and colourless. This is further differentiated by using suitable selective medium which supports good growth of only hybrid cells.

In case of albino mutants one parent protoplast was nuclear albino (chlorophyll-deficient) and the other one was cytoplasmic albino. Both of them are not able to grow into green somatic hybrids. But fusion product of both was green and develops into the well developed somatic hybrids. Here both cytoplasmic and nuclear components help together to develop chlorophyll, but alone is not sufficient for the chlorophyll development.

Flow Cytometric Method

Various research workers are using flow cytometry and fluorescent activated cell sorting for the viable protoplasts. These technique have also been applied for the sorting and selective enrichment of heterokaryons. The hybrid calli derived from this sorted material are known to regenerate hybrid plants.

23.6 REGENERATION OF PROTOPLAST

Various steps in regeneration of protoplasts are as follows:

1. The development of a cell wall around the membrane of protoplasts (**cell wall regeneration**): The development of a cell wall around the membrane of protoplasts is must for regeneration. Protoplasts produced should be protected from rupture of protoplasts (due to osmosis) until they regenerate strong cell wall. Cell wall formation of the protoplasts starts after a few hours

of culturing and completed in two to several days. The more concentration of sucrose is added to the medium and less concentration of osmatic stabilizer is helpful in cell wall formation. Once the cell wall is formed the spherical shape of the protoplasts will be lost.

2. Induction of divisions in cells derived from the protoplasts to form callus tissue (**cell division**): Once the cell develops wall layer it was known to increase in its mass and directly undergoes cell divisions. There is a direct relationship between wall formation and its division. Poorly developed wall in protoplasts always shows improper cell divisions and after few divisions it forms small colonies of cells. Further divisions lead to formation of macroscopic callus after 2-3 weeks.
3. Induction of callus to regenerate whole plant (**plantlet formation**): The callus tissue is transferred onto osmotic free medium to develop good callus. Then it was induced to undergo differentiation to form shoot or embryo. This was further grown into well developed plants.

23.7 CYBRIDS (CYTOPLASMIC HYBRID)

Cybrids are cells or plants containing nucleus of one species and cytoplasm of both the parental species. These are generally produced during protoplast fusion in variable frequencies. Cybrid formation may result by fusion of normal protoplasts of one species with enucleated protoplasts (cytoplasm), elimination of the nucleus of one species from a normal heterokaryon, gradual elimination of the chromosomes of one species from a hybrid cell during the further mitotic divisions (Fig. 23.4). The cybrids can be produced in high frequencies by irradiation of one parental protoplast before fusion in order to inactivate the nuclei or by preparing enucleate protoplast of one species and fusing them with normal protoplast of other species.

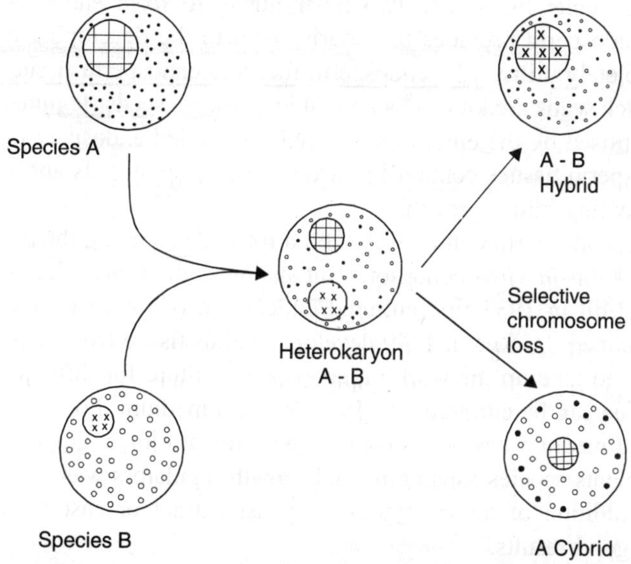

Fig. 23.4. Some fusion products resulting from protoplast culture. The fusion of protoplasts A and B results in a binucleate heterokaryon containing the cytoplasmic contents of the two original protoplasts. Fusion of the two nuclei results in a tetraploid hybrid cell or synkaryocyte. If one of the nuclei degenerates, a cybrid or heteroplast is produced.

23.8 PRACTICAL APPLICATION OF SOMATIC HYBRIDIZATION (HYBRID vs. CYBRIDS)

Genetic recombination in asexual or sterile plants: Somatic hybridization is the only means of genetic recombination in plants that cannot reproduce sexually or which are completely sterile. The sterile plants are made to produce fertile diploids and polyploids by protoplast fusion of the parents.

Overcoming sexual incompatibility: Sexual crossing between two different species or two different genus fail to produce hybrids due to incompatibility barriers. This can be overcome by somatic cell fusion of two different species or genus.

Transfer of cytoplasmic characters: Cytoplasmic characters such as cytoplasmic male sterility, rate of photosynthesis, resistance to diseases or herbicides and low or high temperature tolerance are generally made to transfer to the other strains. To recover recombinants of mitochondrial or chloroplast DNA.

Mitochondria of one species can be combined with chloroplasts of another species. This may be very important in some cases and is not possible by sexual methods even between easily crossable species.

Recombinant organelle genomes, especially of mitochondria are generated in somatic hybrids and cybrids. Some of these recombinant genomes contain useful characters.

23.9 ENDOSPERM CULTURE AND PRODUCTION OF TRIPLOIDS

Endosperm is the nourishing tissue in seeds. During the process of fertilization in angiosperms the two male gamete/nuclei fuse separately with two different female nuclei and this process is known as **double fertilization**. The one male nuclei fuses with egg and forms the zygote (diploid tissue), whereas other male nuclei fuses with two polar nuclei to form endosperm (triploid tissue) in the embryo sac. This endosperm provides the nourishment to the developing embryo in the seed. In the seeds embryo completely utilises the endosperm then the endosperm tissue will not be present in the mature seeds and such seeds are known as non-endospermous seeds (legumes, cucurbits). If endosperm is not completely utilised by the embryo such seeds are called endospermous seeds (coconut, coffee). In such seeds endosperm tissues contain lot of reserve food materials and was consumed during seed germination and seedling initial growth.

In the beginning endosperm culture was started for understanding the potentiality of the endosperm growth and regeneration *in vitro* condition. But later its importance has been realised in the tissue culture. Lamp and Mill in 1933 first cultured endosperm of maize on the nutrient medium but full success was not obtained. La Rue in 1949 developed callus tissue from the endosperm culture and this made other workers to take up the work on endosperm culture for different species of plants.

Endosperm tissue can be cultured on MS or White's medium. For the induction of callus from an endosperm the basal medium has been amended with tomato juice, grape juice, maize juice and yeast extract. For some plants species asparagine and growth regulators was added in the basal medium. In some other cases addition of auxin, cytokinin, yeast extract or casein hydolysate and sucrose are known to produce good results.

Generally mature endosperm is considered not a good material for explant. Because endosperms undergo some changes after (12 days) pollination which makes it unsuitable for culturing. Thus immediately after a few (3-10) days of pollination endosperm was selected for this purpose. The entire seed or kernel was first surface sterilized and endosperm tissue was excised under sterile conditions. The excised endosperm was placed on nutrient medium. For the induction of callus from the endosperm culture should be incubated in darkness or diffuse light at 25 °C. But to induce shoot or other organ, the culture should be kept in bright light of intensity 2000-4000 lux. Sometimes association of embryo is found to be better for the response of endosperm in the cultural conditions.

23.10 APPLICATIONS OF TRIPLOIDS

Triploid plants are generally self-sterile and seedless. This character is very important in case of edible fruits such as grapes, mango, apple, banana, mulberry, watermelon and others. In case of timber or oil yielding plants triploids are showing better performance than the diploid or tetraploids counterparts.

By breeding programme triploids are produced by crossing tetraploids and diploids plants. But in majority of the plants it is not possible and successful, thus endosperm culture was the better alternative and feasible method.

PRACTICAL APPLICATION OF TISSUE OR ORGAN CULTURE (SUMMARY)

In crop improvement methods transfer of gene and manipulation of genes in case of plants was done by conventional breeding. The importance was given to useful characters such as resistance to pathogenic diseases, insects and stress, high yielding varieties, nutritionally important varieties, long term storage capacity. This was a lengthy process and even results were not 100% successful. Nowadays by enormous development in the field of chemistry, biochemistry, genetics, tissue culture, molecular biology and genetic engineering made it possible to genetic manipulation and transfer genes in plants is very easy.

- Plant cells which are in the culture condition is convenient to handle and use according to the requirement. Cells of plants grown in cultural conditions are similar to the plants grown in the natural conditions. Based on this plant cells which are known to be resistant to herbicides in culture are used for crop improvement of commercial varieties. This method is used for developing herbicide resistance in tomato, tobacco and carrot.
- Variation found in the cells or tissues in cultured condition can show heritable changes. Which accumulates in the somatic cells and expresses in the *in vitro* regenerated plants. This somaclonal variation found in plants was considered for development of disease resistance, high nutritional content, male sterility, quantity and quality of plants and pigmentation.
- Somatic cell hybridization helps in the fusion of desired cytoplasm and protoplasm. This is possible only by the cultured cells. This helps in the varied combination of cytoplasmic genes and nuclear genes. The fusion of interspecific or intergeneric protoplasts is possible in production of new characters in the desired plants.
- By culturing hybrid embryos it is possible to obtain the plants with the useful characters. This helps in solving the post-fertilization development which prevents normal seed development.
- Through ovule or ovary culture it is possible to overcome the problems involved in the embryo development.
- Culture system of plants provides the production of haploids by culturing anthers or ovary cells.
- The protoplasts offer good material for uptake of manipulated DNA directly or microinjection of desired genes by micropipettes or gene gun or bacterial or viral mediated gene transfer system.

Cultured cells are considered to be good for the infection by *Agrobacterium tumefaciens* and *A. rhizogenes* which successfully transfer the desired genes into the cells.

Culture of tissue also helps in the production of trioploids, polyploids in all the types of crops.

EXERCISE

1. Define somatic hybridization.
2. Explain different methods of isolation of protoplast. (BU Oct. 2001)
3. Describe various tests used for testing viability of protoplasts. (BU Oct. 2001)
4. Briefly explain protoplast fusion. (BU April 2001)
5. Give an account on hybrid cells selection. (BU April 2001)
6. What do you mean by cybrids?
7. Write a note on application of somatic hybridization. (BU Oct. 2001)
8. Explain how triploids are produced.
9. Give a detailed account on practical application of tissue or organ culture.
10. What is somatic embryogenesis? Give an account of its induction and importance. (BU Oct. 2001)

Chapter

24

Use of Plant Cell, Protoplasts and Tissue Culture for Genetic Manipulation of Plants

24.1 INTRODUCTION

Modern advances made in the gene technology made it possible to genetically manipulate the plants for the crop improvements. This genetic manipulation can be done by introduction of foreign genes (from both prokaryotic or eukaryotic organisms) into plant cells, protoplasts and tissues of both monocot and dicot plants. Today a large number of plants are genetically modified to produce transgenic plants in many important agricultural and commercial crops. Besides the selectable marker genes, a number of other foreign genes conferring insect resistance, herbicide resistance, viral, bacterial and fungal resistance, human genes and genes related to increase shelf life of plants or plant products have been transferred to plants from a wide range of plants, bacteria, viruses and humans. In these introduction of genes in plants resulted in expression and inherited in stable manner without any detrimental effect to the progenies. These transgenic plants also showed equal type of character in field condition similar to the laboratory condition.

24.2 AGROBACTERIUM TUMEFACIENS

In nature there are varieties of organisms especially microorganisms which have the natural ability to change the composition of the genome of their host organisms for their survival. This offers greater amount of chance for their survival in a new host environment. One such organism is *Agrobacterium tumefaciens,* which changes the genome of the host organisms after infection by integrating the DNA segment of its own. Hence this microorganism is known as **nature's genetic engineer**. *A. tumefaciens* is an aerobic, rod shaped, gram negative bacterium discovered by Smith and Townsend in 1907 as an infective agent of a widespread crown gall disease of plants. This bacterium forms overproduction of infected tissues to form tumor (mass of undifferentiated cells). They enter the plant cells at wound site generally around the crown of roots below the soil surface and hence this is known as crown gall tumor. *A. tumefaciens* contains an autoreplicative, extra-chromosomal DNA segment known as plasmid Zaena *et al.* (1974). This plasmid consists of genes responsible for transfer of host cells into tumor in plants called **T-DNA (Transferred DNA)**, which gets transferred from the bacterial plasmid to plant cells genome (natural genetic engineering). The plasmid is named as **tumor inducing plasmid (Ti-plasmid)** by van Larenbecke and others in 1975. This T-DNA expresses itself along with the normal host DNA.

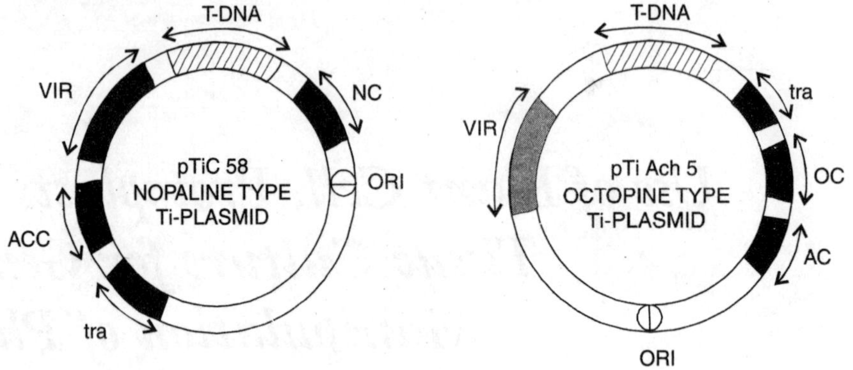

Fig. 24.1. The general structure of nopaline type and octopine type Ti-plasmids of *Agrobacterium*. VIR, virulence region; NC, nopaline catabolism; ACC, agrocinopine catabolism; *tra*, conjugative transfer; OC, octopine catabolism; AC, agropine catabolism; ORI, origin of replication in *Agrobacterium*.

Tumor inducing plasmid (Ti-plasmid) is a megaplasmid of about 200 kb and consists of T-DNA, *vir* region, *ori* sites, *tra* site (Fig. 24.1). Ti-plasmid also contains genes for the synthesis and catabolism of a special type of nitrogenous compounds known as opines (e.g., nopaline, octopine, leucinopine, mannopine, succinamopine, agropine and agrocinopine), auxins and cytokinins. Based on the type of opines synthesized and DNA homology Ti-plasmid can be of three major types namely, nopaline Ti-plasmid, octopine Ti-plasmid and agropine Ti-plasmid. The structural formulae of a number of opines are represented in Fig. 24.3. Opines are utilized by *Agrobacterium* present in the intercellular spaces of the tumor as their carbon and nitrogen sources. Normal plant cells do not synthesizes opines in the absence of the plasmid. Auxins and cytokinin synthesized by plasmid genes is responsible for tumorous growth of the plant cells. *Vir* region regulates the transfer of T-DNA into plant cells. *Ori* regions are responsible for the origin of replication for the plasmid multiplication. The *tra* part is helpful in conjugative transfer of the plasmid.

24.3 STRUCTURE OF TRANSFERRED DNA (T-DNA)

The transferred DNA (T-DNA) portion is approximately 23 kb in length and is transferred into the plant genome in tumor cells from the Ti-plasmid after infection of *Agrobacterium* (Fig. 24.2). It consists of 24 bp direct repeat sequence on the border known as TL (left T-DNA) and TR (right T-DNA). In between these segments there are genes responsible for tumor formation (oncogenes-three genes) and opine synthesis. The *iaa M* and *iaa H* codes for tryptophan-2-monooxygenase and indoleacetamide hydrolase respectively and helps together in synthesis of an auxin, indole acetic acid from the tryptophan precursor. The *ipt* gene codes for isopentenyl transferase helps in synthesis of a cytokinin, isopentenyl adenine. The *iaa M* and *iaa H* together constitute the *tms* (tumor with shoot) locus and *ipt* constitute *tmr* (tumor with root) locus. When mutation in *iaa M* and *iaa H* gene results in the absence of auxin and forms the shoots in tumor because of cytokinin. Whereas mutation in *ipt* gene results in lack to cytokinin and forms the roots because of auxin. One more locus tml (tumor large) mutation results in the production of large tumor. The opine biosynthesis gene are present near the right (TR) border of T-DNA.

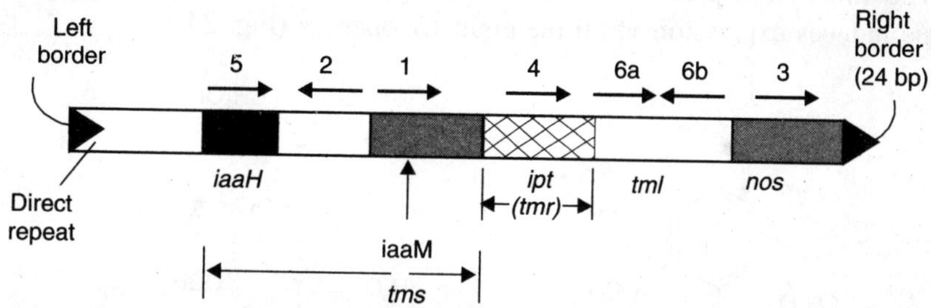

Fig. 24.2. Structural organisation of a nopaline type Ti plasmid T-DNA of *Agrobacterium*. Arrows represent the direction of transcription and numbers indicate the transcriptional units. *iaaM,* encodes tryptophan-2-monooxygesase; *iaaH,* encodes indoleacetamide hydrolase [both involved in auxin biosynthesis, together constitute the *tms* (shooty tumour) locus]; *ipt,* endodes isopentenyl transferase (cytokinin biosynthesis); *tml,* large tumour locus; *nos,* encodes nopaline synthase.

The virulence (*vir*) genes are involved in the transfer of T-DNA into plant genomes, but they will not be transferred themselves. These *vir* genes only mediate the T-DNA transfer and consist of eight operons (*vir A, vir B, vir C, vir D, vir E, vir F, vir G and vir H*) having about 40 kb length (Fig. 24.3). The eight operons consist of about 25 genes of which *vir A* and *vir G* only code for only one protein each and other codes for many proteins. The *vir A* and *vir G* are constitutive genes and regulate the expression of all other genes in *vir* operons. The genes in T-DNA are not required for the transfer of the DNA.

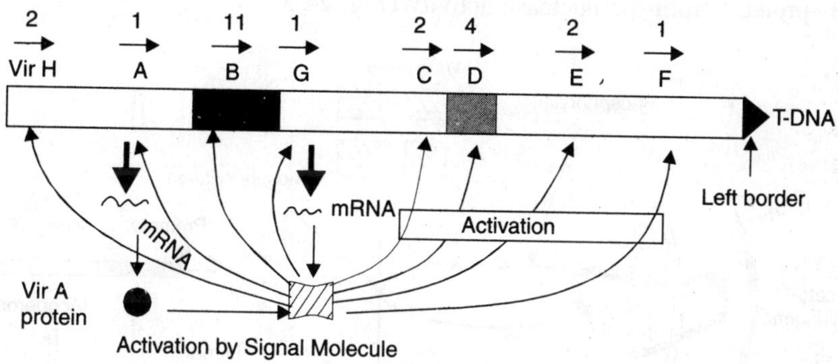

Fig. 24.3. Structural organisation of the *vir* region of an octopine type Ti-plasmid of *Agrobacterium*. Operons *vir A* and *vir G* show low levels of constitutive expression. In response to plant phenolic signal molecules, *Vir A* protein phosphorylates itself and *Vir G* protein, which then induces all the 8 operons of *vir* regulon. The arrows, above the letters shows the operons, indicate the direction of transcription, and the numbers above the arrows represent the total number of genes in different operon.

The proper mechanism of T-DNA transfer is not completely understood. The transfer is regulated by the *vir* genes products especially *vir A* and *vir G* proteins. The *vir A* protein is activated by the phenicolic signal molecules acetosyringone (AS) and α-hydroxyacetosyringone (OH-AS) (Fig. 24.4) which are produced by the all dicots plant tissues during wounding. These signaling molecules were discovered by Stachel *et al.* in 1985. *Vir A* protein acts as receptor for phenolic signal molecules binding and work along with *Chv E* protein present in inner membrane of the bacterium. Binding of phenolic signal molecules activates from its own autokinase activity and phosphorylates itself. This phosphorylated *vir A* protein further activates *vir G* protein by phosphorylation and results in dimerization. The dimer *vir G* protein is a DNA binding protein which requires phosphorylation for its binding and this induces expression of all the eight *vir* operons (Fig. 24.5).

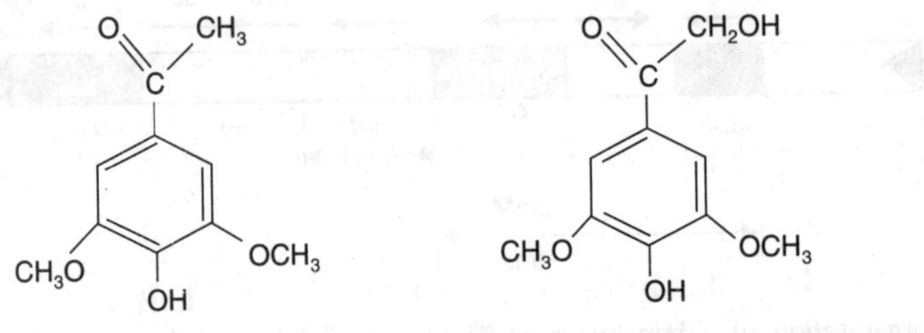

Fig. 24.4. Structures of phenolic signal compounds; acetosyringone and α-hydroxyacetosyringone. These signal molecules are produced by wounded plant tissues, and constitute the wound response. These compounds activate the *vir* operons whose function is essential for T-DNA transfer into plant cells chromosomes.

The *vir* D1 gene product, *vir* D1 protein shows topoisomerase activity and binds to the right border sequence of T-DNA which results in relaxation of supercoiling in DNA. This allows *vir* D2 proteins to show its endonuclease activity and results in the nick in the DNA with 3′ open end. The vir D2 protein remains binding at 5′ end and helps in transport. The open 3′ end at the nick site acts as primer for new DNA synthesis (by DNA polymerase) enzyme in 5′-3′ direction which displaces the old strand from the duplex. This later nicked at left border to give rise to free T-DNA binded to *vir* E2 single stranded protein and thus protects from the nuclease activity (Fig. 24.5).

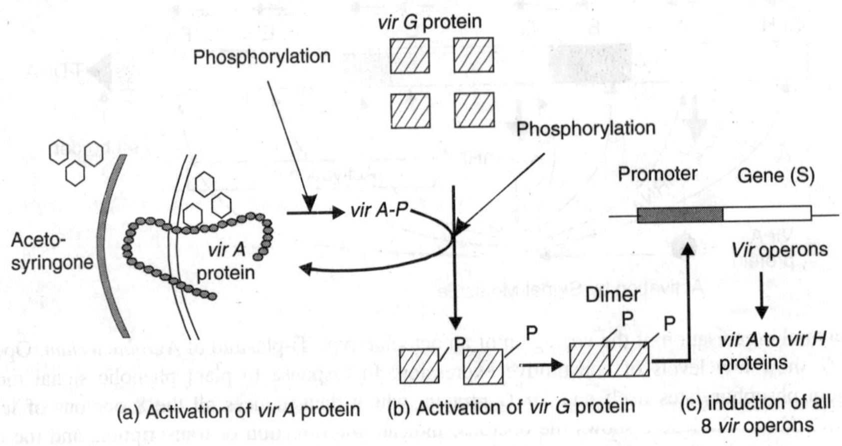

Fig. 24.5. Process of activation of *vir* operons by plant phenolic signal molecules acetosyringone/α-hydroxyacetosyringone that leads to expression of 8 operons. (a) *Vir A* protein is activated by the phenolics to autophosphorylate (*Vir-P*); (b) Phosphorylated *Vir A* protein phosphorylates *Vir G* protein, and *Vir G* forms dimers; (c) Activated *Vir G* induces the 8 operons of *vir* regulon to yield proteins from *Vir A* to *Vir H* operons.

Various *vir B* proteins (nearly 11 genes product) found in the membrane along with *vir* D4 protein forms the channel for the transport of T-DNA from bacterium to the plant cells. Energy is given by

ATPase of *vir B11* protein. Then T-DNA is finally transported into the nucleus through nuclear pore. The single stranded T-DNA converted into double stranded DNA and integrates to chromosomal DNA at random sites by homologous recombination (Fig. 24.6). Then T-DNA starts expressing along with the plant genes and this remains as most stable part in the plant cells.

Agrobacterium nucleoid ('chromosomal') genes are also involved in the attachment of *Agrobacterium* to specific host plant cells. These genes include; the *chvA* and *chvB* are two linked loci that synthesize and excrete β-1,2-glucan, the *cel* locus synthesizes cellulose fibrils, the *pscA* (*exoC*) affects both cyclic glucan and acidic succinoglycan synthesis and the *att* locus affects cell surface proteins.

24.4 TUMOR FORMATION ON PLANTS USING *A. TUMEFACIENS* (MONOCOTS vs. DICOTS)

Tumor can be induced in any dicotyledonous plants by making the wound on the surface sterilized stem or root or any other plant segments. The *A. tumefaciens* was made to contain recombinant gene in the Ti-plasmid before induction of tumor. This recombinant gene should be generally a beneficial gene which has to express in the plants continuously (Table 24.1). The removal of oncogenes involved in synthesis of auxin and cytokinin from T-DNA in Ti-plasmid is called **disarming.** This disarming was not affecting transforming ability of the T-DNA from the bacterium. But cells did not show tumorous growth even with T-DNA in the cells. Now it was clear that only RT and LT is required for such transformation and not the entire T-DNA. This gave a idea of using Ti-plasmid as vector for the transferring any gene to the plant cells without any difficulty.

Table 24.1: List of Plant Species that are Genetically Transformed by Using *Agrobacterium tumefaciens*

Plant Species	*Material Used*	*Vector*	*Result*
Alfalfa *(Medicago saliva)*	Stem segments	pTiBO542	Transformed callus and plants
Asparagus *(Asparagus officinalis)*	Stem segments	pTiC58	Transformed callus
Cotton *(Gossypium hirsutum)*	Hypocotyl segments	pCMC204	Transformed plants
Potato *(Solanum tuberosum)*	Stem or leaf segments	pTiBO542/ pGA 472	Transformed callus
Soyabean *(Glycine max)*	Hypocotyl, embryonic axes	pNon 9749 pTV1100	Transformed plants
Tobacco *(Nicotiana tabacum)*	Leaf discs	pTiBO542/ pGA472	Transformed callus and plants
Tomato *(Lycopersicon esculentum)*	Leaf discs	pMON200SE	Transformed callus and plants
Walnut *(Juglans regia)*	Somatic embryo	pCEN594	Transformed plant

The required gene along with marker gene (antibiotic-kanamycin resistance) was inserted within T-DNA region of disarmed Ti-plasmid of *Agrobacterium tumefaciens.* This recombinant *A. tumefaciens*

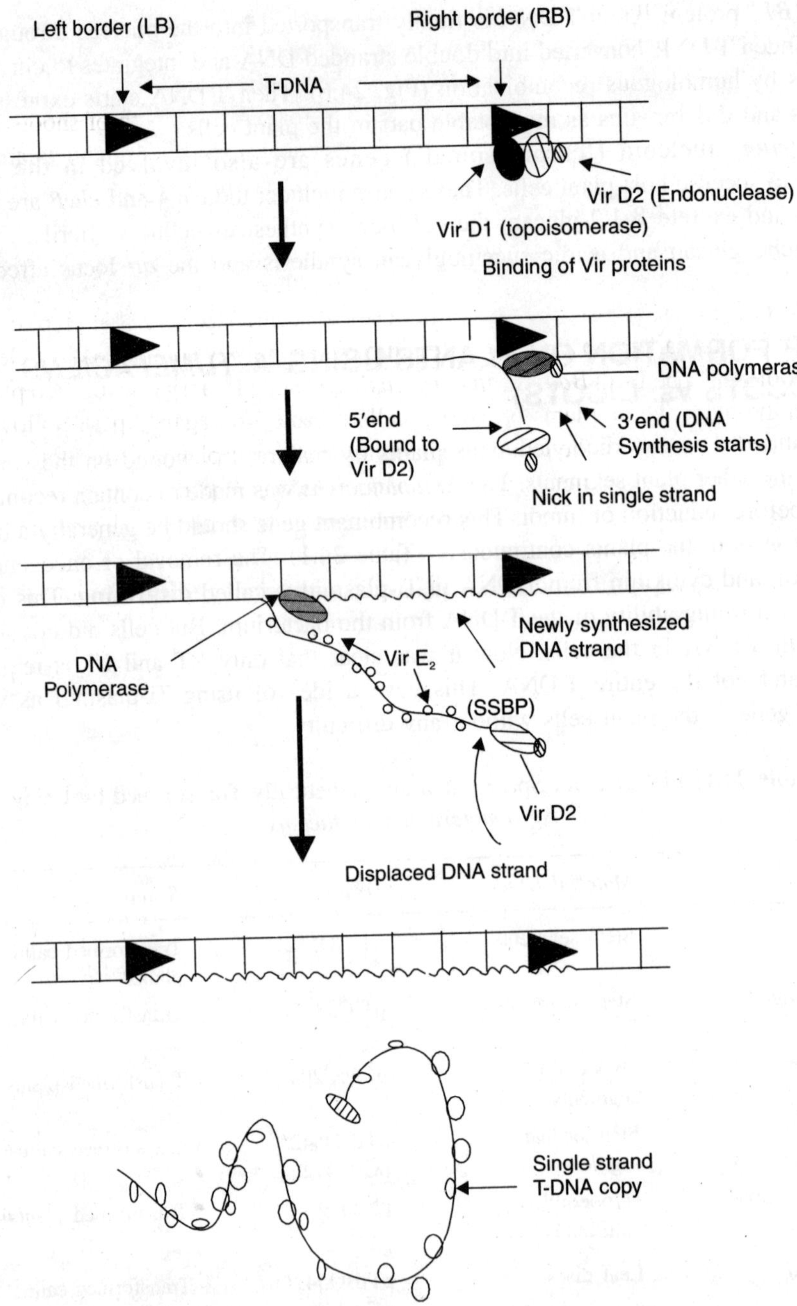

Fig. 24.6. Diagrammatic representation of the generation of T-DNA copy (single-stranded) necessary for transfer into plant cells genome. LB and RB, left and right borders of T-DNA; SSBP, single-strand binding proteins.

was cultured in suspension culture along with the wounded plant segments for two days. Generally circular leaf discs (a few mm diameter) are taken out from surface sterilized leaves. This forms the co-culture of *A. tumefaciens* with plant tissues (Fig. 24.7). At incubation period *A. tumefaciens* infects the leaf discs through wound done during disc preparation. These discs are transferred onto the fresh shoot induction agar medium containing suitable concentration of kanamycin and cerbenicillin. In this medium only transformed plant cells regenerate to shoots because kanamycin present in the medium will not allow normal plant cells to grow. The antibiotic carbenicillin inhibits the growth of *A. tumefaciens.* The regenerated shoots are subcultured onto fresh medium for proper growth of shoots and later roots are induced in rooting medium. These recombinant or genetically modified plantlets are finally acclimatized to the soil condition through series of hardening process.

In the early 1980s number of workers in Europe and U.S.A. used this bacterium to insert foreign genes into plant cells or tissues, they were then able to recover whole plant from transformed cells. The resulting plants were found to incorporate foreign genes into their own genomes and thus had acquired new, stable and inherited characters. This led to the use of bacterial genes for antibiotic resistance and for the production of unusual amino acids. The Belgium biotechnology company, Plant Genetic Systems in 1987 using endotoxin gene from *Bacillus thuringiensis* developed transgenic tobacco plant expressing the toxic gene in the plant. These plants provide excellent protection against pest. Followed by this all over the world scientists reported the genetic manipulation of crop plants.

A. tumefaciens normally does not infect the monocotyledonous plants and does not form tumors. This is due to the fact that wound response is not present in such plants. The wound responses are necessary to initiate *A. tumerfaciens* through induction of *vir* genes.

Especially the signaling molecules are not synthesized which triggers the infection process. However in few cases they are known to case infection with tumor formation e.g., *Asparagus officinalis,*

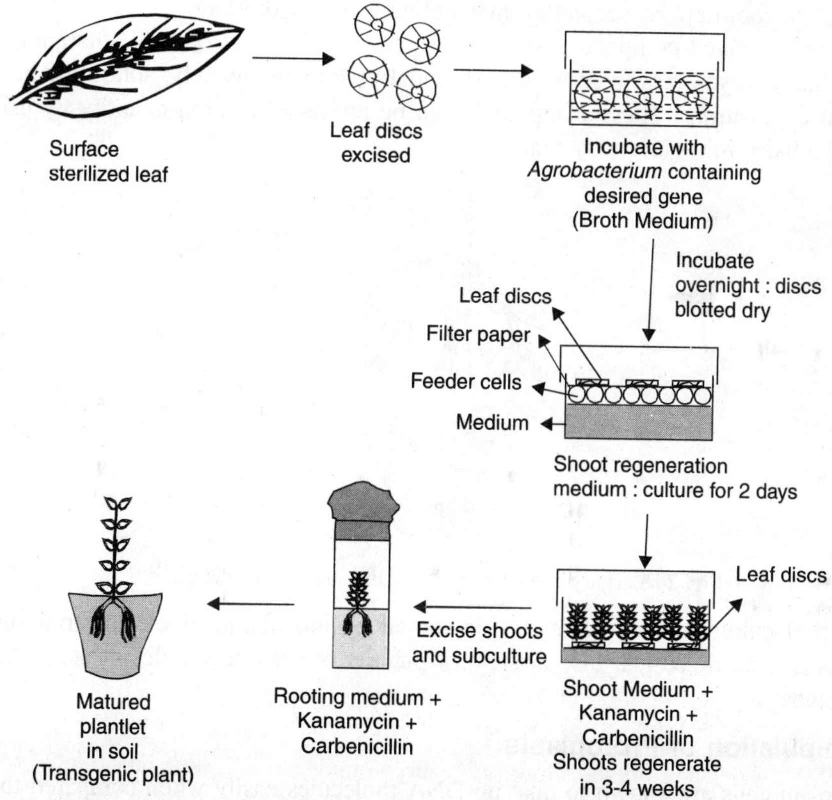

Fig. 24.7. A schematic representation of the procedure for *Agrobacterium-mediated* gene transfer into leaf tissues to obtain genetic transformation.

Chlorophylum, Narcissus, Dioscorea bulbifera and *Oryza sativa.* This infection process can be increased with effective transformation by adding external acetosyringone or wound exudates during the co-culture of plant cells with *A. tumefaciens.* Some monocot plants produce compounds which suppress the *vir* gene induction mechanism. This can be overcome by supplying excess levels of acetosyringone. By this mechanism transformation was found to be successful in case of barley, wheat, maize and rice.

24.5 ROOT FORMATION USING *AGROBACTERIUM RHIZOGENES*

Agrobacterium rhizogenes is a soil bacterium known to infect the plant belonging to dicots. This was discovered by Riker in 1930. It is similar to *A. tumefaciens* in having tumor inducing plasmid, but this plasmid is known as root inducing plasmid (Ri-plasmid) because it induces the infected site to proliferate rapidly to form large number of roots and is known as hairy roots (Fig. 24.8). This change results from insertion of T-DNA from Ri-plasmid into plant genome. The T-DNA codes for auxin synthesis and other rhizogenic functions. The transformed region after infection with *A. rhizogenes* produces high degree of lateral branching, increased root hairs and absence of geotropism. All these characteristics make suitable for the production of required root specific chemicals which are useful in the commercial market. As these infected roots can be removed from the parent plant and grown continuously in simple defined medium free of growth hormones (auxins). But normal root cells requires external supply of auxins and growth hormones. These hairy root cultures are having great potentiality in the production of root-derived secondary metabolites from dicot plant.

Number of root derived compounds such as tropane alkaloids, atropine, catharanthin, betalanin, nicotine, anatabine, hyscyamine, shikonin and steroidal precursors, namely, solasidine having significant commercial applications. These compounds can be produced in high quantity by using the specific plants in the hairy root culture system.

Fig. 24.8. Hairy roots formation at infected site of the plant.

The hairy root culture of any plant can be started by inoculating thick bacterial suspension by wounding of surface sterilized leaf disc or stem of plantlet or seed or petioles or stem segments from green house plants.

Genetic Manipulation of Protoplasts

Protoplasts of plant cells are known to take up DNA molecules easily when compared to intact plant cells with cell wall. This makes protoplasts as good materials for introduction of new traits from distant source to any plant cells. Now reports are available to indicate this type of transformation in case of tomato protoplasts. These protoplasts are taken in lambda phage DNA. By this method of gene transfer host plant can be made to change their genetic characters.

Protoplasts are also considered as very good source for the incorporation of cell organelles or prokaryotic cells such as nuclei (*Nicotiana tabacum, Petunia hybrida* and *Zea mays*), chloroplast (*Petunia, Daulcus carota*), cyanobacteria (cereals) and *Rhizobium*. After incorporation of these foreign cells or cell organelles the host plant cells show change of genetic characters such as conversion of albino forms to chlorophyllous plant, disease resistant plant, nitrogen fixing cereal plant and so on.

Practical Application of Genetic Transformation

- By using the genetic transformation of plants it is possible to obtain the plants with generally insecticidal activity. This reduces the application of chemical pesticides to the crop plants and this method acts as ecofriendly approach.
- It is possible to produce plants which are resistance to herbicides, salt and disease resistance. The herbicide resistance (phosphinothricin) plants such as tobacco, tomato and potato are produced by using acetyl phosphinothricin transferase gene from *Streptomyces hygroscopicus*. Through the expressing genes for the chitinase and other fungal cell wall degrading enzymes in tomato and tobacco plants protection against fungal pathogen such as *Alternaria longipes, Rhizoctonia solani, Phytophthora infestans* was achieved.
- Antisense RNA is a RNA molecule having complementary sequences for the mRNA. Thus, it inhibits the expression of such specific proteins from the mRNA. By using this antisense sequence in plants we can control the expression of particular enzymes which is involved in ripening processes. For example polygalacturonase is responsible for the ripening of tomato fruits and leads to spoilage. If we have antisense RNA for this enzyme in the plants the ripening process in prolonged.
- Genetic manipulation also carried out in crops for making desirable storage proteins, vitamins, amino acids and for the suppression of antinutritional proteins synthesis.
- Through the genetic manipulation plants are made to produce the antigenic proteins of the pathogenic organism and thus can be used as edible vaccines. Such efforts are made in number of plants including potato, tomato, tobacco for *Vibrio cholerae,* Hepatitis B virus and others.
- Plants are made to transform into hairy roots characters to produce high concentration of secondary metabolites having high commercial value.

EXERCISE

1. Write a note on *Agrobacterium tumefaciens*.
2. Which organism is known as nature's genetic engineer?
3. Define Ti-plasmid.
4. Name different types of Ti-plasmid.
5. Explain structure of T-DNA.
6. Describe mechanism involved in T-DNA transfer to plants genome. (BU April 2003)
7. Write a note on the importance of plants genome *Agrobacterium rhizogenes*. (BU Oct. 2002)
8. Give an account of practical application of genetic transformation.
9. Write a short note on Ti-plasmid. (BU Oct. 2001)
10. Elucidate practical applications of genetic manipulation.
11. Describe role played by signal molecules in plants infection by *Agrobacterium*.
12. Briefly explain activation of VIR operons in *Agrobacterium*.
13. **Name the plants where genetic transformation was done by using *Agrobacterium*.**

Glossary

A—Adenine residue, in either DNA or RNA.

Ab—Antibody.

Abiotic stress—The effect of non-living factors which can harm living organisms. These non-living factors include drought, extreme temperatures, pollutants, etc.

Abscisic acid—A plant growth regulator involved in abscission, dormancy, stomatal opening/closure, and inhibition of seed germination. It also affects the regulation of somatic cell embryogenesis in some plant species.

Absciss; abscissa—The horizontal axis of a graph. *cf* ordinate.

Absorb (L. *ab*, away + *sorbere*, to suck in)—To suck up, or to take in. In the cell, materials are taken in (absorbed) from a solution. *cf* adsorb.

Absorption—In general: the process of absorbing; taking up of water and nutrients by assimilation or imbibition. The taking up by capillary, osmotic, chemical or solvent action, such as the taking up of a gas by a solid or liquid, or taking up of a liquid by a solid. *cf* adsorption. In biology: the movement of a fluid or a dissolved substance across a cell membrane. In plants: water and mineral salts are absorbed from the soil by roots. In animals: solubulized food material is absorbed into the circulatory system through cells lining the alimentary canal.

Abzyme—Catalytic antibody.

Acaricide—A pesticide used to kill or control mites or ticks.

Accessory bud—Lateral bud occurring at the base of a terminal bud or at the side of an axillary bud.

Acclimatization—The adaptation of a living organism (plant, animal or micro-organism) to a changed environment that subjects it to physiological stress. Acclimatization should not be confused with adaptation (q.v.). *cf* acquired.

Acellular—Describing tissues or organisms that are not made up of separate cells but often have more than one nucleus. *cf* syncytium.

Acentric chromosome—Chromosome fragment lacking a centromere.

Acetyl co-enzyme A; acetyl CoA—A compound formed in the mitochondria when an acetyl group (CH_3CO—) derived from breakdown of fats, proteins, or carbohydrates-combines with the thiol group (-SH) of co-enzyme A.

Acquired—Developed in response to the environment, not inherited, such as a character trait (acquired characteristic) resulting from environmental effect(s). *cf* acclimatization.

Acridine dyes—A class of positively charged polycyclic molecules that intercalate into DNA and induce frameshift mutations.

Acrocentric—A chromosome that has its centromere near the end.

Acropetal—Developing or blooming in succession towards the apex, such as leaves or flowers developing acropetally. Transport or movement of substances towards the apex, such as the movement of water through the plant. The opposite tendency is termed basipetal.

Acrosome—An apical organelle in the head of a spermatozoon, q.v.

Acrylamide gels—*See* polyacrylamide gels.

Actin—One of the two contractile proteins in muscle (the other being myosin). Actin is also found in the microfilaments that form part of the cytoskeleton of all cells.

Activated charcoal; activated carbon—Charcoal which has been treated to remove hydrocarbons and to increase its adsorptive properties. It acts by condensing and holding a gas or solute onto its surface; thus inhibitory substances in nutrient medium may be adsorbed to charcoal included in the medium. Rooting factors such as phenolamines present as contaminants in charcoal may stimulate growth *in vitro*. Its addition to rooting medium may stimulate root initiation in some plant species. Activated charcoal may differ in origin and in composition. *cf* charcoal; phenolic oxidation.

Active collection—In PGR: Defined in the International Undertaking on Plant Genetic Resources (FAO, 1983) as a collection which complements a base collection (q.v.) and is a collection from which seed samples are drawn for distribution, exchange and other purposes such as multiplication and evaluation.

Activator—A substance or physical agent that stimulates transcription of a specific gene or operon. A compound that, by binding to an allosteric site on an enzyme, enables the active site of the enzyme to bind to the substrate. *See* gene expression.

Active site—A site on the surface of a catalyst at which activity occurs. The site on the surface of an enzyme molecule that binds the substrate molecule.

Adaptation (L. *ad*, to + *aptare*, to fit)—Adjustment of a population to changed environment over generations, associated (at least in part) with genetic changes resulting from selection imposed by the changed environment. *Not* acclimatization, q.v.

Adaptation traits—In AnGR: The complex of traits related to reproduction and survival of the individual in a particular production environment. Adaptation traits contribute to individual fitness; they are the traits subjected to selection during the evolution of animal genetic resources. By definition, these traits are also important to the ability of the animal genetic resource to be sustained in the production environment. (Based on FAO, 1999)

Adaptive radiation—The evolution of new forms, sub-species or species from one species of plant or animal in order to exploit new habitats or food sources. a.k.a. divergent evolution.

Adaptor—A synthetic double-stranded oligonucleotide that has a blunt end, while the other end has a nucleotide extension that can base pair with a cohesive end created by cleavage of a DNA molecule with a specific type II restriction endonuclease. The blunt end of the adaptor can be ligated to the ends of a target DNA molecule and the construct can be cloned into a vector by using the cohesive ends of the adaptor. A synthetic single-stranded oligonucleotide that, after self-hybridization, produces a molecule with cohesive ends and an internal restriction endonuclease site. When the adaptor is inserted into a cloning vector by means of the cohesive ends, the internal sequence provides a new restriction endonuclease site.

Addendum (pl: addenda)—In formulation of tissue culture media: an item or a constituent substance to be added.

Additive allelic effects—Effects of alleles at a locus, where the heterozygote is exactly intermediate between the two homozygotes.

Additive gene effects—Additive allelic effects summed across all the loci that contribute to genetic variation in a quantitative trait.

Adenilate cyclase—The enzyme that catalyses the formation of cyclic AMP.

Adenine—($C_5H_5N_5$; f.w. 135.14) (symbol: A) A white crystalline purine base. A constituent of DNA and RNA and nucleotides such as ADP and ATP. A B-group vitamin (B_4) generally available as $C_5H_5N_5.3H_2O$, m.w. 189.13. It is added to some tissue culture media, as adenine sulphate, to promote shoot formation and for its weak cytokinin effect. It is present in plant tissues combined with aminoamide, phosphoric acids and D-ribose.

Adenovirus—A group of DNA viruses which cause diseases in animals. In man, they produce acute respiratory tract infections with symptoms resembling common cold. They are used in gene cloning, as vectors for expressing large amounts of recombinant proteins in animal cells. They are also used to make live-virus vaccines against more dangerous pathogens. *See* viral vaccines.

ADEPT (antibody-directed enzyme pro-drug therapy)—A way to target a drug to a specific tissue. The drug is administered as an inactive pro-drug, and then converted into an active drug by an enzyme administered with a second injection. The enzyme is coupled to an antibody that concentrates it in the target tissue. When the enzyme arrives at the target tissue, the pro-drug is activated to form the active drug, while elsewhere it remains inactive. *See* drug delivery; targeted drug delivery.

Adhesion (L. *adhaerere*, to stick to)—The attraction of dissimilar molecules for each other. A sticking together of unlike substances, such as soil and water.

A-DNA—A right-handed DNA double helix that has 11 base pairs per turn. DNA exists in this form when partially dehydrated.

Adsorbent—Noun: A substance to which compounds adhere. In tissue culture, an adsorbent is added to the culture medium to adsorb compounds released by cultured cells or tissues, thus minimizing any adverse effect on the subsequent growth in culture. A common adsorbent in tissue culture is *activated charcoal*, q.v.

Adsorption—The formation of a layer of gas, liquid or solid on the surface of a solid. *cf* absorption.

Adult cloning—The creation of identical copies of an adult animal by nuclear transfer (q.v.) from differentiated adult tissue. *See also* cloning; Dolly.

Advanced—Applied to an organism or a part thereof, implying considerable development from the ancestral stage or from the explant stage.

Adventitious—(L. *adventitius*, not properly belonging to) A structure arising at sites other than the usual ones, *e.g.*, shoots from roots or leaves, and embryos from any cell other than a zygote.

Aerate—To supply with or mix with air or gas. The process is aeration.

Aerobe—A micro-organism that grows in the presence of oxygen. Opposite: anaerobe.

Aerobic bacteria—Bacteria that can live in the presence of oxygen.

Aerobic respiration—A type of respiration in which foodstuffs are completely oxidized to carbon dioxide and water, with the release of chemical energy, in a process requiring atmospheric oxygen.

Aerobic—Active in the presence of free oxygen.

Affinity chromatography—A method for separating molecules by exploiting their ability to bind specifically to other molecules. There are several types of biological affinity chromatography. A biological molecule can be immobilized and a smaller molecule (ligand, q.v.) to which it is to bind can be stuck to it, or the smaller ligand can be immobilized and the macromolecule stuck to it. A variant is to use an antibody as the immobilized molecule and use it to 'capture' its antigen: this is often called **immuno-affinity chromatography**. A variation is **pseudo-affinity chromatography**, in which a compound which is like a biological ligand is immobilized on a solid material, and enzymes or other proteins are bound to it. Other techniques include **metal affinity chromatography**, where a metal ion is immobilized on a solid support: metal ions bind tightly and specifically to many biomolecules. The metal ion is bound to a chelator or chelating group, a chemical group that binds specifically and extremely tightly to that metal.

Affinity tag; purification tag—An amino acid sequence that has been engineered into a protein to make its purification easier. These can work in a number of ways. The tag could be another protein, which binds to some other material very tightly and thus allowing the protein to be purified by affinity chromatography (q.v.). The tag could be a short amino acid sequence, which is recognized by an antibody. The antibody would then bind to the protein whereas it would not have done so before. One such short peptide, called FLAG, has been designed so that it is particularly easy to make antibodies against it. The tag could be a few amino acids, which are then used as a chemical tag on the protein. For example, a string of positively charged amino acids will bind very strongly to a negatively charged filter: this could be used as the basis of a separation system. Some amino acids bind metals very strongly, especially in pairs: this chemical property can be exploited by using a filter with metal atoms chemically linked onto it to pull a protein out of a mixture of proteins. *cf* affinity chromatography.

Aflatoxin—Toxic compounds, produced by moulds (fungi) of the *Aspergillus flavus* group, that bind to DNA and prevent replication and transcription. Aflatoxins can cause acute liver damage and cancer. Animals may be poisoned by eating stored food or feed contaminated with the mould.

Agar (Malay, *agar-agar*)—A polysaccharide solidifying agent used in nutrient media preparations and obtained from certain types of red algae (*Rhodophyta*). Both the type of agar and its concentration can affect the growth and appearance of cultured explants.

Agarose—The main constituent of agar.

Agarose gel electrophoresis—A process in which a matrix composed of a highly purified form of agar is used to separate larger DNA and RNA molecules. *See* electrophoresis.

Aggregate—1. A clump or mass formed by gathering or collecting units.
2. A body of loosely associated cells, such as a friable callus or cell suspension.
3. Coarse inert material, such as gravel, that is mixed with soil to increase its porosity.
4. A serological reaction (aggregation) in which the antibody and antigen react and precipitate out of solution.

Agonist—A drug, hormone or transmitter substance that forms a complex with a receptor site that is capable of triggering an active response from a cell.

Agricultural biological diversity—*See* agrobiodiversity.

Agrobacterium—A genus of bacteria that includes several plant pathogenic species, causing tumour-like symptoms. *See Agrobacterium tumefaciens;* crown gall; hairy root culture; Ri plasmid; Ti plasmid.

Agrobacterium tumefaciens—A bacterium that causes crown gall disease in some plants. The bacterium infects a wound, and injects a short stretch of DNA into some of the cells around the wound. The DNA comes from a large plasmid—the Ti (tumour induction) plasmid—a short region of which (called T-DNA, = transferred DNA) is transferred to the plant cell, where it causes the cell to grow into a tumour-like structure. The T-DNA contains genes which *inter alia* allows the infected plant cells to make two unusual compounds, nopaline and octopine, that are characteristic of transformed cells. The cells form a gall, which hosts the bacterium. This DNA-transfer mechanism is exploited in the genetic engineering of plants. The Ti plasmid is modified so that a foreign gene is transferred into the plant cell along with or instead of the nopaline synthesis genes. When the bacterium is cultured with isolated plant cells or with wounded plant tissues, the 'new' gene is injected into the cells and ends up integrated into the chromosomes of the plant.

***Agrobacterium tumefaciens*-mediated transformation**—A naturally occurring process of DNA transfer from the bacterium *A. tumefaciens* to plants.

Agrobiodiversity; agricultural biological diversity—That component of biodiversity that is relevant to food and agriculture production. The term agrobiodiversity encompasses within-species, species and ecosystem diversity. (Based on FAO, 1999)

AIDS—(acquired immunodeficiency syndrome) The usually fatal human disease in which the immune system is destroyed by a retrovirus (Human Immunodeficiency Virus, HIV). The virus infects and destroys helper T-cells, which are essential for combating infections.

Airlift fermenter—A cylindrical fermentation vessel in which the cells are mixed by air introduced at the base of the vessel and that rises through the column of culture medium. The cell suspension circulates around the column as a consequence of the gradient of air bubbles in different parts of the reactor.

Albinism—Hereditary absence of pigment in an organism. Albino animals have no colour in their skin, hair and eyes. The term is also used for absence of chlorophyll in plants.

Albino—An organism lacking pigmentation, due to genetic factors. The condition is *albinism*, q.v. A conspicuous plastome (plastid) mutant involving loss of chlorophyll.

Aleurone—The outermost layer of the endosperm in a seed.

Algal biomass—Single-celled plants, such as *Chlorella* spp. and *Spirulina* spp., are grown commercially in ponds to make feed materials. Chlorella is grown commercially to make into fish food: it is fed to zooplankton, and these in turn are harvested as feed for fish farms. This is a means of converting sunlight into food in a way more convenient and controllable than normal farming.

Alginate—Polysaccharide gelling agent.

Alkylating agents—Chemicals that transfer alkyl (methyl, ethyl, etc.) groups to the bases in DNA.

Allele (Gr. *allelon*, of one another, mutually each other); **allelomorph** (adj. allelic, allelomorphic)—One of a pair, or series, of variant forms of a gene that occur at a given locus in a chromosome. Alleles are symbolized with the same basic symbol (*e.g.*, *B* for dominant and *b* for recessive); B_1, B_2, ..., B_n *for n* additive alleles at a locus). In a normal diploid cell there are two alleles of any one gene (one from each parent), which occupy the same relative position (locus) on homologous chromosomes. Within a population there may be more than two alleles of a gene. *See* multiple alleles.

Allele frequency—The number of copies of an allele in a population, expressed as a proportion of the total number of copies of all alleles at a locus in a population.

Allele-specific amplification (ASA)—The use of polymerase chain reaction (PCR) at a sufficiently high stringency that only a primer with exactly the same sequence as the target DNA will be amplified. A powerful means of genotyping for single-locus disorders that have been characterized at the molecular level.

Allelic exclusion—A phenomenon whereby only one functional allele of an antibody gene can be assembled in a given B lymphocyte. The 'allele' on the other homologous chromosome in a diploid mammalian cell cannot undergo a functional re-arrangement, which would result in the production of two different antibodies by a single plasma cell.

Allelomorph—*See* allele.

Allelopathy—The phenomenon by which the secretion of chemicals, such as phenolic and terpenoid compounds, by a plant inhibits the growth or reproduction of other plant species with which it is competing.

Allergen—An antigen that provokes an immune response.

Allogamy—Cross fertilization in plants. *See* fertilization.

Allometric—When the growth rate of one part of an organism differs from that of another part or of the rest of the body.

Allopatric speciation—Speciation occurring at least in part because of geographic isolation.

Allopolyploid (Gr. *allos*, other, + polyploidy)—A polyploid organism (usually a plant) having multiple sets of chromosomes derived from different species. Hybrids are usually sterile, because they do not have sets of homologous chromosomes and therefore pairing cannot take place. However, if doubling of the chromosome number occurs in a hybrid derived from two diploid (2*n*) species, the resulting tetraploid (4*n*) is a fertile plant, since it contains two sets of homologous chromosome and pairing may occur; this tetraploid is an allotetraploid.

Allosteric control—Allosteric regulation.

Allosteric enzyme—An enzyme that has two structurally distinct forms, one of which is active and the other inactive. Active forms of allosteric enzymes tend to catalyse the initial step in a pathway leading to the synthesis of molecules. The end product of this synthesis can act as a feedback inhibitor, converting the enzyme to the inactive form, thus controlling the amount of product synthesized.

Allosteric regulation—A catalysis-regulating process in which the binding of a small effector molecule to one site on an enzyme affects the activity at another site.

Allosteric transition—A reversible interaction of a small molecule with a protein molecule, resulting in a change in the shape of the protein and a consequent alteration of the interaction of that protein with a third molecule.

Allotetraploid—An organism with four genomes derived from hybridization of different species. Usually, in forms that become established, two of the four genomes are from one species and two are from another species. *See* allopolyploid.

Allozygote—A diploid individual that is homozygous at a locus in which the two genes are not identical by descent from a common ancestor.

Allozyme—*See* allosteric enzyme.

Alphalactalbumin—Protein component of milk.

Alternative mRNA splicing—The inclusion or exclusion of different exons to form different mRNA transcripts.

Alu sequences—A family of 300-bp sequences occurring nearly a million times in the human genome.

Ambient temperature—Air temperature at a given time and place; not radiant temperature.

Amino acid (Gr. *Ammon*, from the Egyptian sun god, in M. L. used in connection with ammonium salts)—An acid containing the group NH_2. In particular, any of 20 basic building blocks of proteins with a free amino (NH_2) and a free carboxyl (COOH) group, and having the basic formula $NH_2—C_R—COOH$. According to the side group R, they are subdivided into: polar or hydrophilic (serine, threonine, tyrosine, asparagine and glutamine); non-polar or hydrophobic (glycine, alanine, valine, leucine, isoleucine, proline, phenylalanine, tryptophan and cysteine); acidic (aspartic acid and glutamic acid) and basics (lysine, arginine, hystidine). The sequence of amino acids determines the shape, properties and the biological role of a protein. Plants and many micro-organisms can synthesize amino acids from simple inorganic compounds, but animals are unable to synthesize some of them, called essential amino acids, so they must be present in the diet.

Aminoacyl site; A-site—One of two sites on ribosomes to which the incoming aminoacyl tRNA binds.

Aminoacyl tRNA synthetase—Enzyme that attaches each amino acid to its specific tRNA molecule.

Amitosis—Cell division (cytokinesis), including nuclear division through constriction of the nucleus, without chromosome differentiation as in mitosis. The maintenance of genetic integrity and diploidy during amitosis is uncertain. This process occurs in the endosperm of flowering plants.

Amniocentesis—A procedure for obtaining amniotic fluid from a pregnant mammal for the diagnosis of some diseases in the unborn foetus. Cells are cultured, and metaphase chromosomes are examined for irregularities (*e.g.*, Down syndrome, spina bifida, etc., in humans).

Amnion—The thin membrane that lines the fluid-filled sac in which the embryo develops in higher vertebrates, reptiles and birds.

Amniotic fluid—Liquid contents of the amniotic sac of higher vertebrates, containing cells of the embryo (not of the mother). Both fluid and cells are used for diagnosis of genetic abnormalities in the embryo or foetus.

Amorph; null mutation—A mutation that obliterates gene function.

Amphidiploid—A species or type of plant derived from doubling the chromosomes in the F_1 hybrid of two species; an allopolyploid. In an amphidiploid the two species are known, whereas in other allopolyploids they may not be known.

Amphimixis—True sexual reproduction involving the fusion of male and female gametes and the formation of a zygote.

Ampicillin (β-lactamase)—A penicillin-derived antibiotic that prevents bacterial growth by interfering with synthesis of the cell wall.

Amplification—Treatment (*e.g.*, use of chloramphenicol) designed to increase the proportion of plasmid DNA relative to that of bacterial (host) DNA.

Amplified fragment length polymorphism (AFLP)—A type of DNA marker, generated by digestion of genomic DNA with two restriction enzymes to create many DNA fragments, ligation of specific sequences of DNA (called adaptors) to the ends of these fragments, amplification of the fragments via PCR (using a set of primers with sequences corresponding to the adapters, plus various random combinations of three additional bases at the end), and visualization of fragments via gel electrophoresis. The PCR will amplify any fragment whose sequence happens to start with any of the three-base sequences in the set of primers. AFLPs have the important advantage that many markers can be generated with relatively little effort. They are a very useful means of quantifying the extent of genetic diversity within and between populations. Their major disadvantage is that they are not specific to a particular locus and, because they are scored as the presence or absence of a band, heterozygotes cannot be distinguished from homozygotes, *i.e.*, they are inherited in a dominant fashion.

Amplify—To increase the number of copies of a DNA sequence, either *in vivo* by inserting into a cloning vector that replicates within a host cell, or *in vitro* by polymerase chain reaction (PCR).

Ampometric—Enzyme electrode.

Amylase—A group of enzymes that degrade starch, glycogen and other polysaccharides, producing a mixture of glucose and maltose. Plants have both α- and β-amylase; animal have only α-amylase.

Amylolytic—The capability of breaking down starch into sugars.

Amylopectin—A polysaccharide comprising highly branched chains of glucose molecules. The water-insoluble portion of starch.

Amylose—A polysaccharide consisting of linear chains of 100 to 1,000 glucose molecules. The water-soluble portion of starch.

Anabolic pathway—A pathway by which a metabolite is synthesized; a biosynthetic pathway.

Anaerobe—An organism that can grow in the absence of oxygen. Opposite: aerobe.

Anaerobic—An environment or condition in which molecular oxygen is not available for chemical, physical or biological processes.

Anaerobic digestion—Digestion of materials in the absence of oxygen. *See* anaerobic respiration.

Anaerobic respiration—Respiration in which foodstuffs are partially oxidized, with the release of chemical energy, in a process not involving atmospheric oxygen, such as alcoholic fermentation, in which one of the end products is ethanol.

Analogous—Features of organisms or molecules that are superficially or functionally similar but have evolved in a different way or contain different compounds.

Anaphase(Gr. *ana*, up + *phais*, appearance)—The stage of mitosis or meiosis during which the daughter chromosomes (sister chromatids) pass from the equatorial plate to opposite poles of the cell (toward the ends of the spindle). Anaphase follows metaphase and precedes telophase.

Anchor gene—A gene that has been positioned on both the physical map and the linkage map of a chromosome.

Androgen—Any hormone that stimulates the development of male secondary sexual characteristics, and contributes to the control of sexual activity in vertebrate animals. Usually synthesized by the testes.

Androgenesis—Male parthenogenesis, *i.e.*, the development of a haploid embryo from a male nucleus. The maternal nucleus is eliminated or inactivated subsequent to fertilization of the ovum, and the haploid individual (referred to as androgenetic) contains in its cells the genome of the male gamete only. Androgenesis is detected by cytological staining. *See* anther culture; gynogenesis; parthenogenesis.

Aneuploid (Gr. *aneu*, without + ploid)—An organism or cell having a chromosome number that is not an exact multiple of the monoploid (x) with one chromosome being present in greater (*e.g.*, trisomic $2n + 1$) or lesser (*e.g.*, monosomic $2n - 1$) number than the normal diploid number.

Animal cell immobilization—Entrapment of animal cells in some solid material in order to produce some natural product or genetically engineered protein. Animal cells have the advantage that they already produce many proteins of pharmacological interest, and that genetically engineered proteins are produced by them with the post-translation modifications normal to animals. However, because animal cells are much more fragile than bacterial ones, they cannot tolerate a commercial fermentation process. Typical materials are hollow fibre membrane bioreactors, or porous carriers made of polysaccharide, protein, plastic or ceramic materials with microscopic holes inside which the cells grow.

Animal genetic resources databank—A databank that contains inventories of farm animal genetic resources and their immediate wild relatives, including any information that helps to characterize these resources.

Animal genome (gene) bank—A planned and managed repository containing animal genetic resources. Repositories include the environment in which the genetic resource has developed, or is now normally found (*in situ*) or facilities elsewhere (*ex situ - in vivo* or *in vitro*). For *in vitro, ex situ* genome bank facilities, germplasm is stored in the form of one or more of the following: semen, ova, embryos and tissue samples.

Anion—A negatively charged ion; opposite: cation.

Anneal—The pairing of complementary DNA or RNA sequences, via hydrogen bonding, to form a double-stranded polynucleotide. Most often used to describe the binding of a short primer or probe.

Annealing—The process of heating (de-naturing step) and slowly cooling (re-naturing step) double-stranded DNA to allow the formation of hybrid DNA or complementary strands of DNA or of DNA and RNA.

Annual (L. *annualis*, within a year) (adj.)—Taking one year, or occurring at intervals of one year.Noun: In botany, a plant that completes its life cycle within one year. During this time the plant germinates, grows, flowers, produces seeds, and dies. *See* biennial, perennial.

Anonymous DNA marker—A DNA marker (q.v.) detectable by virtue of variation in its sequence, irrespective of whether or not it actually occurs in or near a coding sequence. Microsatellites are typical anonymous DNA markers.

Antagonism—An interaction between two organisms (*e.g.*, moulds or bacteria) in which the growth of one is inhibited by the other. *cf* synergism.

Antagonist—A compound that inhibits the effect of an agonist in such a way that the combined biological effect of the two becomes smaller than the sum of their individual effects.

Anther culture—The aseptic culture of anthers for the production of haploid plants from microspores. *See* androgenesis; gynogenesis; parthenogenesis.

Anther—Microsporangium bearing microspores which develop into pollen (microgametophytes). The upper part of a stamen, containing pollen sacs within which are numerous pollen grains.

Anthesis—The flowering period or efflorescence. Anthesis is the time of full bloom, which lasts till fruit set.

Anthocyanin—Water-soluble blue, purple and red flavonoid pigments found in vacuoles of cells.

Antiauxin—A chemical that interferes with the auxin response. Antiauxin may or may not involve prevention of auxin transport or movement in plants. Some antiauxins are said to promote morphogenesis *in vitro,* such as 2,3,5-tri-iodobenzoate (TIBA; f.w. 499.81), or 2, 4, 5-trichlorophenoxyacetate (2, 4, 5-T; m.w. 255.49), which stimulate the growth of some cultures.

Antibiosis—The prevention of growth or development of an organism by a substance or another organism.

Antibiotic—A class of natural and synthetic compounds that inhibit the growth of or kill some micro-organisms. Antibiotics such as penicillin are often used to control (to some extent kill) contaminating organisms. However, resistance to particular antibiotics can be acquired through mutations. Some contaminating organisms are only suppressed or their metabolism slowed to an insignificant level. *See* antibiotic resistance; bactericide; bacteriostat.

Antibiotic resistance—The ability of a micro-organism to produce a protein that disables an antibiotic or prevents transport of the antibiotic into the cell.

Antibody (Gr. *anti*, against + body)—An immunological protein (called an immunoglobulin, Ig) produced by certain white blood cells (lymphocytes) of the immune system of an organism in response to a contact with a foreign substance (antigen). Such an immunological protein has the ability of specifically binding with the foreign substance and rendering it harmless. The basic immunoglobulin molecule consists of two identical heavy and two identical light chains. *See* monoclonal antibodies; polyclonal antibodies.

Antibody class—The class to which an antibody belongs, depending on the type of heavy chain present. In mammals, there are five classes of antibodies: IgA, IgD, IgE, IgG, and IgM.

Antibody structure—Antibodies have a well-defined structure. Each antibody has two identical 'light' chains and two identical 'heavy' chains. Each chain comprises a constant region, *i.e.*, a region that is the same between antibodies of the same class and sub-class, and a variable region that differs between. The antigen-binding region or binding site - *complementarity determining region* - is in the variable region. The antibody can be cut by proteases into several fragments, known as Fab, Fab', and Fc.

Antibody-mediated (humoral) immune response—The synthesis of antibodies by B cells in response to an encounter of the cells of the immune system with a foreign immunogen.

Anticlinal—The plane of cell wall orientation or cell division perpendicular (at right angles) to the surface of an organ. *See* tunica; periclinal.

Anticoding strand—The strand of the DNA double helix that is actually transcribed. Also known as the antisense or template strand.

Anticodon—A triplet of nucleotides in a tRNA molecule that pairs with a complementary triplet of nucleotides, or codon, in an mRNA molecule during translation. *See* codon; mRNA; RNA; tRNA.

Antigen; immunogen—A compound that elicits an immune response by stimulating the production of antibodies. The antigen, usually a protein, when introduced into a vertebrate organism is bound by the antibody or a T cell receptor. *See* antigenic determinant; antigenic switching.

Antigenic determinant—A surface feature of a micro-organism or macromolecule, such as a glycoprotein, that elicits an immune response. *See* epitope.

Antigenic switching—The altering of a micro-organism's surface antigens through genetic re-arrangement, to elude detection by the host's immune system.

Antihaemophilic globulin—Blood globulin that reduces the clotting time of haemophilic blood.

Anti-idiotype antibodies—Antibodies which recognize the binding sites of other antibodies. Their binding sites are complementary to the binding sites of another immunoglobulin. When an animal becomes immune to something, it not only acquires antibodies against that something, it also acquires antibodies against those antibodies. This forms a network of antibodies which can all bind to each other to various degrees, helping to regulate the immune response. Some allergic responses are in part due to the breakdown of this sort of regulation.

Antimicrobial agent—Any chemical or biological agent that harms the growth of micro-organisms.

Antinutrients—Compounds that inhibit normal uptake of nutrients.

Anti-oncogene—A gene whose product prevents the normal growth of tissue. *cf* recessive oncogene.

Antioxidant solution—Pre-treatment solution (*e.g.*, Vitamin C; citric acid) that retards senescence and browning of tissue. It is employed to incubate explants prior to surface sterilization.

Antioxidant—Compound that slows the rate of oxidation reactions.

Antiparallel orientation—The normal arrangement of the two strands of a DNA molecule, and of other nucleic-acid duplexes (DNA-RNA, RNA-RNA), in which the two strands are oriented in opposite directions so that the 5´-phosphate end of one strand is aligned with the 3´-hydroxyl end of the complementary strand.

Antisense DNA—The strand of chromosomal DNA that is transcribed. A DNA sequence that is complementary to all or part of an mRNA molecule.

Antisense gene—A gene that produces a transcript (mRNA) that is complementary to the pre-mRNA or mRNA of a normal gene (usually constructed by inverting the coding region relative to the promoter).

Antisense RNA—An RNA sequence that is complementary to all or part of a functional mRNA molecule, to which it binds, blocking its translation. *See* RNA.

Antisense therapy—The *in vivo* treatment of a genetic disease by blocking translation of a protein with a DNA or an RNA sequence that is complementary to a specific mRNA.

Antiseptic—Any substance that kills or inhibits the growth of disease-causing micro-organism (a micro-organism capable of causing sepsis), but is essentially non-toxic to cells of the body.

Antiserum—The fluid portion of the blood of an animal (after coagulation of the blood), containing antibodies

Anti-terminator—A type of protein which enables RNA polymerase to ignore certain transcriptional stop or termination signals and read through them to produce longer mRNA transcripts.

Antitranspirant—A compound designed to reduce transpiration when sprayed or painted on leaves of newly transplanted trees, shrubs or vines, or used as a dip for cuttings in lieu of misting; may interfere with photosynthesis and respiration if the coating is too thick or unbroken.

Apex—(L. *apex*, a tip, point, or extremity; pl. apices)—The tip, point or angular summit. The tip of a leaf; the portion of a root or shoot containing apical and primary meristems. Usually used to designate the apical tip of the meristem.

Apical cell—A meristematic initial in the apical meristem of shoots or roots of plants. As this cell divides, new tissues are formed.

Apical dominance—The phenomenon of inhibition of growth of lateral (axillary) buds in a plant by the presence of the terminal (apical) bud on the branch, due to auxins produced by the apical bud.

Apical meristem—A region of the tip of each shoot and root of a plant in which cell division is continually occurring to produce new stem and root tissue, respectively. Two regions are visible in the apical meristem: (i) An outer 1-4-cell layered region (called the tunica), where cell divisions are anticlinal, *i.e.*, perpendicular to the surface; and below the tunica, (ii) the corpus, where the cells divide in all directions, giving them an increase in volume.

Apoenzyme—Inactive enzyme that has to be associated with a specific organic molecule called a co-enzyme in order to function. The apoenzyme/co-enzyme complex is called a holoenzyme.

Apomixis—(Gr. *apo*, away from + *mixis*, a mingling; adj. apomictic) The asexual production of diploid offspring without the fusion of gametes. The embryo develops by mitotic division of the maternal or paternal gamete, or, in the case of plants, by mitotic division of a diploid cell of the ovule. *cf* androgenesis; gynogenesis; panmixis; parthenogenesis.

Apoptosis—The process of cell death, which occurs naturally as a part of normal development, maintenance and renewal of tissue in an organism. Apoptosis differs from necrosis, in which cell death is caused by a toxic substance.

Aquaculture—Growing of water plants and animals, rather than harvesting them from wherever they happen to grow in rivers or seas. Usually aquaculture uses fresh water; when it uses sea water it can be called mariculture. It is considered to be a part of biotechnology (although peripheral) because it is a new commercial development, and because it often involves growing organisms in large volumes of water, which has similarities to growing large volumes of yeast or bacteria. Biotechnology also provides clean, well-aerated water for the animals to grow in; food, such as krill or powdered synthetic food; and food additives, such as astaxanthins to ensure that fish and prawns have the right colour. Aquaculture has also been used to mass-produce macro- and micro-algae for chemicals, vitamins and pigments. For both animals and plants, biotechnologists have been using genetic methods to produce triploid and tetraploid organisms, and hybrid algae through plant cell fusions. Triploid trout, for example, are sterile, and can be used for biocontrol of weeds without the threat of their being able to breed themselves.

Arabidopsis—A genus of flowering plants in the Cruciferae. *A. thaliana* is used in research as a model plant because it has a small genome (5 pairs of chromosomes; $2n = 10$) and can be cultured easily, with a generation time of two months.

ARS (autonomous replicating sequence)—Any eukaryotic DNA sequence that initiates and supports chromosomal replication; they have been isolated in yeast cells. Also called autonomous(ly) replicating segment.

Artificial inembryonation—Non-surgical transfer of embryo(s) to a recipient female. As *in vitro* embryo technology develops, artificial inembryonation will gradually replace artificial insemination.

Artificial insemination—The deposition of semen, using a syringe, at the mouth of the uterus to make conception possible. It is used in the breeding of domestic animals.

Artificial medium—*See* culture medium.

Artificial seed—Encapsulated or coated somatic embryos (embryoids) that are planted and treated like seed.

Artificial selection—The practice of choosing individuals from a population for reproduction, usually because these individuals possess one or more desirable traits.

Ascorbic acid; vitamin C—($C_6H_8O_6$; f.w. 176.12)—A water-soluble vitamin present naturally in some plants, and also synthetically produced. Aside from its role as a vitamin, it is used as an antioxidant in plant tissue culture; and included in disinfection solutions.

Ascospore—One of the spores contained in the ascus (q.v.) of certain fungi.

Ascus (pl. asci)—Reproductive sac in the sexual stage of a type of fungi (*Ascomycetes*) in which ascospores are produced.

Aseptic—Asepsis or sterile. The state of being free of contaminating organisms (bacteria, fungi, algae and all micro-organisms except viruses) but not necessarily free of internal symbionts. *See* axenic.

Asexual (Gr. *a*, without + L. *sexualis*, sexual)—Any type of reproduction not involving meiosis or the union of gametes.

Asexual embryogenesis—The sequence of events whereby embryos develop from somatic cells. a.k.a. somatic cell embryogenesis.

Asexual propagation—Vegetative, somatic, non-sexual reproduction of a plant without fertilization. *cf* apomixis.

Asexual reproduction—Reproduction that does not involve the formation and union of gametes from the different sexes or mating types. It occurs mainly in lower animals, micro-organisms and plants. In plants, asexual reproduction is by vegetative propagation (*e.g.*, bulbs, tubers, corms) and by formation of spores.

A-site—*See* aminoacyl site.

Asn—*See* asparagine.

Asparagine (abbr. Asn; $C_4H_8N_2O_3$; f.w. 132.12)—One of the 20 essential amino acids. It is occasionally included in plant tissue culture media as a source of reduced nitrogen.

Aspartic acid (abbr. Asp; $C_4H_7NO_4$; f.w. 132.12)—An amino acid necessary for nucleotide synthesis and occasionally included in plant tissue culture media.

Assay—1. To test or evaluate.

2. The procedure for measuring the quantity of a given substance in a sample (chemically or by other means).

3. The substance to be analysed.

Assortative mating—Mating in which the partners are chosen on the basis of phenotypic similarity.

Assortment—*See* segregation.

Asynapsis—The failure or partial failure in the pairing of homologous chromosomes during the meiotic prophase.

ATP (adenosine triphosphate)—A nucleotide of fundamental importance as a carrier of chemical energy in all living organisms. It consists of adenosine with three phosphate groups, linked together linearly. The phosphates are attached to adenosine through its ribose (sugar) portion. Upon hydrolysis, bonds yield either one molecule of ADP (adenosine diphosphate) and an inorganic phosphate, or one molecule of AMP (adenosine monophosphate) and pyrophosphate; in both cases releasing energy that is used to power biological processes. ATP is regenerated by rephosphorilation of AMP and ADP, using chemical energy derived from the oxidation of food.

ATPase—An enzyme that brings about the hydrolysis of ATP, by the cleavage of either one phosphate group with the formation of ADP and inorganic phosphate, or of two phosphate groups, with the formation of AMP and pyrophosphate.

Attenuated vaccine—A virulent organism that has been modified to produce a less virulent form, but nevertheless retains the ability to elicit antibodies against the virulent form.

Attenuation—A mechanism for controlling gene expression in prokaryotes that involves premature termination of transcription.

Attenuator—A nucleotide sequence in the 5′ region of a prokaryotic gene (or in its RNA) that causes premature termination of transcription, possibly by forming a secondary structure.

Authentic protein—A recombinant protein that has all the properties —including any post-translational modifications—of its naturally occurring counterpart.

Autocatalysis—Catalysis in which one of the products of the reaction is a catalyst for the reaction. Usually the catalysis starts slowly and increases as the quantity of the catalyst increases, falling off as the product is used up.

Autocatalytic reaction—*See* autocatalysis.

Autoclave—An enclosed chamber in which substances are heated under pressure to sterilize utensils, liquids, glassware, etc., using steam. The routine method uses steam pressure of 103.4×10^3 Pa at 121°C for 15 minutes, or longer to allow large volumes to reach the critical temperature. A pressure cooker used to sterilize growth medium and instruments for tissue culture work.

Auto-immune disease—Disorder in which the immune systems of affected individuals produce antibodies against molecules that are normally produced by those individuals (called self antigens).

Auto-immunity—A disorder in the body's defence mechanism in which an immune response is elicited against its own (self) tissues.

Autologous cells—Cells that are taken from an individual, cultured (or stored), and, possibly, genetically manipulated before being infused back into the original donor.

Autolysis—The process of self destruction of a cell, cell organelle, or tissue. It occurs by the action of lysosomic enzymes.

Autonomous(ly) replicating segment (or sequence)—

Autonomous—A term applied to any biological unit that can function on its own, *i.e.*, without the help of another unit, such as a transposable element that encodes an enzyme for its own transposition.

Autopolyploid—A polyploid that has multiple and identical or nearly identical sets of chromosomes (genomes) all derived from the same species. A polyploid species with genomes derived from the same original species.

Autoradiograph—A picture prepared by labelling a substance such as DNA with a radioactive material such as tritiated thymidine and allowing the image produced by decay radiation to develop on a film over a period of time.

Autoradiography—A technique that captures the image formed in a photographic emulsion as a result of the emission of either light or radioactivity from a labelled component that is placed next to unexposed film. The technique is used for detecting the location of an isotope in a tissue, cell or molecule. The sample is placed in contact with a photographic emulsion, usually an X-ray film. The emission of β-particles from the sample activates the silver halide grains in the emulsion and allows them to reduce to metallic silver when the film is developed. In genetic engineering, autoradiography is most commonly used to detect the hybridization of a radioactive DNA (probe) molecule to denatured DNA in either the Southern transfer or colony hybridization procedures.

Autosome—A chromosome that is not involved in sex determination.

Autotrophic—Self-nourishing organisms capable of utilizing carbon dioxide or carbonates as the sole source of carbon and obtaining energy for life processes from radiant energy or from the oxidation of inorganic elements, or compounds such as iron, sulphur, hydrogen, ammonium and nitrites. *See* heterotrophic.

Autotrophy—Autotrophy is the capacity of an organism to use light as the sole energy source in the synthesis of organic material from inorganic elements or compounds. Autotrophic organisms include green photosynthesizing plants and some photosynthetic bacteria. *cf* heterotrophy.

Auxin (Gr. *auxein*, to increase)—A group of plant growth regulators (natural or synthetic) which stimulate cell division, enlargement, apical dominance, root initiation, and flowering. One naturally produced auxin is indole-acetic acid (IAA).

Auxin-cytokinin ratio—The relative proportion of auxin to cytokinin present in plant-tissue-culture media. Varying the relative amounts of these two hormone groups in tissue culture formulae affects the proportional growth of shoots and roots *in vitro*. As the ratio is increased (increased auxin or decreased cytokinin), roots are more likely to be produced, and as it is decreased root growth declines and shoot initiation and growth are promoted. This relationship was first recognized by C.O. Miller and F. Skoog in the 1950s.

Auxotroph—(Gr. *auxein*, to increase + *trophe*, nourishment)—A mutant cell or micro-organism lacking the capacity to form an enzyme or metabolite present in the parental strain, and that consequently will not grow on a minimal medium, but requires the addition of some compound—such as an amino acid or a vitamin—for growth.

Availability—A reflection of the form and location of nutritional elements and their suitability for absorption. In tissue culture media this is related to the abundance of each nutritional element, the osmotic concentration and pH of the medium, the stability and solubility of the item in question, the presence of absorbing agents in the media, and other factors.

Axenic culture—Free of external contaminants and internal symbionts; generally not possible with surface sterilization alone, and incorrectly used to indicate *aseptic culture*, q.v.

Axillary bud—A bud found at the axil of a leaf (synonymous with lateral bud).

Axillary bud proliferation—Propagation in culture using protocols and media which promote axillary (lateral shoot) growth. This is a technique for mass production (micropropagation) of plantlets in culture, achieved primarily through hormonal inhibition of apical dominance and stimulation of lateral branching.

Bacillus thuringiensis **(Bt)**—A bacterium that kills insects; a major component of the microbial pesticide industry.

Back mutation—A second mutation at the same site in a gene as the original mutation. The second mutation restores the wild-type nucleotide sequence.

Backcross—Crossing an organism with one of its parents or with the genetically equivalent organism. The offspring of such a cross are referred to as the backcross generation or backcross progeny. *See* testcross.

Bacteria—Plural of bacterium, q.v.

Bacterial toxin—A toxin produced by a bacterium, such as Bt toxin by *Bacillus thuringiensis*.

Bacteriocide; bactericide—A chemical or drug that kills bacterial cells.

Bacteriocin—A protein produced by bacteria of one strain and active against those of a closely related strain.

Bacteriophage—A virus that infects bacteria. Also called simply phage. Altered forms are used in DNA cloning work, where they are convenient vectors. The bacteriophages most used are derived from two 'wild' phages, called M13 and lambda (λ). Lambda phages are used to clone segments of DNA in the range of around 10-20 kb. They are lytic phages, *i.e.*, they replicate by lysing their host cell and releasing more phages. On a bacteriological plate, this results in a small clear zone - a plaque. Some lambda vectors have also been developed which are expression vectors (q.v.). The M13 system can grow inside a bacterium, so that it does not destroy the cell it infects but causes it to make new phages continuously. It is a single-stranded DNA phage, and is used for the Sanger di-deoxy DNA sequencing method (*see* DNA sequencing). Both of these phages grow on *Escherichia coli* as a host bacterium.

Bacteriostat—A substance that inhibits or slows down growth and reproduction of bacteria.

Bacterium (Gr. *bakterion*, a stick; pl. bacteria)—Common name for the class Schizomycetes: minute (0.5-5 μm), unicellular organisms, without a distinct nucleus. Bacteria are prokaryotes, and most of them are identified by means of Gram staining (q.v.). They are classified on the basis of their oxygen requirement (aerobic vs anaerobic) and shape (spherical = coccus; rodlike = bacillus; spiral = spirillum; comma-shaped = vibrio; corkscrew-shaped = spirochaete; filamentous). Bacteria usually reproduce asexually, by simple cell division, although a few undergo a form of sexual reproduction, termed *conjugation*. A few bacteria can photosynthesize (including green-blue cyanobacteria), some are saprophytes and others are parasites and can cause diseases. They are major agents of fermentation, putrefaction and decay, and frequently a source of contamination in tissue culture. In plant pathology, strains of bacteria causing disease in specific plant cultivars are called pathovars (q.v.).

Baculovirus—Baculoviruses are a class of insect virus which have been used to make DNA cloning vectors for gene expression in eukaryotic cells. Baculoviruses have a gene which is expressed at very high levels late in their infection cycle, filling the nucleus of the cell with many-sided bodies full of a protein which is not needed to produce more viruses, but is necessary for the virus's spread in the wild. In a vector cloning system, this gene is replaced by one that the biotechnologist wants expressed. Production of the protein can be up to 50% of the cells' protein content, and several proteins can be made simultaneously, so that multi-sub-unit enzymes can be made by this system. Being an animal expression system, baculoviruses produce proteins that are glycosylated (addition of carbohydrates) like the proteins in animals, making it an attractive option for the production of biopharmaceuticals. In addition, baculoviruses are non-infective and non-pathogenic to vertebrates.

Balanced lethal system—A system for maintaining a recessive lethal allele at each of two loci on the same pair of chromosomes. In a closed population with no crossing-over between the loci, only the double heterozygotes for the lethal mutations survive.

Balanced polymorphism—Two or more types of individuals maintained in the same breeding population.

Bank—*See* gene bank.

Bar—A unit used for pressure of fluid. 1 bar = 10^5 Pa. 1 bar is approximately equivalent to 1 atmosphere.

Barr body—A condensed mass of chromatin found in the nuclei of female mammals. It is a late-replicating, inactive X-chromosome. Named after its discoverer, Murray Barr (1908).

Basal—Located at the base of a plant or a plant organ. A fundamental formulation of a tissue culture medium.

Basal body—Small granule to which a cilium or flagellum is attached. *cf* kinetosome.

Base—A cyclic, nitrogen-containing compound that is one of the essential components of nucleic acids. Exists in five main forms (adenine, A; guanine, G; thymine, T; cytosine, C; uracil, U). A and G have a similar structure and are called purines; T, C and U have a similar structure and are called pyrimidines. A base joined to a ribose sugar joined to a phosphate group is a nucleotide - the building block of nucleic acids.

Base analogues—Unnatural purine or pyrimidine bases that differ slightly in structure from the normal bases, but can be incorporated into nucleic acids. They are often mutagenic.

Base collection—In PGR: Defined in the International Undertaking on Plant Genetic Resources (FAO, 1983) as a collection of seed stock or vegetative propagating material (ranging from tissue cultures to whole plants) held for long-term security in order to preserve the genetic variation for scientific purposes and as a basis for plant breeding as multiplication and evaluation. *cf* active collection.

Base pair (bp)—The two strands that constitute DNA are held together by specific hydrogen bonding between purines and pyrimidines (A pairs with T; and G pairs with C). The size of a nucleic acid molecule is often described in terms of the number of base pairs (symbol: bp) or thousand base pairs (kilobase pairs; symbol: kb; a more convenient unit) it contains.

Base substitution—Replacement of one base by another in a DNA molecule. *See* transition; transversion.

Basophil—A type of white blood cell (leucocyte), produced by stem cells in the red bone marrow.

Batch culture—A suspension culture in which cells grow in a finite volume of liquid nutrient medium and follow a sigmoid pattern of growth. *cf* continuous culture; batch fermentation.

Batch fermentation—A process in which cells or micro-organisms are grown for a limited time. At the beginning of the fermentation, an inoculum is introduced into fresh medium, with no addition or removal of medium for the duration of the process.

B cells—An important class of white blood cells that mature in bone marrow and produce antibodies. They are largely responsible for the antibody-mediated or humoral immune response; they give rise to the antibody-producing plasma cells and some other cells of the immune system. *See* B lymphocytes.

β-DNA—The normal form of DNA found in biological systems. It exists as a right-handed helix.

Bench-scale process—A small- or laboratory-scale process; commonly used in connection with fermentation.

β-galactosidase—An enzyme that catalyses the formation of glucose and galactose from lactose.

Biennial (L. *biennium*, a period of two years)—In botany, a plant which completes its life cycle within two years and then dies. For most biennial plants, the two growing seasons have to be separated by a period of cold temperature sufficient to induce flowering and fruit formation.

Bifunctional vector—*See* shuttle vector.

Binary vector system—A two-plasmid system in *Agrobacterium tumefaciens* for transferring into plant cells a segment of T-DNA that carries cloned genes. One plasmid contains the virulence gene (responsible for transfer of the T-DNA), and another plasmid contains the T-DNA borders, the selectable marker and the DNA to be transferred. *See also* cDNA; carrier DNA; plasmid; vector.

Binding—The ability of molecules to stick to each other because of the exact shape and chemical nature of parts of their surfaces. Many biological molecules bind extremely tightly and specifically to other molecules: enzymes to their substrates; antibodies to their antigens; DNA strands to their complementary strands; and so on. Binding can be characterized by a *binding constant* or *association constant* (K_a), or its inverse, the *dissociation constant* (K_d).

Binomial nomenclature—In biology, each species is generally identified by two terms: the first is the genus to which it belongs, and the second is the specific epithet that distinguishes it from others in that genus (*e.g.*, *Quercus suber*, cork oak). The genus name always has an initial capital; the specific epithet is never capitalized, even though it may be derived from a proper name (*e.g.*, keranda nut, *Elaeocarpus bancroftii*). Both terms in the binomial are italicized. Based on the system of classification developed by Carolus Linnaeus.

Binomial expansion—The probability that an event will occur 0, 1, 2, ..., n times out of n is given by the successive terms of the expression $(p + q)^n$, where p is the probability of the event occurring, and $q = 1 - p$.

Bio—A prefix derived from *bios* and used in scientific words to associate the concept of 'living organisms.' Usually written with a hyphen before vowels, for emphasis or in neologisms; otherwise usually without a hyphen.

Bio-accumulation—In an organism, concentration of materials which are not components critical for that organism's survival. Usually it refers to the accumulation of metals or other compounds (*e.g.*, DDT). Many organisms—plants, fungi, protists, bacteria, etc. accumulate metals when grown in a solution of them, either as part of their defence mechanism against the poisonous effect of those compounds, or as a side-effect of the chemistry of their cell walls. Bio-accumulation is important as part of the microbial mining cycle (q.v.), removing toxic metals from wastewater, as a purification (bioremediation) process, etc. *See also* biosorption; microbial mining.

Bio-assay—A procedure for the assessment of a substance by measuring its effect in living cells or on organisms. Animals have been used extensively in drug research in bio-assays for the pharmacological activity of drugs. However, bio-assays are now usually developed using bacteria or animal or plant cells, as these are usually much easier to handle than whole animals or plants, are cheaper to make and keep, and avoid the ethical problems associated with testing of animals. Sometimes used to detect minute amounts of substances that influence or are essential to growth.

Bio-augmentation—Increasing the activity of bacteria that decompose pollutants; a technique used in bioremediation.

Biocatalysis; biocatalyst—Use of enzymes to catalyse chemical reactions. *See* biotransformation.

Biocontrol—The control of living organisms (especially pests) by biological means. Any process using deliberately introduced living organisms to restrain the growth and development of other, very often pathogenic, organisms, such as the use of spider mites to control cassava mealy bug, or the introduction of myxomatosis into Australia to control rabbits. The term also applies to use of disease-resistant crop cultivars. Biotechnology approaches biocontrol in various ways, such as using fungi, viruses or bacteria which are known to attack an insect or weed pest.

Bioconversion—Conversion of one chemical into another by living organisms, as opposed to their conversion by enzymes (which is biotransformation) or by chemical processes. The usefulness of bioconversion is much the same as that of biotransformation - in particular its extreme specificity and ability to work in moderate conditions. However, bioconversion has several other properties, including the possibility of having several chemical steps. A major commercial application is in the manufacture of steroids. The 'basic' steroid molecule, often isolated from plants, is itself a very complicated molecule, and not one that is easy to modify by normal chemical means to produce the very specific molecules needed for drug use. However, a particular type of bioconversion that attacks only specific bits of the molecule can be used. Bioconversion is particularly useful for introducing chemical changes at specific points in large, complex molecules.

Biodegradable—*See* biodegradation.

Biodegradation—The breakdown by living organisms of a compound to its chemical constituents. Materials that can be easily biodegraded are colloquially termed *biodegradable.*

Biodiversity—The variety of species (species diversity) or other taxa of animals, micro-organisms and plants in a natural community or habitat, or of communities in a particular environment (ecological diversity), or of genetic variation in a species (genetic diversity, q.v.). The maintenance of a high level of biodiversity is important for the stability of ecosystems. The variety of life in all its forms, levels and combinations, encompassing genetic diversity, species diversity and ecosystem diversity. *cf* agrobiodiversity.

Bio-energetics—The study of the flow and the transformations of energy that occur in living organisms.

Bio-engineering—The use of artificial tissues, organs and organ components to replace parts of the body that are damaged, lost or malfunctioning.

Bio-enrichment—Adding nutrients or oxygen to increase microbial breakdown of pollutants.

Bio-ethics—The branch of ethics that deals with the life sciences and their potential impact on society. At one extreme, it can be enormously useful in focusing attention on problems that need to be confronted; at the other extreme, it can become a name-calling argument between the 'pro-biotechnology' and 'anti-biotechnology' schools of thought, which, as it reduces discussion to epithets and clichés, can make better sound bites.

Biofuel—A gaseous, liquid or solid fuel that contains energy derived from a biological source. For example, rapeseed oil or fish liver oil can be used in place of diesel fuel in modified engines. A commercial application is the use of modified rapeseed oil, which as rapeseed methyl ester (RME)—can be used in modified diesel engines, and is sometimes named bio-diesel. *cf* biogas.

Biogas—A mixture of methane and carbon dioxide resulting from the anaerobic decomposition of waste such as domestic, industrial and agricultural sewage. a.k.a. *gobar.*

Biogenesis—The principle that a living organism can only arise from other living organisms similar to itself and can never originate from non-living material.

Bio-informatics —The use and organization of information of biological interest. In particular, it is concerned with organizing bio-molecular databases, in getting useful information out of such databases, in utilizing powerful computers for analysing such information, and in integrating information from disparate biological sources.

Biolistics (from **bio**logical + bal**listics**)—A technique to insert DNA into cells. The DNA is mixed with small metal particles—usually tungsten or gold - a fraction of a micrometre across. These are then fired into a cell at very high speed. They puncture the cell and carry the DNA into the cell. Biolistics has an advantage over transfection, transduction, etc., because it can apply to any cell, or indeed to parts of a cell. Thus use of biolistics has inserted DNA into animal, plant and fungal cells, and into mitochondria inside cells. a.k.a. microprojectile bombardment.

Biological ageing—*See* senescence.

Biological control—*See* biocontrol.

Biological containment—Restricting the movement of (genetically engineered) organisms by arranging barriers to prevent them from growing outside the laboratory. Biological containment can take two forms: making the organism unable to survive in the outside environment, or making the outside environment inhospitable to the organism. The latter is rarely suitable for bacteria, which, in principle, could survive almost anywhere. Thus for bacteria and yeasts, the favoured approach is to mutate the genes in the organism so that they require a supply of a specific nutrient that is usually available only in the laboratory. If they get out, they then cannot grow. Making the environment unfriendly to the organism is partly a biological control, partly a physical one. Thus, some of the first genetically engineered rice strains were developed in England (which is too cold for rice to grow) and tried in the field in Arizona (where it is too dry). Biological containment may also involve the use of vector molecules and host organisms which have been genetically disabled such that they can survive only in the peculiar conditions provided by the experimenter and which are unavailable outside the laboratory.

Biotinylated-DNA—A DNA molecule labelled with biotin by incorporation of biotinylated-dUTP into a DNA molecule. It is used as a non-radioactive probe in hybridization experiments, such as Southern transfer. The detection of the labelled DNA is achieved by complexing it with streptavidin (an antibiotic with a high affinity for biotin) to which is attached a colour-generating agent such as horseradish peroxidase that gives a fluorescent green colour upon reaction with various organic reagents.

Biotransformation—The conversion of one chemical or material into another using a biological catalyst: a near synonym is biocatalysis, and hence the catalyst used is called a biocatalyst. Usually the catalyst is an enzyme, or a whole, dead micro-organism that contains an enzyme or several enzymes.

Biotope—A small habitat in a large community.

Biotoxin—A naturally produced toxic compound which shows pronounced biological activity and presumably has some adaptive significance to the organism which produces it.

Bivalent—A pair of synapsed or associated homologous chromosomes (one of maternal origin; the other of paternal origin) that have each undergone duplication. Each duplicated chromosome comprises two chromatids. Thus a bivalent comprises four chromatids.

β-lactamase—An ampicillin resistance gene. *See* selectable marker.

Blastocyst (also blastocist)—A mammalian embryo (fertilized ovum) in the early stages of development, approximately up to the time of implantation. It consists of a hollow ball of cells.

Blastomere—Any one of the cells formed from the first few cleavages in animal embryology. The embryo usually divides into two, then four, then eight blastomeres, and so on.

Blastula—In animals, an early embryo form that follows the morula stage; typically, a single-layered sheet (blastoderm) or ball of cells (blastocyst).

B lymphocytes; B cells—An important class of lymphocytes that mature in bone marrow (in mammals) and the Bursa of Fabricius (in birds), that are largely responsible for the antibody-mediated or humoral immune response; they give rise to the antibody-producing plasma cells and some other cells of the immune system.

Bovine somatotrophin (BST) (also bovine somatotropin)—a.k.a. bovine growth hormone, this protein is found naturally in cattle, and is the bovine counterpart of human growth hormone, one of the earliest biopharmaceutical products. It has been cloned, using recombinant DNA technology, expressed in large amounts and marketed as an agricultural product to improve the growth rate and protein:fat ratios in farm cattle, and to improve milk yield. Its use is banned in some countries.

Bovine spongiform encephalopathy (BSE)—a.k.a. mad cow disease. *See* proteinaceous infectious particle.

Bp—Abbreviation for *base pair,* q.v.

Bract—A modified leaf that subtends flowers or inflorescences and may appear to be a petal.

Breed—Noun: In AnGR, either (i) a sub-specific group of domestic livestock with definable and identifiable external characteristics that enable it to be separated by visual appraisal from other similarly defined groups within the same species or (ii) a group of domestic livestock for which geographical and/or cultural separation from phenotypically similar groups has led to acceptance of its separate identity. *cf* breed at risk; breed not at risk; critical breed; critical-maintained breed; endangered-maintained breed.

Breed at risk—Any breed that may become extinct if the factors causing its decline in numbers are not eliminated or mitigated. Breeds may be in danger of becoming extinct for a variety of reasons. Risk of extinction may result from, *inter alia,* low population size; direct and indirect impacts of policy at the farm, country or international levels; lack of proper breed organization; or lack of adaptation to market demands. Breeds are categorized as to their risk status on the basis of, *inter alia,* the actual numbers of male and/or female breeding individuals and the percentage of pure-bred females. FAO has established categories of risk status: critical, endangered, critical-maintained, endangered-maintained, and not-at-risk.

Breeding—The process of sexual reproduction and production of offspring.

Breeding value—In quantitative genetics, the part of the deviation of an individual phenotype from the population mean that is due to the additive effects of alleles. In practical terms: if an animal is mated with a random sample of animals from a population, that animal's breeding value for a certain trait is twice the average deviation of its offspring from the population mean for that trait.

Breed not at risk—In AnGR: A breed where the total number of breeding females and males is greater than 1,000 and 20 respectively; or the population size approaches 1,000 and the percentage of pure-bred females is close to 100%, and the overall population size is increasing. *cf* breed at risk.

Brewing—The process by which beer is made. In the first stage the barley grain is soaked in water and allowed to germinate (malting), during which the natural enzymes of the grain convert the seed starch to maltose, and then to glucose. Grain is then dried, crushed, and added to water at a specific temperature (steeping) and any remaining starch is converted to sugar. The resulting liquid (wort) is the raw material to which yeast is added to convert sugar to alcohol. Hops (female flowers of *Humulus lupulus*) are added during this process to give a characteristic flavour.

Brewer's yeast—Strains of yeast, often *Saccharomyces cerevisiae,* that are used in the production of beer.

Bridge—A filter paper or other substrate used as a wick and support structure for a plant tissue in culture when a liquid medium is used.

Broad-host-range plasmid—A plasmid that can replicate in a number of different bacterial species.

Broad-sense heritability—In quantitative genetics, the proportion of the total phenotypic variation due to genetic variation.

Browning—Discoloration due to phenolic oxidation of freshly cut surfaces of explant tissue. In later stages of culture, such discoloration may indicate a nutritional or pathogenic problem, generally leading to necrosis.

Brucellosis—Disease caused by infection with organisms of the genus *Brucella.*

BSE—Bovine spongiform encephalopathy. *See* proteinaceous infectious particle.

BST—*See* bovine somatotrophin.

Bt—*See Bacillus thuringiensis.*

Bubble column fermenter—A fermentation vessel, or bioreactor, in which the cells or micro-organisms are kept suspended in a tall cylinder by rising air, which is introduced at the base of the vessel.

Bud—A region of meristematic tissue with the potential for developing into leaves, shoots, flowers or combinations; generally protected by modified scale leaves. A terminal (or apical) bud exists at the tip of a stem or branch, while axillary (or lateral) buds develop in the axils of leaves.

Bud scar—A scar left on a shoot when the bud or bud scales drop.

Bud sport—A somatic mutation arising in a bud and producing a genetically different shoot. Bud sports includes changes due to gene mutation, somatic reduction, chromosome deletion or polyploidy.

Budding—A method of asexual reproduction in which a new individual is derived from an outgrowth (bud) that becomes detached from the body of the parent.

Buffer—A solution that resists change in pH when an acid or alkali is added, or when solutions are diluted.

Buoyant density—The intrinsic density which a molecule, virus or sub-cellular particle has when suspended in an aqueous solution of a salt, such as CsCl, or a sugar, such as sucrose. DNA from different species has a characteristic buoyant density, which reflects the proportion of G=C base pairs. The greater the proportion of G=C, the greater the buoyant density of the DNA.

CAAT box (also CAT box)—A conserved sequence found within the promoter region of the protein-encoding genes of many eukaryotic organisms. It has the consensus sequence GGCCAATCT; it occurs around 75 bases prior to the transcription initiation site; and it is one of several sites for recognition and binding of regulatory proteins called transcription factors.

Cabinet—*See* growth cabinet.

Calf scours—A watery diarrhoea in calves.

Callus (L. *callum*, thick skin; pl. calluses or calli)—A protective tissue, consisting of parenchyma cells, that develops over a cut or damaged plant surface. Mass of unorganized, thin-walled parenchyma cells induced by hormone treatment. Actively dividing non-organized masses of undifferentiated and differentiated cells often developing from injury (wounding) or in tissue culture in the presence of growth regulators.

Callus culture—A technique of tissue culture; it is usually on solidified medium and initiated by inoculation of small explants or sections from established organ or other cultures (the inocula). Callus culture is used as the basis for organogenic (shoot, root) cultures, cell cultures or proliferation of embryoids. Callus cultures can be indefinitely maintained through regular sub-culturing.

Calorie (abbr. cal)—Equivalent to the amount of heat required to raise the temperature of 1 gram of water from 14.5°C to 15.5°C (≈4.19 J). *cf* kilocalorie.

Calyx—(Gr. *kalyx*, a husk, cup)—All the sepals of a flower considered collectively. The outermost whorl of flower parts.

Cambial zone—Region in stems and roots consisting of the cambium and its recent derivatives.

Cambium—(L. *cambium*, one of the alimentary body fluids supposed to nourish the body organs; pl. cambia)—A layer, usually regarded as one or two cells thick, of persistently meristematic tissue between the xylem and phloem tissues, and which gives rise to secondary tissues, thus resulting in an increase in diameter. The two most important cambia are the vascular (fascicular) cambium and the cork cambium.

Cancer—Uncontrolled growth of the cells of a tissue or an organ in a multicellular organism. *cf* oncogenesis.

Candidate gene—A gene whose function suggests that it may be involved in the genetic variation observed for a particular trait, *e.g.*, the gene for growth hormone is a candidate gene for body weight.

Candidate-gene strategy—An experimental approach in which knowledge of the biochemistry and/or physiology of a trait is used to draw up a list of genes whose protein products could be involved in the trait.

Canine—Pertaining to dogs.

Canola—Any of several cultivars of oilseed rape (more fully: canola oil); the vegetable oil high in mono-unsaturated fatty acid obtained from these cultivars.

Cap —The structure found on the 5′-end of eukaryotic mRNA, and consisting of an inverted, methylated guanosine residue. *See* G cap.

Cap site—The site in a gene where translation.

Catalysis—The process of changing the rate of a chemical reaction by use of a catalyst.

Catabolism—The metabolic breakdown of large molecules in living organism, with accompanying release of energy.

Catabolite repression—Glucose-mediated reduction in the rates of transcription of operons that encode enzymes involved in catabolic pathways (such as the *lac* operon).

Catalyst (Gr. *katalyein*, to dissolve)—A substance that promotes a chemical reaction by lowering the activation energy of a chemical reaction, without itself undergoing any permanent chemical change. The process is catalysis. *cf* catalytic antibody; catalytic RNA.

Catalytic antibody (= abzyme)—An antibody selected for its ability to catalyse a chemical reaction by binding to and stabilizing the transition-state intermediate.

Catalytic RNA (= ribozyme; gene shears)—A natural or synthetic RNA molecule that cuts an RNA substrate.

Cation—A positively charged ion; opposite is *anion*.

Caulogenesis—Stem organogenesis; induction of shoot development from callus.

CD molecules; cluster of differentiation molecules—Any group of antigens that is associated with a specific sub-population of T cells. There are designations for surface molecules on various cells of the immune system, *e.g.*, CD4 is present on the surface of helper T cells.

CDNA; complementary DNA—The double-stranded DNA complement of an mRNA sequence; synthesized *in vitro* from a mature RNA template using reverse transcriptase (to create a single strand of DNA from the RNA template) and DNA polymerase (to create the double-stranded DNA). Preparation of cDNAs is often the first step in cloning DNA sequences of interest. Used as specific and sensitive probes in hybridization studies, because cDNAs usually do not include regulatory or other controlling sequences, and so they can be used to identify (probe) and isolate genes and their associated sequences from genomic DNA. *See* binary vector; carrier DNA.

CDNA clone—A double-stranded DNA molecule that is carried in a vector and was synthesized *in vitro* from an mRNA sequence by using reverse transcriptase and DNA polymerase.

CDNA cloning—A method of cloning the coding sequence of a gene, starting with its mRNA transcript. It is normally used to clone a DNA copy of a eukaryotic mRNA. The cDNA copy, being a copy of a mature messenger molecule, will not contain any intron sequences and may be readily expressed in any host organism if attached to a suitable promoter sequence within the cloning vector.

CDNA library—A collection of cDNA clones that were generated *in vitro* from the mRNA sequences isolated from an organism or a specific tissue or cell type or population of an organism. *cf* library .

CDR—(complementarity-determining regions)—These are regions of the variable (V) regions of light and heavy antibody chains that make contact with the antigen. The primary amino acid sequences of these regions are highly variable among antibodies of the same class.

Cell—(L. *cella*, small room)—The smallest structural unit of living matter capable of functioning independently; a microscopic mass of protoplasm surrounded by a semi-permeable membrane, usually including one or more nuclei and various non-living products, capable - either alone or by interacting with other cells - of performing all the fundamental functions of life.

Cell culture—The *in vitro* growth of cells derived from multi-cellular organisms. The cells are usually of one type.

Cell cycle—Sequence of stages that a cell passes through between one division and the next. The cell cycle oscillates between mitosis and the interphase, which is divided into G, S, and G_2. In the G phase there is a high rate of biosynthesis and growth; in the S phase there is the doubling of the DNA content as a consequence of chromosome replication; in the G_2 phase the final preparations for cell division (cytokinesis) are made.

Cell differentiation—Continuous loss of physiological and cytological characters of young cells, resulting in getting the characters of adult cells. The unspecialized cells become modified and specialized for the performance of specific functions. Differentiation results from the controlled activation and de-activation of genes.

Cell division—Formation of two or more daughter cells from a single mother cell. The nucleus divides first, followed by the formation of a cell membrane between the daughter nuclei. Division of cytoplasm and nucleus into two or more parts by formation of a cell plate.

Cell fusion—Formation of a single hybrid cell from two cells of different species, cultured *in vitro*. The cells fuse and coalesce, but their nuclei may remain separated. During subsequent cell division, a single spindle is formed so that each daughter cell has a single nucleus containing sets of chromosomes from each parental line. Subsequent divisions often result in the loss of chromosomes and therefore of genes. The cell fusion technique can be used to determine the control of specific genes and their assignment to chromosomes. *cf* cell hybridization.

Cell generation time—The interval between the beginning of consecutive divisions of a cell. The time that it takes for a population of single-celled organisms to double its cell number. Successive generations of cells or organisms within a population are separated by a time interval called generation time. The cell regeneration time can be determined with the aid of time-lapse microcinematography.

Cell hybridization—The fusion of two or more dissimilar cells leading to the formation of a somatic hybrid. *cf* cell fusion.

Cell line—A cell lineage that can be maintained in culture. A cell line arises from a primary culture. It implies that cultures from it consist of several lineages of the cells originally present in the primary culture.

Cell-mediated immune response—The activation of T cells of the immune system in response to the presence of a foreign antigen.

Cell membrane—The membrane that separates the cell wall and the cytoplasm, and regulates the flow of material into and out of the cell. *See* plasmalemma.

Cell number—The number of cells per unit volume of a culture.

Cell plate—The precursor of the cell wall, formed as cytokinesis starts during cell division. The cell plate develops in the region of the equatorial plate and arises from membranes in the cytoplasm.

Cell sap—Water and dissolved substances, sugar, amino acids, waste substances, etc., in the plant cell vacuole.

Cell selection—The process of selecting cells within a group of genetically different cells. Select cells or cell lines are sub-cultured onto fresh medium for continued selection and often are exposed to an increased level of the selection agent. The final objective is to regenerate plants exhibiting the traits selected for at the cellular level.

Cell strain—A strain of cells having specific properties or markers derived from a primary culture of a cell line by selection or cloning. The selected properties must persist during subsequent cultivation. a.k.a. single-cell line.

Cell suspension—Cells in culture in moving or shaking liquid medium, often used to describe suspension cultures of single cells and cell aggregates. *See* suspension culture.

Cellular immune response—*See* T-cell-mediated (cellular) immune response.

Cellular oncogene (proto-oncogene)—A normal mammalian or avian gene that when mutated or improperly expressed contributes to the development of cancer. *See* oncogene.

Cellulase—Enzyme catalysing the breakdown of cellulose.

Cellulose—(cell + ose, a suffix indicating a carbohydrate)—A complex carbohydrate composed of long, unbranched chains of beta-glucose ((1.4)-linked-β-d-glucose) molecules, which contribute to the structural framework of plant cell walls. It comprises 40% to 55% by weight of the plant cell wall.

Cellulose nitrate—*See* nitrocellulose.

Cellulosome—A multi-protein aggregate that is present in some cellulolytic micro-organisms and contains multiple copies of all the enzymes required to completely break down cellulose. This complex is often found on the outer surface of cellulolytic micro-organisms.

Cell wall—A rigid external coat which surrounds plant cells. It is formed outside the plasmalemma and consists primarily of cellulose.

CentiMorgan (cM)—One percent recombination between two loci. *See* map distance; crossing-over unit.

Central dogma—The basic concept that, in nature, genetic information generally can flow only from DNA to RNA to protein. It is now known, however, that information contained in RNA molecules of certain viruses (called retroviruses) can also flow back to DNA.

Central mother cell—A subsurface cell located in a plant apical meristem and characterized by a large vacuole.

Centres of origin—The locations in the world where particular domesticated plants originated. These areas show the highest variation, and are rich in wild alleles.

Centrifugation—Separating molecules by size or density using centrifugal forces generated by a spinning rotor. G-forces of several hundred thousand times gravity are generated in ultracentrifugation. *See* density gradient centrifugation.

Centrifuge—A device in which solid or

Chimera (or chimaera)—From chimera, a mythological creature with the head of a lion, the body of a goat and the tail of a serpent. An organism whose cells are not all derived from the same zygote. Animal. An individual exhibiting two or more genotypes in patches derived from two or more embryos. An individual derived from two embryos by experimental intervention. Plant. Part of a plant with a genetically different constitution as compared with other parts of the same plant. It may result from different zygotes that grow together, or from artificial fusion (grafting); it may either be periclinal chimera, in which one tissue lies over another as a glove fits a hand; mericlinal chimera, where the outer tissue does not completely cover the inner tissue; and sectoral chimera, in which the tissues lie side by side. A recombinant DNA molecule that contains sequences from different organisms.

Chimeric DNA—A recombinant DNA molecule containing unrelated genes.

Chimeric gene—A semi-synthetic gene, consisting of the coding sequence from one organism, fused to promoter and other sequences derived from a different gene. Most genes used in transformation are chimeric. *See* carrier DNA; binary vector; plasmid; transformation; vector.

Chimeric protein—*See* fusion protein.

Chimeric selectable marker gene—A gene that is constructed from parts of two or more different genes and allows the host cell to survive under conditions where it would otherwise die.

Chi-squared test—(χ^2 test)—A significance test used to statistically assess the goodness of fit of observed data to a prediction.

Chitin—A nitrogenous polysaccharide occurring as skeletal material in many invertebrates and fungi.

Chitinase—An enzyme which breaks down chitin.

Chloramphenicol—An antibiotic that interferes with protein synthesis.

Chlorenchyma (Gr. *chloros*, green + *enchyma*, a suffix meaning tissue)—Tissue containing chloroplasts, including leaf mesophyll and other parenchyma cells.

Chlorophyll—(Gr. *chloros*, green + *phyllon*, leaf)—One of the two pigments responsible for the green colour of most plants. It is essential in the absorption of light energy for photosynthesis.

Chloroplast—(Gr. *chloros*, green + *plastos*, formed)—Specialized cytoplasmic organelle that contains chlorophyll. Lens-shaped and bounded by a double membrane, chloroplasts contain membranous structures (thylakoids) piled up into stacks, surrounded by a gel-like matrix (stroma). They are the site of solar energy transfer and important reactions of starch or sugar synthesis. Chloroplasts have their own DNA and are inherited cytoplasmically, independent of nuclear genes.

Chloroplastid—*See* chloroplast.

Chlorosis—(Gr. *chloros*, green + *osis*, diseased state)—Failure of chlorophyll development, and appearance of yellow colour in plants, because of a nutritional disturbance or because of an infection by a virus, bacteria or fungus.

Chromatid—(chromosome + *id*, L. suffix meaning 'daughters of')—Each of the two daughter strands comprising a duplicated chromosome. The term remains in use while the two chromatids are still joined at the centromere. As soon as the centromere divides, setting the two chromatids adrift (during anaphase of mitosis; and during anaphase II of meiosis), they are called chromosomes.

Chromatin—(Gr. *chroma*, colour)—Substance of which eukaryotic chromosomes are composed. It consists of primarily DNAm with some proteins (mainly histones), and small amounts of RNA. Originally named because of the readiness with which it stains with certain dyes (chromaticity).

Chromatin fibres—A basic organizational unit of eukaryotic chromosomes, consisting of DNA and associated proteins assembled into strands of 30 nm average diameter.

Chromatography (Gr. *chroma*, colour + *graphein*, meaning to draw or write)— A method for separating and identifying the components of mixtures of molecules having similar chemical and physical properties. The term originally used by Mikhail Tswett (1906) to describe the separation of a mixture of leaf pigments on a calcium carbonate column.

Chromocentre—Body produced by fusion of the heterochromatic regions of the chromosomes in the polytene tissues (*e.g.*, the salivary glands) of certain *Diptera*.

Chromogenic substrate—A compound or substance that contains a colour-forming group.

Chromomeres—Small bodies, described by J. Belling, that are identified by their characteristic size and linear arrangement along a chromosome.

Chromonema (pl. chromonemata)—An optically single thread forming an axial structure within each chromosome.

Chromoplast—Plastid containing pigments, such a chloroplast, or one in which carotenoids predominate.

Chromosomal aberration—Any change in chromosome structure or number. Although it can be a mechanism for enhancing genetic diversity, such alterations are usually fatal or ill-adaptive, especially in animals.

Chromosomal integration site—A chromosomal location where foreign DNA can be integrated, often without impairing any essential function in the host organism.

Chromosomal polymorphism—The occurrence of one to several chromosomes in two or more alternative structural forms within a population; the structurally changed chromosomes are the result of chromosome mutations (*i.e.*, any structural change involving the gain, loss or re-location of chromosome segments).

Chromosome—(Gr. *chroma*, colour + *soma*, body)— A single DNA molecule, a tightly coiled strand of DNA, condensed into a compact structure *in vivo* by complexing with accessory histones or histone-like proteins. A group of nuclear bodies containing genes which are largely responsible for the differentiation and activity of a eukaryotic cell; one of the bodies into which the nucleus resolves itself at the beginning of mitosis and from which it is derived at the end of mitosis. Chromosomes contain most of the cell's DNA. Chromosomes exist in pairs in eukaryotes—one paternal (from the male parent) and one maternal (from the female parent). Each eukaryotic species has a characteristic number of chromosomes. Bacterial and viral cells contain only a single chromosome, consisting of a single or double strand of DNA or, in some viruses, RNA, without histones.

Chromosome aberration—Abnormal structure or number of chromosomes; includes deficiency, duplication, inversion, translocation, aneuploidy, polyploidy, or any other change from the normal pattern.

Chromosome banding—Staining of chromosomes in such a way that light and dark areas occur along the length of the chromosomes in repeatable patterns. Lateral comparisons identify pairs. Each chromosome can be identified by its banding pattern.

Chromosome mutation—A change in the gross structure of a chromosome, usually causing severely deleterious effects in the organism. They are often due to an error in pairing during the crossing-over stage of meiosis. The main types of chromosome mutation are translocation, duplication, deletion and inversion.

Chromosome jumping—A technique that allows two segments of duplex DNA that are separated by thousands of base pairs (about 200 kb) to be cloned together. After sub-cloning, each segment can be used as a probe to identify cloned DNA sequences that, at the chromosome level, are roughly 200 kb apart. *See* positional cloning.

Chromosome theory of inheritance—The theory that chromosomes carry the genetic information and that their behaviour during meiosis provides the physical basis for segregation and independent assortment.

Chromosome walking—A technique that identifies overlapping cloned DNA fragments that form one continuous segment of a chromosome. These fragments can be generated either by random shearing or by partial digestion with a four-base-pair cutter such as *Sau*3A. A series of colony hybridizations is then carried out, starting with some cloned fragment which has already been identified and which is known to be in the region encompassed by the overlapping clones. This identified fragment is used as a probe to pick out clones containing adjacent sequences. These are then used as probes themselves to identify clones carrying sequences adjacent to them and so on. At each round of hybridization one 'walks' further along the chromosome from the initial fragment. *See* positional cloning.

Chymosin—An enzyme that clots milk; it is used in the manufacture of cheese

Clonal propagation—Asexual propagation of many new plants (ramets) from an individual (ortet); all have the same genotype.

Clonal selection—The production of a population of plasma cells all producing the same antibody in response to the interaction between a B lymphocyte producing that specific antibody and the antigen bound by that antibody.

Clone (Gr. *klon*, a twig or slip)—A group of cells or organisms that are genetically identical as a result of asexual reproduction, breeding of completely inbred organisms, or forming genetically identical organisms by nuclear transplantation.

Cloning vector—A small, self-replicating DNA molecule—usually a plasmid or viral DNA chromosome - into which foreign DNA is inserted in the process of cloning genes or other DNA sequences of interest. It can carry inserted DNA and be perpetuated in a host cell. Also called a cloning vehicle, vector, or vehicle.

Co-integrate vector system—A two-plasmid system for transferring cloned genes to plant cells. The cloning vector has a T-DNA segment that contains cloned genes. After introduction into *Agrobacterium tumiefasciens*, the cloning vector DNA undergoes homologous recombination with a resident disarmed Ti plasmid to form a single plasmid carrying the genetic information for transferring the genetically engineered T-DNA region to plant cells.

Co-integrate—A DNA molecule formed by the fusion of two different DNA molecules, usually mediated by a transposable element.

Copy number—The average number of molecules of a plasmid or gene per genome contained in a cell.

Co-repressor—An effector molecule that forms a complex with a repressor and turns off the expression of a gene or set of genes.

Cryoprotectant—Compound preventing cell damage during freezing and thawing processes. Cryoprotectants are agents with high water solubility and low toxicity. Two types of cryoprotectant agent are commonly used: permeating (glycerol and DMSO (q.v.)) and non-permeating (sugars, dextran, ethylene glycol, polyvinyl pyrolidone and hydroxyethyl starch).

Cybrid—A cytoplasmic hybrid, originating from the fusion of a cytoplast (the cytoplasm without nucleus) with a whole cell, as in nuclear transfer (although the term is not used in that context). Note that the nucleus and cytoplasm of the fused cell products are from different genetic sources.

Cyclic AMP (cyclic adenosine monophosphate)—A 'messenger' that regulates many intracellular reactions by transducing signals from extracellular growth factors to cellular metabolic pathways.

Death phase—The final growth phase, during which nutrients have been depleted and cell number decreases. *See* growth phases.

Deceleration phase—The phase of declining growth rate, following the linear phase and preceding the stationary phase in most batch-suspension cultures. *See* growth phases.

De-differentiation—The process by which cells lose their specialization and proliferate by cell division to form a mass of cells which, in response to appropriate stimuli, may later differentiate again to form either the same cell type or a different one. De-differentiation occurs in response to wounding and in tissue cultures. *See* re-differentiation.

Deficiency—Insufficiency or absence of one or more usable forms of enzymatic, nutritional or environmental requirements, so that development, growth or physiological functions are affected.

Diabetes—A disease associated with the absence or reduced levels of insulin, which is a hormone essential for the transport of glucose to cells.

Diagnostic procedure—A test or assay used to determine the presence of an organism, substance or nucleic acid sequence alteration.

Diakinesis—A stage of meiosis just before metaphase I, in which the separation of homologous chromosomes is almost completed.

Diazotroph—An organism that can fix atmospheric nitrogen.

Dicentric chromosome—A chromosome having two centromeres.

Dichogamy—The condition in which the male and the female reproductive organs of a flower mature at different times, thereby making self-fertilization improbable or impossible.

Direct embryogenesis—Embryoids form directly in culture, without an intervening callus phase, on the surface of zygotic or somatic embryos or on explant tissues (leaf section, root tip, etc.).

Direct organogenesis—Formation of organs directly on the surface of cultured intact explants. The process does not involve callus formation.

DNA micro-array—A small glass surface to which has been fixed an array of DNA fragments, each with a defined location. A typical DNA chip would contain 10,000 discrete spots (each containing a different DNA fragment) in an area of just a few square centimetres. When a solution of fluorescently labelled DNA fragments is hybridized to the chip, spots to which hybridization occurs are visible as fluorescence. If the spots on the chip are genes (expressed sequence tags, q.v.), hybridization with cDNA from a particular tissue shows which genes are expressed in that tissue. If the spots are short, synthesized oligonucleotides (approximately 25 bases) corresponding to that part of a gene containing a single nucleotide polymorphisms (SNP)(q.v.), with a separate spot for each of the 4 possible bases at that site, hybridization with genomic DNA from an individual plant or animal enables that individual to be genotyped at as many SNP loci as are represented on the chip. The big advantage of DNA chips is the extent to which the process of genotyping can be automated, thereby enabling huge numbers of plants or animals to be genotyped for a huge number of loci.

DNA polymerase—An enzyme that catalyses the synthesis of double-stranded DNA, using single-stranded DNA as a template. *See* polymerase.

DNA sequencing—Procedures for determining the nucleotide sequence of a DNA fragment. There are two common methods for doing this: the Maxam and Gilbert technique (chemical degradation), that uses different chemicals to break the DNA into fragments at specific bases; or the Sanger technique (called the di-deoxy or chain-terminating method) uses DNA polymerase to make new DNA chains, with di-deoxy nucleotides (chain terminators) to stop the chain randomly as it grows.In both cases, the DNA fragments are separated according to length by polyacrylamide gel electrophoresis, enabling the sequence to be read directly from the gel.

Drug delivery—Method by which a therapeutic agent is delivered to its site of action. For traditional therapeutic agents this is another name for *formulation*. However, biotechnology has allowed the development of a range of new therapeutic-agent delivery systems, such as liposomes and other encapsulation techniques, and a range of mechanisms that target a therapeutic agent to a particular cell or tissue. *See* therapeutic agent.

Dry weight—The moisture-free weight of tissue obtained by drying at high (oven-drying) or low (freeze-drying) temperatures for an interval sufficient to remove all water.

Dual culture—A culture made of a plant tissue and one organism (such as a nematode) or an obligate parasite/ micro-organism (such as a fungus). Dual culture techniques are used for a variety of purposes, including assessing host-parasite interactions and the production of axenic cultures.

Duplex DNA—Double-stranded DNA.

Duplication—The occurrence of a segment more than once in the same chromosome or genome.

Electromagnetic radiation—Electromagnetic waves, including ultraviolet (UV), X-rays, and gamma radiation (γ rays). Electromagnetic radiation is used to produce mutant cells or organisms, or, in the case of UV, disinfestation and sterilization, in tissue culture.

Electromagnetic spectrum—The range of wavelengths or frequencies over which electromagnetic radiation extends.

Electron microscope—A microscope that uses an electron beam focused by magnetic 'lenses.' *See* scanning electron microscope; transmission electron microscope.

Electrophoresis—A technique that separates charged molecules such as DNA, RNA or protein—on the basis of relative migration in an appropriate matrix (such as agarose gel or polyacrylamide gel) subjected to an electric field. *See* agarose gel electrophoresis; polyacrylamide gel electrophoresis (PAGE); pulsed-field gel electrophoresis (PFGE).

Electroporation—An electrical treatment of cells that induces transient pores, through which DNA can enter the cell. The introduction of DNA or RNA into protoplasts or other cells by the momentary disruption of the cell membrane through exposure to an intense electric field. Note: Although the precise mechanism of electroporation is poorly understood, pores are thought to form by the local polarization of the cell membrane when it is exposed to a high electric potential. These openings persist for a variable amount of time, depending upon the temperature at which the cell is treated. Macro-molecules, such as DNA or RNA, enter through these openings either through diffusion or through electrophoretic movement. The membrane openings then re-seal, capturing introduced DNA and preventing escape of the cell contents.

ELISA (enzyme-linked immunosorbent assay)—A sensitive technique for accurately determining specific molecules in a mixed sample. The amount of protein or other antigen in a given sample is determined by means of an enzyme-catalysed colour change, avoiding both the hazards and expense of radioactive techniques.It takes various forms. In the most common form, two antibody preparations are used in ELISA. An antibody (primary) specific to the test protein is adsorbed onto a solid substrate, and a known amount of the sample is added; all molecules of the test protein in the sample are bound by the antibody. A second antibody specific for a second site on the test protein is added; this is conjugated with an enzyme, which catalyses a colour change in the fourth reagent, added finally. The colour change is measured photometrically and compared against a standard curve to give the concentration of protein in the sample. ELISA is widely used for diagnostic and other purposes.

Elongation factors—Soluble proteins that are required for polypeptide chain elongation.

Embryo (Gr. *en*, in + *bryein*, to swell)—An immature organism in the early stages of development. In mammals, this occurs in the first months in the uterus. In plants, it is the structure that develops in the megagametophyte, as result of the fertilization of an egg cell, or without fertilization. In aseptic cultures, adventitious embryos show polarization, followed by the growth of a shoot from one end and a root from the other end.

Embryo cloning—The creation of identical copies of an embryo by embryo splitting (q.v.) or by nuclear transfer (q.v.) from undifferentiated embryonic cells.

Embryo culture—The culture of embryos on nutrient media.

Embryogenesis (General)—Development of an embryo. (In plants) *In vitro* formation of plants from plant tissues, through a pathway closely resembling normal embryogeny from the zygote; if this development in culture involves somatic cells and not the zygote, it can be indicated by using the term *adventitious embryogenesis* or *somatic embryogenesis*. The generation of embryos has two stages: initiation and maturation. Initiation needs a high level of the group of plant hormones called auxins; maturation needs a lower level. Other chemicals have to be at suitable levels. The procedure involves the explanting of a piece of plant tissue and putting it on a high-auxin medium, where the cells grow into a mass of callus. This is then transferred to a maturation medium, where the callus starts to initialize organs, ultimately growing a root and a shoot.

Embryo multiplication and transfer (EMT)—The cloning of animal embryos and their subsequent transfer to recipients (via artificial inembryonation (q.v.)). The cloned embryos can be clones of an embryo or of an adult.

Embryo sac—A large thin-walled space within the ovule of the seed plant in which the egg and, after fertilization, the embryo develop; the mature female gametophyte in angiosperms. Generally a seven-celled structure. The seven cells are two synergids, one egg cell, three antipodal cells (each with a single haploid nucleus) and one endosperm mother cell with two haploid nuclei.

Embryo sexing—The determination of the sex of an embryo, typically by means of PCR (q.v.) involving amplification from a small sample of embryonic tissue, using primers specific for a locus on the Y chromosome.

Embryo splitting—The splitting of young embryos into several sections, each of which develops into an animal. A form of animal cloning, *i.e.*, of producing animals that are genetically identical. In practice, the number of identicals (identical organisms) that can be produced from a single embryo is less than 10.

Embryo technology—Generic name for any modification of mammalian embryos. It encompasses embryo cloning, embryo splitting, *in vitro* fertilization, and embryo storage.

Embryo transfer (ET)—*See* multiple ovulation and embryo transfer.

Embryoid—An embryo-like body developing *in vitro*. It forms a complete, self-contained platelet with no vascular connection with the callus. The term embryoid is no longer commonly used. *See* embryo.

Embryonic stem cells—Cells of the early embryo that can give rise to all differentiated cells, including germ line cells.

End-labelling—The introduction of a radioactive atom at the end of a DNA or RNA molecule. A commonly used method is to use T4 polynucleotide kinase to introduce a ^{32}P atom onto the end of a DNA molecule.

Endocrine gland—Any gland in an animal that manufactures hormones and secretes them directly into the bloodstream to act at distant sites in the body, known as target organs or cells. *cf* exocrine gland. *See* gland.

Endocrine interference—Interference with the normal balance hormones.

Endocytosis—The process by which materials enter a cell without passing through the cell membrane. The membrane folds around material outside the cell, resulting in the formation of a saclike vesicle into which the material is incorporated. This vesicle is then pinched off from the cell surface so that it lies within the cell. *See* phagocytosis; pinocytosis.

Endoderm—The internal layer of cells of the gastrula, which will develop into the alimentary canal (gut) and digestive glands of the adult.

Endodermis (Gr. *endon*, within + *derma*, skin)—The layer of living cells, with various characteristically thickened walls and no intercellular spaces, which surrounds the vascular tissue of certain plants and occurs in nearly all roots and certain stems and leaves. The endodermis separates the cortical cells from cells of the pericycle.

Endogamy—The fusion of reproductive cells from closely related parents, *i.e.*, inbreeding. *cf* exogamy.

Endogenote—The part of the bacterial chromosome that is homologous to a genome fragment (exogenote) transferred from the donor to the recipient cell in the formation of a merozygote.

Endogenous (Gr. *endon*, within, + *genos*, race, kind)—Developed or added from within the cell or organism.

Endomitosis—Duplication of chromosomes without division of the nucleus, resulting in increased chromosome number within a cell. Chromosome strands separate, but the cell does not divide.

Endonuclease—An enzyme that breaks strands of DNA at internal positions; these enzymes are important tools in recombinant DNA technology. *See* nuclease.

Endophyte—An organism that lives inside a plant.

Endoplasmic reticulum—(Gr. *endon*, within + *plasma*, anything formed or moulded; L. *reticulum*, a small net) —A cytoplasmic net of membranes, adjacent to the nucleus, made visible by the electron microscope. Any system of paired membranes within the cytoplasm. Frequently abbreviated to ER. They are sites of protein synthesis.

Endopolyploidy—The result of nuclear divisions without subsequent cytoplasmic division (cytokinesis); the polyploids so obtained are called endopolyploids. *See* polyploidy.

Endoprotease—An enzyme that cleaves the peptide bonds between amino acids within a protein. Cleavage is usually at one or more specific sites.

Endoreduplication—Chromosome reproduction during interphase. 4-chromatid chromosomes (diplochromosomes) are seen during this phase.

Endosperm (Gr. *endon*, within + *sperma*, seed)—Nutritive tissue that develops in the embryo sac of most angiosperms. It usually forms after the fertilization of the two fused primary endosperm nuclei of the embryo sac with one of the two male gamete nuclei. In most diploid plants, *e.g.*, cereals, the endosperm is triploid ($3n$), but in some (*e.g.*, lily) it is often consumed as the seed matures.

Endosperm mother cell—One of the seven cells of the mature embryo sac, containing the two polar nuclei and, after reception of a sperm cell, gives rise to the primary endosperm cell from which the endosperm develops.

Endotoxin—A component of the cell wall of Gram-negative bacteria that elicits, in mammals, an inflammatory response and fever.

End-product inhibition—The inhibition of an enzyme by a metabolite. The enzyme is sometimes the first enzyme in a biosynthetic pathway, and the metabolite is generally the product of the last step in the pathway. *See* feedback inhibition.

Enhancer—A substance or object that increases a chemical activity or a physiological process.—A DNA sequence that increases the transcription of a eukaryotic gene when they are both on the same DNA molecule. a.k.a. enhancer element; enhancer sequence.

Enhancer element; enhancer sequence—A sequence found in eukaryotes and certain eukaryotic viruses which can increase transcription of a gene when located (in either orientation) up to several kilobases from the gene concerned. These sequences usually act as enhancers when on the 5′ side (upstream) of the gene in question. However, some enhancers are active when placed on the 3′ side (downstream) of the gene. In some cases enhancer elements can activate transcription of a gene with no (known) promoter.

Epigenesis—Describes the developmental process whereby each successive stage of normal development is built up on the foundations created by the preceding stages of development; an embryo is built up from a zygote, a seedling from an embryo, and so on.

Epigenetic variation—Non-hereditary and reversible variation; often the result of a change in gene expression.

Epigenetic—A term referring to the non-genetic causes of a phenotype.

Epinasty—A process by which the growth of branches or petioles is abnormally pointing downward. This phenomenon is caused by the more rapid growth of the upper side. Epinasty may result from either nutritional deficiencies or irregularities at the plant growth regulator level. Not to be confused with wilting, as epinastic tissues are turgid.

Epiphyte—A plant that grows upon another plant, but is neither parasitic on it nor rooted in the ground.

Episome—A genetic extrachromosomal element (*e.g.*, the fertility factor (F) in *Escherichia coli*) which replicates within a cell independently of the chromosome and is able to integrate into the host chromosome. The step of integration may be governed by a variety of factors and so the term episome has lost favour and been superseded by the wider term *plasmid.* Plasmids and F factors are episomes.

Epistasis—Interaction between genes at different loci, *e.g.*, one gene suppresses the effect of another gene that is situated at a different locus. Suppressed genes are said to be hypostatic. Dominance is associated with members of allelic pairs, whereas epistasis is interaction among products of non-alleles.

Epitope—A specific chemical domain on an antigen that stimulates the production of, and is recognized by, an antibody. Each epitope on a molecule such as protein elicits the synthesis of a different antibody. a.k.a. antigenic determinant.

Ethylenediamine tetraacetic acid (EDTA)—A chelating compound. In tissue culture it is used to keep nutrients, such as iron, bound in a form that leaves them still available to the plant but which prevents them from precipitating out.

Etiolation—An abnormal increase in stem elongation, accompanied by poor or absent leaf development. Physiological etiolation is caused by a lack of chlorophyll, and is typical of plants growing under low light intensity or in complete darkness. It can also be caused by disease.

Exon—A segment of a eukaryotic gene that is transcribed as part of the primary transcript and is retained, after processing, with other exons to form a functional mRNA molecule. *See* DNA; intron; split gene; splicing.

Exon amplification—A procedure that is used to amplify exons.

Exonuclease—An enzyme that digests DNA or RNA, beginning at the ends of strands. It requires a free end in order to degrade a DNA or RNA molecule. 5′ exonucleases require a free 5′ end and degrade the molecule in the 5′ → 3′ direction. 3′ exonucleases require a free 3′ end and degrade the molecule in the opposite direction. *cf* endonuclease.

Exonuclease III—An *E. coli* enzyme that removes nucleotides from the 3′-hydroxyl ends of double stranded DNA. a.k.a. exo III; exodeoxyribonuclease III.

Exopolysaccharide—A high-molecular-weight polymer that is composed of sugar residues and is secreted by a micro-organism into the surrounding environment.

Exotoxin—A toxin released by a bacterium into the medium in which it grows.

Explant—A plant part aseptically excised and prepared for culture in a nutrient medium. *cf* donor plant.

Explantation—The removal of cells, tissues or organs of animals and plants for observation of their growth and development in appropriate culture media.

Explant donor—The source plant or mother plant from which is taken the explant used to initiate a culture. *cf* donor plant.

Exponential phase—The growth stage where cells undergo their maximum rate of cell division. The exponential phase follows the lag phase, and precedes the linear growth phase. *See* growth phases. a.k.a logarithmic phase.

Export—To transport a compound out of a cell. a.k.a. to secrete; to excrete.

Express—To transcribe and translate a gene's message into a peptide product.

Expressed sequence tag (EST)—Short cDNA sequence. So-called because it represents part of the sequence (*i.e.*, the 'tag' of a sequence) of an expressed gene.

Expression library—A population of different DNA molecules encoding peptides, that has been cloned into one kind of expression vector. *See* library.

Expression system—Combination of host and vector which provides a genetic context for making a cloned gene function, *i.e.*, produce peptide, in the host cell.

Expression vector—A cloning vector that has been constructed in such a way that, after insertion of a DNA molecule, its coding sequence is properly transcribed and the RNA is translated. The cloned gene is put under the control of a promoter sequence for the initiation of transcription, and often also has a transcription termination sequence at its end. Such promoters are termed *high level*; examples include P1 (the leftward promoter of phage l) and the promoter of the yeast PGK (phosphoglycerate kinase) gene.

Expressivity—Degree of expression of a trait controlled by a gene. A particular gene may show different degrees of expression in different individuals. *See* variable expressivity.

***Ex situ* conservation**—A conservation method which entails the actual removal of germplasm resources (seeds, pollen, sperm, individual organisms) from the original habitat or natural environment. *cf* gene bank; cryobiological preservation; *in situ* conservation.

***Ex situ* conservation of farm animal genetic diversity**—In AnGR: All conservation of genetic material *in vivo,* but out of the environment in which it developed, and *in vitro* including, *inter alia,* the cryoconservation of semen, oocytes, embryos, cells or tissues. Note that *ex situ* conservation and *ex situ* preservation are considered here to be synonymous. (Source: FAO, 1999)

***Ex situ* preservation** *See ex situ* conservation.

Extension—Single-stranded DNA region consisting of one or more nucleotides at the end of a strand of duplex DNA. a.k.a. protruding end; sticky end; overhang; cohesive end. *See* extension.

3′-extension—A short single-stranded nucleotide sequence on the 3′-hydroxyl end of a double-stranded DNA molecule. a.k.a. 3′ protruding end; 3′ sticky end; 3′ overhang.

5′-extension—A short single-stranded nucleotide sequence on the 5′-hydroxyl end of a double-stranded DNA molecule. a.k.a. 5′ protruding end; 5′ sticky end; 5′ overhang.

Extinct breed—In AnGR: A breed where it is no longer possible to recreate the breed population. Extinction is absolute when there are no breeding males (semen), breeding females (oocytes), nor embryos remaining.

Extinction—The irreversible condition of a species or other group of organisms of having no living representatives in the wild, which follows the death of the last surviving individual of that species or group. Extinction may occur on a local or global level; it can result from various human activities, including the destruction of habitats or the overexploitation of species that are hunted or harvested as a resource.

Extrachromosomal—Things that are not part of the chromosomes; DNA units in the cytoplasm that control cytoplasmic inheritance.

Extrachromosomes—Self-replicative genetic elements separate from main chromosome(s) of a cell. This definition usually excludes viruses, but the division is somewhat arbitrary. In bacteria, plasmids are the principal extrachromosomes; they encode functions which are not essential to the growth and division of the host cell. In eukaryotes, extrachromosomes may be either essential or dispensable. They may inhabit (i) the nucleus, *e.g.*, extrachromosomal rDNA molecules, yeast 2 μm plasmid; (ii) the cytosol, *e.g.*, dsRNA molecules in fungi; or (iii) the cytoplasmic organelles, *e.g.*, mitochondrial DNA, chloroplast DNA. Eukaryotic extrachromosomal elements may be recognized genetically by their failure to show segregation at meiosis.

Exude—Slowly discharge; leak liquid material (exudate such as tannins or oxidized polyphenols) through pores or cuts, or by diffusion into the medium. In some woody plant species, exudation is associated with a lethal browning of explants.

Ex vitro (L. 'from glass')—Organisms removed from tissue culture and transplanted; generally plants to soil or potting mixture.

***Ex vivo* gene therapy**—The delivery of a gene or genes to the isolated cells of an individual. After culturing, the transformed cells are introduced back into the individual by transfusion, infusion or injection, to alleviate a genetic disorder.

Fermentation—The anaerobic breakdown by micro-organisms of complex organic substances, especially carbohydrates like glucose. The process is energy-yielding. Fermentation is often misused to describe large-scale aerobic cell culture in specialized vessels (fermenters, bioreactors) for secondary product synthesis.

Fermentation substrates—Materials used as food for growing micro-organism. The fermentation substrates and the trace materials needed, together with chemicals added to make the fermentation easier, form the culture medium.

Fermenter—*See* bioreactor.

Fertile—Of an organism: Capable of breeding and reproduction.

Fertilization—(L. *fertilis*, capable of producing fruit)—The union of two gametes from opposite sexes to form a zygote; it involves the fusion of nuclei of gametes (karyogamy) and the fusion of cytoplasm (plasmogamy). Typically, each gamete contains a haploid set of chromosomes. Hence, after fusion of the nuclei, the resulting nucleus contains a diploid set of chromosomes.

Flocculant—Agent that causes small particles to aggregate (flocculate).

Fluorescence *in situ* hybridization (FISH)—Hybridization of cloned DNA to intact chromosomes, where the cloned DNA has been labelled with a fluorescent dye. This is the major method of physical mapping of cloned DNA fragments on chromosomes.

Fluorescence-activated cell sorting (FACS)—The use of laser beams to detect differences in fluorescence between different types of cells in a mixture, and the subsequent deflection of cells into separate bins corresponding to each type of cell in the mixture. One of the popular uses of this technology is in sperm sexing (q.v.).

Fluorescent probe—Probe whose response is based on the fluorescence intensity of individual cells or cell components.

Flow cytometry—A technique used to sort cells or other biological materials by means of flow through apertures of defined size or by laser sorting.

Flower—The structure in angiosperms (flowering plants) that bears the organs for sexual reproduction.

Follicle stimulating hormone (FSH)—A hormone, secreted by the anterior pituitary gland in mammals, that stimulates, in female mammals, ripening of specialized structures in the ovary (Graafian follicles) that produce ova and, in males, the formation of sperm in the testis. It is a major constituent of fertility drugs.

Founder animal—In transgenesis research, an organism that carries a transgene in its germ line and can be used in matings to establish a pure-breeding transgenic line, or one that acts as a breeding stock for transgenic animals.

Founder principle—The possibility that a new, small, isolated population may be genetically different from the 'parent' population, because the founding individuals (being a small, random sample from the large, 'parent' population) could be quite different from typical members of the 'parent' population.

Fusion biopharmaceuticals—Biopharmaceutical proteins formed as a result of fusion proteins.

Fusion gene—A hybrid gene created by joining portions of two different genes (to produce a new protein) or by joining a gene to a different promoter (to alter or regulate gene transcription).

Fusion protein—A polypeptide made from a recombinant gene that contains portions of two or more different genes. The different genes are joined so that their coding sequences are in the same reading frame: the genetic apparatus reads the gene fusion as a single gene, and so produces a fusion protein. Also known as hybrid protein or chimeric protein.

Gametoclone—A plant regenerated from a tissue culture originating from gametic tissue.

Gametogenesis—The process of the formation of gametes.

Gas transfer—The rate at which gases are transferred from gas into solution. It is an important parameter in fermentation systems because it controls the rate at which the organism can metabolize. Gas transfer can be done by several methods, including use of small gas bubbles, that diffuse faster than larger ones, thanks to their larger surface area per unit of volume; or spreading the liquid out, for example in a thin sheet, or in a thin permeable tube, as in hollow fibre bioreactor.

Gastrula—An early animal embryo consisting of two layers of cells; an embryological stage following the blastula.

GC island—A segment of DNA that is rich in G=C base pairs and often precedes a transcribed gene in the genomes of vertebrate organisms.

Gel electrophoresis—An analytical method for separating molecules according to their size. Samples are put at one end of a slab of polymer gel; an electric field across the gel pulls the molecules through it; the smaller molecules pass more easily and so move towards the other end faster; the various sizes of molecules end up at different positions according their size.

Gene cloning—The process of synthesizing multiple copies of a particular DNA sequence using a bacteria cell or another organism as a host. The gene of interest is inserted into a self-replicating DNA molecule (DNA vector, often a plasmid) and the resulting recombinant DNA molecule is amplified in an appropriate host cell. Used in genetic engineering. a.k.a. molecular cloning; cloning. *See* DNA; host.

Gene conservation; genetic resources conservation—The conservation of species, populations, individuals or parts of individuals, by *in situ* or *ex situ* methods, to provide a diversity of genetic materials for present and future generations.

Gene conversion—A process, often associated with recombination, during which one allele is replicated at the expense of another, leading to non-Mendelian segregation ratios. In whole tetrads, for example, the ratio may be 6:2 or 5:3 instead of the expected 4:4.

Gene expression—The process by which a gene produces RNA and protein, and hence exerts its effects on the phenotype of an organism.

Gene flow—The spread of genes from one breeding population to another (usually) related populations by migration, possibly leading to changes in allele frequency.

Gene library—Collection of cloned DNA fragments that ideally includes all the genetic information of an organism. If the original source of the DNA is the genomic DNA from an organism, then it is called a genomic library; if the DNA is from cDNA made by enzymatic copying of RNA, then the library includes representative fragments from all genes that were being expressed at the time the RNA was sampled, and would be called a cDNA library. a.k.a. gene bank; library; clone bank; bank; library.

Gene probe—A single-stranded DNA or RNA fragment used in genetic engineering to search for a particular gene or other DNA sequence. The probe has a base sequence complementary to the target sequence and will thus attach to it by basepairing. By labelling the probe, it can be identified after subsequent separation and purification.

Generally Regarded as Safe (GRAS)—Designation given to foods, drugs, and other materials that have been used for a considerable period of time and have a history of not causing illness to humans, even though extensive toxicity testing has not been conducted. More recently, certain host organisms for recombinant DNA experimentation have been given this status.

Generate—To propagate or (mass) proliferate. The process is generation or regeneration.

Generative nucleous—One of the two male gametes in the pollen tube of angiosperms.

Gene recombination—The appearance of gene combinations in the progeny that differ from the combinations present in the parents.

Gene replacement—The incorporation of a transgene into a chromosome at its normal location by homologous recombination, thus replacing the copy of the gene originally present at the locus.

Gene sequencing—The process of elucidating the nucleotide sequence of a gene. *See* DNA sequencing.

Gene splicing—A stage in the processing of mRNA, occurring only in eukaryote cells, in which intervening sequences (introns) are removed from the primary RNA transcript (hnRNA), and the coding exons are joined together to form the mature mRNA molecule.

Gene therapy—The treatment of inherited diseases by introducing into the cells of affected individuals the wild-type copies of the defective gene causing the disorder. If reproductive cells are modified, the procedure is called germ-line or heritable gene therapy. If cells other than reproductive cells are modified, the procedure is called somatic-cell or non-inheritable gene therapy.

Genetically engineered organism (GEO)—Sometimes used for genetically modified organism, q.v.

Genetically modified organism (GMO)—An organism that has been modified by the application of recombinant DNA technology.

Genetic engineering—Changes in the genetic constitution of cells (apart from selective breeding) resulting from the introduction or elimination of specific genes through modern molecular biology techniques. This technology is based on the use of a vector for transferring useful genetic information from a donor organism into a cell or organism that does not possess it. *See* gene cloning. A broader definition of genetic engineering also includes selective breeding and other means of artificial selection.

Genetic equilibrium—Condition in a group of interbreeding organisms in which the allele frequencies remain constant over time.

Genetic fingerprinting—A technique in which an individual's DNA is analysed to reveal the pattern of repetition of particular nucleotide sequences throughout the genome. *See* DNA fingerprint.

Genetic heterogeneity—The situation in which different mutant genes produce the same phenotype.

Genetic immunization—Delivery to a host organism of a cloned gene that encodes an antigen. After the cloned gene is expressed, it elicits an antibody response that protects the organism from infection by a virus, bacterium or other disease-causing organism.

Genetic information—Information contained in a nucleotide base sequence in chromosomal DNA or RNA.

Genetic map—The linear array of genes on a chromosome, based on recombination frequencies (linkage map) or physical location (physical or chromosomal map). *See* mapping.

Genetic mapping—Determining the linear order of genes and/or DNA markers along a chromosome. a.k.a. mapping.

Genetic marker—A DNA sequence used to 'mark' or track a particular location (locus) on a particular chromosome. *cf* marker gene.

Genetic resources conservation—*See* **gene conservation.**

Genetics—The science of heredity and variation.

Genetic selection—The process of selecting genes, cells, clones, etc., within populations or between populations or species. Genetic selection usually results in differential success rates of different genotypes, reflecting many variables, including selection pressure and genetic variability in populations.

Genetic transformation—The transfer of extracellular DNA among and between species by using bacterial or viral vectors.

Genetic variation—Differences between individuals attributable to differences in genotypes.

Gene tracking—Following the inheritance of a particular gene from generation to generation.

Gene translocation—The movement of a gene from one chromosomal location to another.

Genome—The entire complement of genetic material (genes + non-coding sequences) present in each cell of an organism, or in a virus or organelle.

GEO—Genetically engineered organism. *See* GMO.

Germ layers; primary germ layers—The layers of cells in an animal embryo at the gastrula stage, from which the various organs of the animal's body will be derived.

Germ line cells—Cells that produce gametes.

Germ line gene therapy—The delivery of a gene or genes to a fertilized egg or an early embryonic cell. The transferred gene(s) is present in all or some of the nuclei of the cells of the mature individual, including possibly the reproductive cells, and alters the phenotype of the individual that develops.

Germ line—A lineage of 'generative' cells (= germ track) ancestral to the gametes (sperm and egg cells proper) which, during the development of an organism (animal or plant), are set aside as potential gamete-forming tissues. These ancestral cells, together with the gametes, are called germ cells, as opposed to somatic cells. Location, nature and time of formation of potential gamete-forming tissues are species specific, and may vary greatly from one species to another.

GLP (Good Laboratory Practice)**; GMP** (Good Manufacturing Practice)—They are codes of practice designed to reduce to a minimum the chance of accidents which could affect a research project or a manufactured product. The GLP and GMP prescriptions are quite voluminous, but boil down to a few key points. The essential point of both GLP and GMP is that everything is recorded, and that only established procedures are used and by people who have been trained to use them.

Glucocorticoid—A steroid hormone that regulates gene expression in higher animals.

Glycolysis —(Gr. *glycos*, sugar (sweet) + *lysis*, dissolution)—Sequence of reactions that converts glucose into pyruvate, with the concomitant production of ATP.

Glycosylation—The covalent addition of sugar or sugar-related molecules to proteins or polynucleotides.

GM food Food that comprises, in whole or in part, material that has been modified by the application of recombinant DNA technology.

Growth factor—Any of various chemicals, particularly polypeptides, that have a variety of important roles in the stimulation of new cell growth and cell maintenance. They bind to the cell surface on receptors. Specific growth factors can cause new cell proliferation.

Growth hormone (GH); **somatotrophin**; **somatotropin**—A hormone, secreted by the mammalian pituitary gland, that stimulates protein synthesis and growth of the long bones in the legs and arms. It also promotes the breakdown and use of fats as an energy source, rather than glucose. Production of growth hormone is greatest during early life. Its secretion is controlled by the opposing actions of two hormones from the hypothalamus: somatocrinin (growth-hormone-releasing hormone), which promotes its release; and somatostatin (growth-hormone-inhibiting hormone), that inhibits the release of growth hormone from the anterior pituitary gland.

Growth inhibitor—Any substance inhibiting the growth of an organism. The inhibitory effect can range from mild inhibition (growth retardation) to severe inhibition or death (toxic reaction). Two plant growth regulators that may act as inhibitors are ethylene and abscisic acid. The concentration of the inhibitor, the length of exposure to it, and the relative susceptibility of the organisms exposed to the inhibitor, are all important factors which determine the extent of the inhibitory effect.

Growth phases; growth phase curve—The characteristic periods in the growth of a bacterial culture, as indicated by the shape of a graph of viable cell number versus time, namely: lag phase; logarithmic (or exponential) phase; stationary phase; death phase.

Growth rate—Increase in mass per unit of time. *See* growth.

Growth regulator—A synthetic or natural compound that at low concentrations elicits and controls growth responses in a manner similar to hormones.

Growth retardant—A chemical that selectively interferes with normal hormonal promotion of growth and other physiological processes, but without appreciable toxic effects.

Gynandromorph—An individual in which one part of the body is female and another part is male; a sex mosaic.

Gynogenesis—Female parthenogenesis; after fertilization of the ovum, the male nucleus is eliminated and the haploid individual (described as gynogenetic) so produced possesses the maternal genome only. *cf* androgenesis; parthenogenesis; anther culture.

Hairy root disease—A disease of broad-leaved plants, where a proliferation of root-like tissue is formed from the stem. Hairy root disease is a tumorous state similar to crown-gall, and is induced by the bacterium *Agrobacterium rhizogenes,* containing an Ri plasmid. *See Agrobacterium*, crown gall.

Halophyte—A plant that can tolerate a high concentration of salt in the growing medium.

Halothane—A volatile anaesthetic.

Hanging droplet technique—*See* microdroplet array.

Haploid (Gr. *haploos*, single + *oides*, like)—A cell or organism containing only one representative from each of the pairs of homologous chromosomes found in the normal diploid cell. Having a single complete set of chromosomes, or referring to an individual or generation containing such a single set of chromosomes per cell. Usually a gamete. *See* monoploid.

Haploid cell—A cell containing only one set, or half the usual (diploid) number, of chromosomes.

Haplotype—A group of alleles, each from a different locus in the same region of a chromosome, that exist in the same double helix.

Helix—Any structure with a spiral shape. The Watson and Crick model of DNA is in the form of a double helix.

Helminths—Parasitic worms, especially internal parasites of man and animals.

Helper cells— T cells that respond to an antigen displayed by a macrophage by stimulating B and T lymphocytes to develop into antibody-producing plasma cells and killer T cells, respectively.

Helper plasmid—A plasmid that provides a function or functions to another plasmid in the same cell.

HEPA filter (high efficiency particulate air filter)—A filter capable of screening out particles larger than 0.3 μm. HEPA filters are used in laminar air flow cabinets (hoods) for sterile transfer work. *See* prefilter.

Herbicide—Any substance that is toxic to plants; usually applied to agrochemicals intended to kill specific unwanted plants, such as weeds.

Herbicide resistance—The ability of a plant to withstand herbicide. Herbicide resistance has been one of the early targets of plant genetic engineering. If a herbicide is sprayed onto a field planted with such resistant crops, then all the plants except the crop would be killed, thus providing an effective method of weed control without having to develop herbicides specific to each weed type. There is substantial concern in some quarters about the widespread use of this technology, which is essentially giving the plant kingdom the ability to evade man's most effective herbicides. The concern are that, firstly, such engineering will lead to increased use of the herbicides, at a time when it is generally accepted that the use of chemicals should be kept as low as possible, and that, secondly, there is the possibility that resistant crop plants will escape to become weeds, or that their resistant genes could be transferred to other species, including weeds.

Heredity—Resemblance among individuals related by descent; transmission of traits from parents to offspring.

Heterogeneous nuclear RNA (hnRNA)—Large RNA molecules, which are unedited mRNA transcripts, or pre-mRNAs found in the nucleous of a eukaryotic cell. *See* RNA.

Heterokaryon—Cell with two or more different nuclei as a result of cell fusion. Opposite: homokaryon.

Heterologous—From a different source, as in heterologous DNA.

Heterologous probe—A DNA probe that is derived from one species and used to screen for a similar DNA sequence from another species.

Heterologous protein—*See* recombinant protein.

Heteroplasmy—A cellular condition in which two genetically different types of organelles are present. *cf* homoplasmy.

Histocompatibility complex; histocompatibility system—The collection of genes coding for peptides present on the surface of nucleated cells; these peptides are responsible for the differences between genetically non-identical individuals that cause rejection of tissue grafts between such individuals. These peptides were originally called histocompatibility antigens. They are now called histoglobulins, reflecting their structural similarity to immunoglobulins.

Homeobox—A DNA sequence found in several genes that are involved in the specification of organs in different body parts in animals; they are characteristic of genes that influence segmentation in animals. The homeobox corresponds to an amino acid sequence in the polypeptide encoded by these genes; this sequence is called the homeodomain.

Homologous—From the same source, or having the same evolutionary function or structure.

Homologous chromosomes—Chromosomes that occur in pairs and are generally similar in size and shape: one comes from the male parent and the other from the female. Such chromosomes contain the same linear array of genes.

Homologous recombination—The exchange of DNA fragments between two DNA molecules or chromatids of paired chromosomes (during crossing over) at the site of identical nucleotide sequences.

Homology—The degree of identity between individuals, or characters. The degree of identity between the nucleotide sequences of two nucleic acid molecules or the amino acid sequences of two protein molecules. Although sequence determination is the ultimate test of homology, useful estimates can be provided by either DNA-DNA or DNA-RNA hybridization.

Hybrid cell—The mononucleate cell which results from the fusion of two different cells, leading to the formation of a synkaryon.

Hybrid dysgenesis—A syndrome of abnormal germ-line traits, including mutation, chromosome breakage, and sterility, which results from activity of transposable elements.

Hybrid released translation—A method used to detect the proteins encoded by cloned DNA. The cloned DNA is bound to a nitrocellulose filter and a crude preparation of mRNA is hybridized to the filter-bound DNA. Only mRNA sequences homologous to the cloned DNA will be retained on the filter. These mRNA molecules can then be removed by high temperature or by using formamide. The purified mRNA is then placed in an *in vitro* translation system and the proteins encoded by the message can be analysed by electrophoresis through a polyacrylamide gel. *See* hybrid arrested translation.

Hybrid selection—The process of choosing individuals possessing desired characteristics from among a hybrid population.

Hybrid vigour; heterosis—The extent to which the performance of a hybrid in one or more traits is better than the average performance of the two parental populations.

Hybridoma—A hybrid cell, derived from a B (antibody producing) lymphocyte fused to a tumour cell, which grows indefinitely in tissue culture and is selected for the secretion of the specific antibody produced by that B cell.

Hydrate—A compound formed by the incorporation of water.

Hydrogen bond—A relatively weak bond formed between a hydrogen atom (which is covalently bound to a nitrogen or oxygen atom) and a nitrogen or oxygen atom with an unshared electron pair. Weak interactions between electro-negative atoms and hydrogen atoms (electro-positive) that are linked to other electro-negative atoms.

Hydrogen-uptake positive (Hup^+)—A term describing a micro-organism that is capable of assimilating (or taking up) hydrogen gas.

Hydrophobic interactions—Association of non-polar groups with each other when present in aqueous solutions because of their insolubility in water.

Hydroponics—The growing of plants in aerated water containing all the essential mineral nutrients, with no soil. Also called soilless gardening or cultivation.

3′-hydroxyl end—The hydroxyl group that is attached to the 3′ carbon atom of the sugar (ribose or deoxyribose) of the terminal nucleotide of a nucleic acid molecule.

Hyperploid—A genetic condition in which a chromosome or a segment of a chromosome is over-represented in the genotype. Opposite: hypoploid.

Hypervariable region—The parts of both the heavy and light chains of an antibody molecule that enable it to bind to a specific site on an antigen.

Hypervariable segment—A region of a protein that varies considerably between strains or individuals.

Hypochlorite—Generic term for aqueous solutions of sodium hypochlorite, potassium hypochlorite or calcium hypochlorite, which are oxidizing agents and used for disinfecting surfaces and surface-sterilizing tissues, and for bleaching.

Hypocotyl (Gr. *hypo*, under + *kotyledon*, a cup-shaped hollow)—Portion of an embryo or seedling below the cotyledons, which is a transitional area between stem and root.

Hypomorph—A mutation that reduces but does not completely abolish gene expression.

Hypoplastic—Reduction in plant growth or development (dwarfing, stunting) resulting from an abnormal condition associated with a disease or nutritional stress.

Hypoploid—A genetic condition in which a chromosome or segment of a chromosome is underrepresented in the genotype. Opposite: hyperploid.

Hypothalamic peptides—Peptides generated in the vertebrate forebrain and concerned with regulating the body's physiological state.

Hypothesis (Gr. *hypothesis*, foundation)—A tentative theory or supposition provisionally adopted to explain certain facts and to guide in the investigation of other facts. Once proven by rigorous scientific investigation, it becomes a theory or a law.

Immediate early gene—A viral gene that is expressed promptly after infection.

Immobilized cells—Cells entrapped in matrices such as alginate, polyacrylamide and agarose designed for use in membrane and filter bioreactors.

Immortalization—The genetic transformation of a cell type into a cell line which can proliferate indefinitely.

Immortalizing oncogene—A gene that upon transfection enables a primary cell to grow indefinitely in culture.

Immune response—The processes, including the synthesis of antibodies, that are used by vertebrates to respond to the presence of a foreign antigen. *See* primary immune response; secondary immune response.

Immunity—The state of relative insusceptibility of an animal or plant to infection by disease-producing organisms or to the harmful effects of their poisons.

Immunization—The production of immunity in an individual by artificial means. Active immunization involves the introduction, either orally or by infection, of specially treated bacteria, viruses or their toxins so as to stimulate the production of antibodies.

Immuno-affinity chromatography—A purification technique in which an antibody is bound to a matrix and is subsequently used to bind and separate a protein from a complex mixture. *cf* affinity chromatography.

Immunoglobulin—One of a group of proteins (globulins) in the body that act as antibodies. They are produced by specialized cells (B lymphocytes) and are present in blood serum and other body fluids.

Immunosuppression—The suppression of immune response. Immunosuppression is necessary following organ transplants in order to prevent the host rejecting the grafted organ.

Immunoassay; immunodiagnostics—An assay system which detects proteins by using an antibody specific to that protein. A positive result is seen as a precipitate of an antibody-protein complex. The antibody can be linked to a radioactive atom or to an enzyme which catalyses an easily monitored reaction such as a change in colour.

Immunochemical control—Use of immune agents to combat infections.

Immunodiagnostics—*See* immunoassay.

Immunogenicity—The ability to elicit an immune response.

Immunosensor—A biosensor having an antibody as biological part.

Immunosuppressor—A substance, an agent or a condition that prevents or diminishes the immune response.

Immunotherapy—The use of an antibody or a fusion protein containing the antigen binding site of an antibody to cure a disease or enhance the well-being of a patient.

Immunotoxin—Protein drugs consisting of an antibody joined to a toxin molecule. Immunotoxins can be made by linking toxin and antibody molecules chemically, or by fusing the genes for the toxin and the antibody. The antibody portion of the molecule facilitates binding to a target molecule or cell, and the toxin inactivates or kills the target molecule or cell.

Impeller—An agitator that is used for mixing the contents of a bioreactor.

Inactivated agent—A virus, bacterium or other organism that has been treated to prevent it from causing a disease.

Inbred line—The product of inbreeding, *i.e.*, the mating of individuals that have ancestors in common; in plants and laboratory animals, it refers to populations resulting from at least 6 generations of selfing or 20 generations of brother-sister mating, that are for all practical purposes, completely homozygous. In farm animals, the term is sometimes used to describe populations that have resulted from several generations of the mating of close relatives, without having reached complete homozygosity.

Inbreeding—Matings between individuals that have one or more ancestors in common, the extreme condition being self-fertilization, which occurs in many plants and some primitive animals.

Inbreeding depression—Reduction in vigour, yield, etc., of a population that is commonly seen as the level of inbreeding increases. The traits that show greatest inbreeding depression are those that are most closely associated with viability and reproductive ability.

Inclusion body—Protein that is overproduced in a recombinant bacterium and forms a crystalline array inside the bacterial cell.

Indirect embryogenesis—Embryo formation on callus tissues derived from zygotic or somatic embryos, seedling plant or other tissues in culture. *cf* direct embryogenesis.

Indirect organogenesis—Organ formation on callus tissues derived from explants.

Inducer—A low-molecular-weight compound or a physical agent that is bound by a repressor so as to produce a complex that can no longer bind to the operator; thus, the presence of the inducer turns on the expression of the gene(s) controlled by the operator.

Inducible—A gene or gene-product is said to be inducible if its transcription or synthesis is increased by exposure of the cells to an effector. Effectors are usually small molecules whose effects are specific to particular operons or groups of genes. *See* constitutive.

Inoculum (pl. inocula)— A small piece of tissue cut from callus, or an explant from a tissue or organ, or a small amount of cell material from a suspension culture, transferred into fresh medium for continued growth of the culture. *See also* minimum inoculum size.

***In situ* colony; *in situ* plaque hybridization**—A procedure for screening colonies or plaques growing on plates or membranes for the presence of specific DNA sequences by the hybridization of nucleic acid probes to the DNA molecules present in these colonies or plaques.

***In situ* conservation**—A conservation method that attempts to preserve the integrity of genetic resources by conserving them within the evolutionary dynamic ecosystems of their original habitat or natural environment. *cf ex situ* conservation.

Insulin—A peptide hormone secreted by the Langerhans islets of the pancreas, and that regulates the level of sugar in the blood.

Integrating vector—A vector that is designed to integrate cloned DNA into the host cell chromosomal DNA.

Integration—The recombination process which inserts a small DNA molecule (usually by homologous recombination) into a larger one. If the molecules are circular, integration involves only a single crossing-over; if linear, then two crossings-over are required. A well known example is the integration of phage λ (lambda) DNA into the *E. coli* genome.

Integration-excision (I/E) **region**—The portion of bacteriophage lambda (λ) DNA that enables bacteriophage lambda (λ) DNA to be inserted into a specific site in the *E. coli* chromosome and excised from this site.

Integument—One of the layers that enclosed the ovule, and is the precursor of the seed coat.

Intensifying screen—A plastic sheet impregnated with a rare-earth compound, such as calcium tungstate, which absorbs β radiation and emits light. When placed on one side of a piece of X-ray film with a radioactive sample on the other side, the intensifying screen will capture some of the β emissions which pass through the film, blackening the X-ray film and so greatly enhancing the sensitivity of the detection. An intensifying screen is used in southern and northern blotting procedures.

Interaction—In statistics, an effect that cannot be explained by the additive action of contributing factors; a departure from strict additivity.

Interference—Crossing over at one point that alters the chance of another crossing-over nearby; detected by studying the pattern of crossings-over with three or more linked genes. Interference is positive or negative depending on whether the chance of another crossing-over nearby is reduced or increased, respectively.

Interferon—A family of small proteins that stimulate viral resistance in cells.

Intergeneric—A cross between two different genera.

Intergenic regions—DNA sequences located between genes; they comprise a large percentage of a genome and have no known function.

Interleukin—A group of proteins that transmit signals between immune cells and are necessary for mounting normal immune responses.

Intracytoplasmic sperm injection (ICSI)—The injection, using micromanipulation, of a single sperm into the cytoplasm of a mature oocyte.

Intrageneric—Within a genus, such as a hybrid resulting from a cross between species within one genus.

Intragenic complementation—Complementation that occurs between two mutant alleles of a gene; common only when the product of the gene functions as a homomultimer.

Intraspecific—Within a species or its populations, including subspecies, such as an intraspecific cross, or variation.

Introgression—The introduction of new gene(s) into a population by crossing between two populations, followed by repeated backcrossing to that population while retaining the new gene(s).

Intron; intervening sequence—A segment of DNA sequence of a eukaryotic gene, not represented in the mature (final) mRNA transcript, because it is spliced out of the primary transcript before it can be translated; a process known as intron splicing. Some genes of higher eukaryotes contain a large number of introns, which make up the bulk of the DNA sequence of the gene. Introns are also found in genes whose RNA transcripts are not translated, namely eukaryotic rRNA and tRNA genes. In these cases the intron sequence does not appear in the functional RNA molecule. *cf* exon.

Invariant—Constant, unchanging; usually referring to the portion of a molecule that is the same across species.

Invasiveness—Ability of a plant to spread beyond its introduction site and become established in new locations, where it may have a deleterious effect on organisms already existing there.

Inversion—A chromosome re-arrangement that reverses the order of a linear array of genes in it.

Inverted repeat—Two regions of a nucleic acid molecule which have the same nucleotide sequence but in an inverted orientation

***In vitro* embryo production** (IVEP)—The combination of ovum pickup (q.v.), *in vitro* maturation (q.v.) of ova, and *in vitro* fertilization (q.v.). A potential means of overcoming the variability between donors in number of ova collected in embryo-transfer programmes.

***In vitro* fertilization** (IVF)—A widely used technique in human and animal science, whereby the egg is fertilized with sperm outside the body. Usually, the fertilized egg is cultured outside the body for a few days (to confirm that fertilization has occurred) before re-implantation into a female.

***In vitro* maturation** (IVM)—Culture of immature ova in the laboratory, usually until they are ready for *in vitro* fertilization.

***In vitro* transcription; cell-free transcription**—The specific and accurate synthesis of RNA in the test tube using purified DNA preparations as a template. So-called 'coupled systems' may be obtained from *E. coli* which carry out both mRNA synthesis and its translation into protein. For eukaryotes, separate cell-free systems have to be set up to demonstrate the activity of the three functionally distinct RNA polymerase complexes.

***In vitro* translation; cell-free translation**—The synthesis of proteins in the test-tube from purified mRNA molecules using cell extracts containing ribosomal sub-units, the necessary protein factors, tRNA molecules and aminoacyl tRNA synthetases, ATP, GTP, amino acids and an enzyme system for re-generating the nucleoside triphosphates. Prokaryotic translation systems are usually prepared from *E. coli* or the thermophilic bacterium, *Bacillus stearothermophilus.* Eukaryotic systems usually employ rabbit reticulocyte lysates or wheat germ.

In vivo (L. for 'in living')—The natural conditions in which organisms reside. Refers to biological processes that take place within a living organism or cell under normal conditions.

***In vivo* gene therapy**—The delivery of a gene or genes to a tissue or organ of an individual to alleviate a genetic disorder.

Ionizing radiation—The portion of the electromagnetic spectrum that results in the production of positive and negative charges (ion pairs) in molecules. X-rays and gamma rays are examples of ionizing radiation.

IPTG (Isopropyl-β-d-thiogalactopyranoside)—An inducer of the *lac* (lactose) operon. In recombinant DNA technology, IPTG is often used to induce cloned genes that are under the control of the *lac* repressor-*lac* promoter system.

Irradiation (L. *in*, into + *radius*, ray)— Exposure to any form of radiation.

IS element (insertion sequence)—A short (800-1,400 nucleotide pairs) DNA sequence found in bacteria that is capable of transposing to a new genomic location; other DNA sequences that are bounded by IS elements may also be transposed.

Isoalleles—Different forms of a gene that produce the same phenotype or very similar phenotypes.

Isochromosome—A chromosome with two identical arms and identical genes. The arms are mirror images of each other.

Isodiametric—Commonly used to describe cells with equal diameters.

Isotope—One of two or more forms of an element that have the same number of protons (atomic number) but differing numbers of neutrons (mass numbers). Radioactive isotopes are commonly used to make DNA probes and metabolic tracers.

Isozyme—A variant of a particular enzyme. In general, all the isozymes of a particular enzyme have the same function and sometimes the same activity, but they differ in amino-acid sequence. With the help of isozyme analysis, based on electrophoresis techniques that can separate the different variants, related species or cultivars can be distinguished.

Juvenile hormone—A hormone secreted by insects from a pair of endocrine glands close to the brain. It inhibits metamorphosis and maintains the larval features.

Kanamycin—An antibiotic of the aminoglycoside family that poisons translation by binding to the ribosomes.

Kanr—Kanamycin-resistance gene. *See* selectable marker.

Kappa chain—One of two classes of antibody light chains. *See* lambda chain.

Karyogamy—The fusion of nuclei or nuclear material that occurs during sexual reproduction. *See* fertilization.

Karyogram—A diagram representing the characteristic features of the chromosomes of a species.

Karyokinesis—The division of a cell nucleus. *See* meiosis; mitosis.

Karyotype—The chromosome constitution of a cell, an individual, or of a related group of individuals, as defined both by the number and the morphology of the chromosomes, usually in mitotic metaphase; chromosomes arranged in order of length and according to position of centromere; also, the abbreviated formula for the chromosome constitution, such as 47, XX + 21 for human trisomy–21 (Down's syndrome).

Kb—Kilobase pairs. *See* base pair; kilobase.

K_{cat}—The catalytic rate constant that characterizes an enzyme-catalysed reaction. The larger the k_{cat} value, the faster the conversion of substrate into product.

K_{cat}/K_m—The catalytic efficiency of an enzyme-catalysed reaction. The greater the value of k_{cat}/K_m, the more rapidly and efficiently the substrate is converted into product.

Killer T cells—T cells that carry T-cell receptors and kill cells displaying the recognized antigens.

Kilobase (kb)—A length unit equal to 1,000 base pairs of a double-stranded nucleic acid molecule. One kilobase of double-stranded DNA has a mass of about 660 kilodalton.

Kilobase pairs (kb)—Used interchangeably with kilobase.

Kilocalorie (kcal)—It is equal to 1,000 cal.

Kilodalton (kDa)—A unit of atomic mass equal to 1,000 daltons.

Kilojoule (kJ)—It is equal to 1 000 J; 1 kcal = 4.184 kJ.

Kinase—An enzyme that can transfer a phosphate from a high energy phosphate such as ATP, to an organic molecule.

Kinetics—Dynamic processes involving motion.

Kinetin—(Gr. *kinetikos*, causing motion)—One of the cytokinins, a group of growth regulators that characteristically promote cell division in plants.

Kinetochores—Two parallel bushlike filaments situated in the centromere of a chromosome.

Kinetosome—Granular cytoplasmic structure which forms the base of a cilium or flagellum. *cf* basal body.

Kinin—A substance promoting cell division. In plant systems, the prefix cyto- has been added (cytokinin) to distinguish it from kinin in animal systems.

Klenow fragment—A product of proteolytic digestion of the DNA polymerase I from *E. coli*; it has both polymerase and 3´-exonuclease activities but not 5´-exonuclease activity.

K_m—A dissociation constant that characterizes the binding of an enzyme to a substrate. The smaller the value of K_m, the tighter the binding of the enzyme to the substrate. Also called the Michaelis constant.

Knockout—An animal resulting from an embryonic stem cell in which a normal functional gene has been replaced by a non-functional form of the gene. This technique is used extensively in mice - much can be learned about the function of a gene by studying the phenotype of animals that lack the peptide product of that gene.

Leaf margin—The edge of a leaf.

Leaf primordium (L. *primordium*, a beginning)—A lateral outgrowth from the apical meristem, which will become a leaf when fully developed and expanded.

Leaf roll—Virus diseases characterized by symptomatic curling of the host's leaves.

Leaf scar—Mark left on a stem after leaf abscission.

Leaflet—Expanded leaflike part of a compound leaf.

Lectin—Any of a group of proteins, derived from plants, that can bind to specific oligosaccharides on the surface of cells, causing cells to clump together.

LINEs (long interspersed nuclear elements)—Families of long (average length = 6,500 bp), moderately repetitive (about 10,000 copies). LINEs are cDNA copies of functional genes present in the same genome; also known as processed pseudo-genes.

Linkage—The tendency of non-allelic genes to be inherited together more than would be expected if they were assorting independently. Linkage exists between two loci when they are located sufficiently close on the same chromosome that some gametes are produced without crossing-over occurring between the two loci.

Linkage map—A linear or circular diagram that shows the relative positions of genes on a chromosome as determined by recombination fraction.

Lipid (Gr. *lipos*, fat)—Any of a group of fats or fat-like compounds insoluble in water and soluble in fat solvents.

Lipofection—Delivery into eukaryotic cells of DNA, RNA or other compounds that have been encapsulated in an artificial phospholipid vesicle.

Lipopolysaccharide (LPS)—A compound containing lipid bound to a polysaccharide; often a component of microbial cells walls.

Liposome—A microscopic artificial membrane vesicle consisting of a spherical phospholipid bilayer. Liposomes can be incorporated into living cells and used to transport relatively toxic drugs into diseased cells, where they can exert their maximum effect. DNA molecules may be entrapped in, or bound to the surface of, the vesicles, and subsequent fusion of the liposome with the cell membrane will deliver the DNA into the cell. Liposomes have been used to develop an efficient transfection procedure for *Streptomyces* bacteria.

Liquefaction—Enzymatic digestion (often by a-amylase) of gelatinized starch to form lower molecular weight polysaccharides.

Liquid media—Media without a solidifying agent.

Liquid membranes—Thin films made up of liquids (as opposed to solids) which are stable in another liquid (usually water). Thus the liquid must not dissolve in the water, but nevertheless must be prevented from collapsing into a lot of small droplets.

Liquid nitrogen—Nitrogen gas condensed to a liquid with a boiling point of about –196°C. Very commonly the medium in which containers of genetic material are stored. *See* cryopreservation.

Litmus—A pH indicator paper (range 4.5-8.3) impregnated with an extracted lichen pigment. It turns red in acidic and blue in alkaline solutions. However, the use of litmus paper as an indicator is not a precise method of pH measurement.

Live vaccine—A living, non-virulent form of a micro-organism or virus that is used to elicit an antibody response that will protect the inoculated organism against infection by a virulent form of the micro-organism or virus. Also a living, non-virulent micro-organism or virus that express a foreign antigenic protein and is used to inoculate humans or animals. The latter organisms are also called a live recombinant vaccines.

Locus (pl: loci)—A site on a chromosome.

Long terminal repeat (LTR)—A string of bases that occurs at each end of the 'genome' of a retrovirus that has become integrated into the host genome. Involved in the integration process.

Luteinizing hormone—A pituitary hormone which causes growth of the yellow body of the ovary and also stimulates activity of the interstitial cells of the testis.

Lymphocyte—A general class of white blood cells that are important components of the immune system of vertebrates.

Lymphokine—Generic name for proteins that are released by lymphocytes to act on other cells involved in the immune response. The term includes interleukins and interferons.

Lymphoma—Cancer originating in the lymph nodes, spleen and other lympho-reticular sites.

Lyase—Any of a class of enzymes that catalyse either the cleavage of a double bond and the addition of new groups to a substrate, or the formation of a double bond.

Lyophilize—Rapid freezing followed by a dehydration under high vacuum. The process is lyophilization. *cf*—freeze-drying.

Lysis (Gr. *lysis*, a losing)—The destruction or breakage of cells either by viruses or by chemical or physical treatment.

Lysogen—A bacterial cell whose chromosome contains integrated viral DNA.

Lysogenic—A type or phase of the virus life cycle during which the virus integrates into the host chromosome of the infected cell, remaining essentially dormant for some period of time. *See* lysogen.

Lysogenic bacteria—Bacteria harbouring temperate (non-virulent, lysogenic) bacteriophages.

Lysogeny—A condition in which a bacteriophage genome (pro-phage) survives within a host bacterium, either as part of the host chromosome or as part of an extrachromosomal element, and does not initiate lytic functions.

Lytic—A phase of the virus life cycle during which the virus replicates within the host cell, releasing a new generation of viruses when the infected cell lyses.

Lytic cycle—The steps in viral production that usually lead to cell lysis.

Lysosome—A membrane-bound sac within the cytoplasm of animal cells that contains enzymes responsible for the digestion of material in food vacuoles, the dissolution of foreign particles entering the cell and, on the death of the cell, the breaking down of all cell structures. The digestive system of the cell.

M13 strand—The single-stranded DNA molecule that is present in the infective form of bacteriophage M13.

Macromolecule—Molecule of large molecular weight, such as proteins, nucleic acids and polysaccharides.

Macronutrient (Gr. *makros*, large + L. *nutrire*, to nourish)—For growth media: an essential element normally required in concentrations >0.5 millimole/l.

Macrophages—Large, white blood cells that ingest foreign substances and display on their surfaces antigens produced from the foreign substances, to be recognized by other cells of the immune system.

Macropropagation—Production of plant clones from growing parts.

Major histocompatibility antigen—A cell-surface macromolecule that allows the immune system to distinguish foreign or 'non-self' from 'self'. A better term is histoglobulin (*See* histocompatibility antigen). These are the antigens that must be matched between donors and recipients during organ and tissue transplants to prevent rejection.

Major histocompatibility complex—The large cluster of genes that encode the major histocompatibility antigens in mammals.

Malignant—Having the properties of cancerous growth.

Malt extract—A mixture of organic compounds from malt, used as a culture medium adjunct. *See* organic complex; undefined.

Malting—A process of generating starch-degrading enzymes in grain by allowing it to germinate in a humid atmosphere. *See* brewing.

Mammary glands—The milk-producing organs of female mammals, which provide food for the young.

Mammary tumours—Tumours of the milk glands.

Management of farm animal genetic resources—In AnGR: The sum total of technical, policy and logistical operations involved in understanding (characterization), using and developing (utilization), maintaining (conservation), accessing, and sharing the benefits of animal genetic resources. (Source: FAO, 1999)

Mannitol ($C_6H_{14}O_6$; f.w. 182.17)—A sugar alcohol widely distributed in plants. Mannitol is commonly used as a nutrient and osmoticum (q.v.) in suspension medium for plant protoplasts.

Marker—An identifiable DNA sequence that facilitates the study of inheritance of a trait or a gene. Such markers are used in mapping the order of genes along chromosomes and in following the inheritance of particular genes: genes closely linked to the marker will generally be inherited with it. Markers must be readily identifiable in the phenotype, for instance by controlling an easily observable feature (such as eye colour) or by being readily detectable by molecular means, *e.g.*, microsatellite markers (q.v.). *See* gene tracking.

Marker-assisted introgression—The use of DNA markers to increase the speed and efficiency of introgression (q.v.) of a new gene or genes into a population. The markers will be closely linked to the gene(s) in question.

Marker-assisted selection (MAS)—The use of DNA markers to increase the response to selection in a population. The markers will be closely linked to one or more quantitative trait loci (q.v.).

Marker gene—A gene of known function and known location on the chromosome. *cf* genetic marker.

Marker peptide—A portion of fusion protein that facilitates its identification or purification.

Metacentric chromosome—A chromosome with the centromere near the middle and, consequently, two arms of about equal length.

Metallothionein—A protective protein that binds heavy metals such as cadmium and lead.

Metaphase (Gr. *meta*, after + *phasis*, appearance)—Stage of mitosis during which the chromosomes, or at least the kinetochores, lie in the central plane of the spindle. It is the stage following prophase and preceding anaphase.

Metastasis—The spread of cancer cells to previously unaffected organs.

Methionine—A sulphur-containing amino acid.

Methylation—The addition of a methyl group (—CH_3) to a macromolecule, such as the addition of a methyl group to specific cytosine and, occasionally, adenine residues in DNA.

Micro-carriers—Small particles used as a support material for cells, and particularly mammalian cells, which are too fragile to be pumped and stirred as bacterial cells are in a large-scale culture.

Microdroplet array; multiple drop array (MDA); hanging droplet technique—Introduced by Kao and Konstabel (1970), this technique is used to evaluate large numbers of media modifications, employing small quantities of medium into which are placed small numbers of cells. Droplets of liquid are arranged on the lid of a Petri dish, inverted over the bottom half of the dish containing a solution with a lower osmotic pressure, and the dish is sealed. The cells or protoplasts form a monolayer at the droplet meniscus and can easily be examined.

Micro-element—An element required in very small quantities.

Micro-encapsulation—A process of enclosing a substance in very small sealed capsules from which material is released by heat, solution or other means.

Micro-environment (Gr. *mikros*, small + O.F. *environ*, about)—The environment close enough to the surface of a living or non-living object to be influenced by it.

Microfibrils (Gr. *mikros*, small + *fibrils*, diminutive of fibre)—Microfibrils are exceedingly small fibres visible only at the high magnification of the electron microscope.

Micropropagation—Miniaturized *in vitro* multiplication and/or regeneration of plant material under aseptic and controlled environmental conditions on specially prepared media that contain substances necessary for growth; used for three general types of tissue: excised embryos (= embryo culture); shoot-tips (= meristem culture or mericloning); and pieces of tissue that range from bits of stems to roots. Four stages of plant tissue culture have been defined by Murashige:

Stage I. Establishment of an aseptic culture.
Stage II. The multiplication of propagules.
Stage III. Preparation of propagules for successful transfer to soil (rooting and hardening).
Stage IV. Establishment in soil.

Microsatellite—A form of VNTR (q.v.). Specifically, a segment of DNA characterized by the occurrence of a variable number of copies (from a few up to 30 or so) of a sequence of around 5 or fewer bases (called a repeat unit, q.v.). A typical microsatellite is the repeat unit AC, which occurs at approximately 1,00,000 different sites in a typical mammalian genome. At any one site (locus), there are usually several different 'alleles,' each identifiable according to the number of repeat units. These alleles can be detected by PCR (q.v.), using primers designed from the unique sequence that is located on either side of the microsatellite. When the PCR product is run on an electrophoretic gel, alleles are seen to differ in length in units equal to the size of the repeat unit, *e.g.*, if the primers correspond to the unique sequence immediately on either side of the microsatellite and are each 20 bases long, and an individual is heterozygous for an AC microsatellite with one allele comprising 5 repeats and the other comprising 6 repeats, the heterozygote will exhibit two bands on the gel, one band being 20 + (2 × 5) + 20 = 50 bases long, and the other allele being 20 + (2 × 6) + 20 = 52 bases long. Microsatellites have been the standard DNA marker: they are easily detectable by PCR, and they tend to be evenly located throughout the genome. Thousands have been mapped in many different species.

Mini-prep—A small-scale (mini-) preparation of plasmid or phage DNA. Used to analyse DNA in a cloning vector after a cloning experiment.

Minisatellite—A form of VNTR (q.v.) in which the repeat units (q.v.) typically range from 10 to 100 bases. They are usually detected by Southern hybridization (q.v.), using a probe comprising a clone of the repeat unit. The first DNA fingerprints (q.v.) were minisatellites detected in this way. Minisatellites tend to be located at the ends of chromosomes and in regions with a high frequency of recombination.

Minitubers—Small tubers (5-15 mm in diameter) formed on shoot cultures or cuttings of tuber-forming crops, such as potato.

Mismatch—The lack of a complementary pair of bases in a double helix of DNA, *e.g.*, A:C, G:T.

Mismatch repair—DNA repair processes that correct mismatched base pairs.

Missense mutation—A mutation that changes a codon for one amino acid into a codon specifying another amino acid.

Mist propagation—Application of fine droplets of water to leafy cuttings in the rooting stage to reduce transpiration. *cf* fog.

Molality—The number of moles of solute per litre of solvent.

Molarity—The number of moles of a substance contained in a kilogram of solution. *See* mole.

Mole (symbol: M)—Amount of substance that has a weight in grams numerically equal to the molecular weight of the substance. Also called gram molecular weight. A mole contains 6.023×10^{23} molecules or atoms of a substance.

Molecular biology—The area of knowledge concerned with the molecular aspects of organisms and their cells.

Molecular cloning—The biological amplification of a specific DNA sequence through mitotic division of a host cell into which it has been transformed or transfected. *See* cloning.

Molecular genetics—The area of knowledge concerned with the genetic aspects of molecular biology, especially with DNA, RNA and protein molecules.

Molecule (L. diminutive of *moles,* a little mass)—A unit of matter, the smallest portion of an element or a compound that retains chemical identity with the substance in mass. The molecule usually consist of a union of two or more atoms; some organic molecules containing a very large number of atoms.

Monoclonal antibody (mAB) A single type of antibody that is directed against a specific epitope (antigen, antigenic determinant) and is produced by a single clone of B cells or a single hybridoma cell line, which is formed by the fusion of a lymphocyte cell with a myeloma cell. Some myeloma cells synthesize single antibodies naturally.

Morphogen—A substance that stimulates the development of form or structure in an organism.

Morphogenesis—The development, through growth and differentiation, of form and structure in an organism.

Morphogenic response—The effect on the developmental history of a plant or its parts exposed to a given set of growth conditions or to a change in the environment.

Morphology (Gr. *morphe,* form + *logos,* discourse)—The science of studying form and its development.

Multiple ovulation and embryo transfer (MOET)—A technology by which a single female that usually produces only one or two offspring can produce a litter of offspring. Involves stimulation of a female to shed large numbers of ova; natural mating or artificial insemination; collection of fertilized ova (either surgically, or non-surgically through the cervix); and transfer (usually non-surgical, through the cervix) of these fertilized ova to recipient females.

Multivalent vaccine—A single vaccine that is designed to elicit an immune response either to more than one infectious agent or to several different epitopes of a molecule.

Mutable genes—Genes with an unusually high mutation rate.

Mutagen—An agent or process which is capable of inducing a mutation, such as UV light. *cf* mutation.

Mutagenesis—Change(s) in the genetic constitution of a cell through alterations to its DNA.

Mutant—An organism or an allele that differs from the wild type because it carries one or more genetic changes in its DNA. A mutant organism may carry mutated gene(s) (= gene mutation); mutated chromosome(s) (= chromosome mutation); or mutated genome(s) (= genome mutation). a.k.a. a variant.

Mutation—(L. *mutare,* to change)—A sudden, heritable change appearing in an individual as the result of a change in the structure of a gene (= gene mutation); changes in the structure of chromosomes (= chromosome mutation); or in the number of chromosomes (= genome mutation). cf genetic diversity; genetic drift.

Mutation pressure—A constant mutation rate that adds mutant genes to a population; repeated occurrences of mutations in a population.

Mycoprotein—Fungal protein.

Mycotoxin—Toxic substance of fungal origin, such as aflatoxin.

Northern blot—A cellulose or nylon membrane to which RNA molecules have been attached by capillary action. The transferred RNA is hybridized to single-stranded DNA probes. northern blot technique is often used to measure expression (transcription) of a gene for which a specific cDNA is available for use as a probe. *See* southern blot, western blot.

Northern blotting—Similar to southern blotting, except that RNA is transferred onto a matrix and the presence of a specific RNA molecule is detected by DNA-RNA hybridization. *See* northern hybridization.

Northern hybridization—Hybridization of a labelled DNA probe to RNA fragments that have been transferred from an agarose gel to a nitrocellulose filter.

Nucellus (L. *nucella*, a small nut)—Tissue composing the chief part of the young ovule in which the embryo sac develops; megasporangium.

Nuclear transfer—A technology by which animals are created by cloning a single diploid somatic cell. It involves taking a single diploid cell from a culture of cells, and inserting it into an enucleated ovum, *i.e.*, an ovum from which the haploid nucleus has been removed. The resultant diploid ovum develops into an embryo that is placed in a recipient female, which gives birth to the cloned animal in the normal manner. Note that the term is somewhat of a misnomer, since it is a whole cell that is transferred, not just the nucleus.

Nuclease—A class of enzymes that degrade DNA or RNA molecules by cleaving the phospho-diester bonds that link adjacent nucleotides. In deoxyribonuclease (DNase), the substrate is DNA. In endonuclease, it cleaves at internal sites in the substrate molecule. Exonuclease progressively cleaves from the end of the substrate molecule. In ribonuclease (RNase), the substrate is RNA. In the S1 nuclease, the substrate is single-stranded DNA or RNA. Nucleases have varying degrees of base-sequence specificity, the most specific being the restriction endonucleases.

Nucleic acid—A macromolecule composed of phosphoric acid, pentose sugar, and organic bases. The two nucleic acids, deoxyribonucleic acid (DNA) and ribonucleic acid (RNA), are made up of long chains of molecules called nucleotides (q.v.). They were first isolated as part of a protein complex in 1871, and were separated from the protein moiety in 1889. *See* DNA; RNA; nucleotides.

Nullisomy (adj: nullisomic)—An otherwise diploid cell or organism lacking both members of a chromosome pair (chromosome formula $2n - 2$). *See also* disomy.

Nurse culture—Planting a cell from a suspension culture on a raft of filter paper above a callus tissue piece (nurse tissue). The filter paper serves to prevent tissue union but allows the flow of essential substances from the nurse to the isolated cell. In such culture, a piece of callus is first placed on nutrient agar. Over this tissue is laid a strip of filter paper.

Nutrient cycle—The passage of a nutrient or element through an ecosystem, including its assimilation and release by various organisms and its transformation into various organic or inorganic chemical forms.

Nutrient deficiency—Absence or insufficiency of some factor needed for normal growth and development.

Nutrient film technique (NFT)—Hydroponic technique used to grow plants. NFT delivers a film of water or nutrient solution either continuously or through on-off cycles (*e.g.*, on 8 minutes and off 7 minutes).

Nutrient gradient—A diffusion gradient of nutrients and gases that develops in tissues where only a portion of the tissue is in contact with the medium. Gradients are less likely to form in liquid media than in callus cultures.

Nutrient medium (pl: nutrient media)—A solid, semi-solid or liquid combination of: major and minor salts; an energy source (sucrose); vitamins; plant growth regulators; and occasionally other defined or undefined supplements. Often made from stock solutions, then sterilized by autoclaving or filtering through a micropore filter.

Oestrogen; estrogen—The generic term for a group of female sex hormones which control the development of sexual characteristics and control oestrus.

Oestrous cycle (from oestrus)—The cycle of reproductive activity shown by most sexually mature non-pregnant female mammals.

Oestrus (adj: oestrous)—In female mammals, the period of sexual excitement and acceptance of the male. a.k.a. rut, heat.

Offset—Young plant produced at the base of a mature plant.

Offshoot—Short, usually horizontal, stem produced near the crown of a plant.

Offspring; progeny (both same in plural)—New individual organisms that result from the process of sexual or asexual reproduction.

Okazaki fragment—Since DNA can be replicated in only one direction, *i.e.*, nucleotides can be added only at the 3′ end, only one of the two strands of a double helix can be replicated continuously. The other strand is replicated in small segments (Okazaki fragments) that are subsequently joined together by DNA ligase. *See* primosome.

Open reading frame (ORF)—A sequence of nucleotides in a DNA molecule that has the potential to encode a peptide or protein: it starts with a start triplet (ATG), is followed by a string of triplets each of which encodes an amino acid, and ends with a stop triplet (TAA, TAG or TGA). This term is often used when, after the sequence of a DNA fragment has been determined, the function of the encoded protein is not known. The existence of open reading frames is usually inferred from the DNA (rather than the RNA) sequence.

Operator—The region of DNA that is upstream from a gene or genes and to which one or more regulatory proteins (repressor or activator) binds to control the expression of the gene(s).

Operon—A functionally integrated genetic unit for the control of gene expression in bacteria. It consists of one or more genes that encode one or more polypeptide(s) and the adjacent site (promoter and operator) that controls their expression by regulating the transcription of the structural genes.

Opine—The condensation product of an amino acid with either a keto-acid or a sugar, produced by the crown gall tissues. Opine synthesis is a unique characteristic of tumour cells.

Ordinate—The vertical axis of a graph. Opposite: absciss; abscissa.

Organ—A tissue or group of tissues that constitute a morphologically and functionally distinct part of an organism.

Organ culture—The aseptic culture of complete living organs of animals and plants outside the body in a suitable culture medium. Animal organs must be small enough to allow the nutrients in the culture medium to penetrate all the cells.

Organellar genes—Genes located on organelles outside the nucleus.

Organelle—A membrane-bounded specialized region within a cell, such as the mitochondrion or dictyosome, that carries out a specialized function in the life of a cell.

Organic—Referring in chemistry to compounds containing carbon, many of which have been in some manner associated with living organisms.

Organic complex—A chemically undefined compound added to nutrient media to stimulate growth, *e.g.*, coconut milk; yeast extract; casein hydrolysate.

Organic co-solvent—A compound used to dissolve some neutral organic substances, such as in media preparation. Organic co-solvents include alcohols (usually ethanol), acetone and dimethylsulphoxide (DMSO).

Organic evolution—The process by which changes in the genetic composition of populations of organisms occur in response to environmental changes.

Organized growth—The development under tissue culture conditions of organized explants (meristem tips or shoot tips, floral buds or organ primordia). *See* unorganized growth.

Organized tissue—Composed of regularly differentiated cells.

Organizer—An inductor; a chemical substance in a living system that determines the fate in development of certain cells or groups of cells.

Overlapping reading frames—Start triplets in different reading frames generate different polypeptides from the same DNA sequence. *See* reading frame.

Ovulation—In mammals, the process of escape of the ovum (egg cell) from the ovary.

Ovule (L. *ovulum*, diminutive of *ovum*, egg)—The part of the reproductive organs in seed plants that consists of the nucellus, the embryo sac and integuments.

Ovum (L. *ovum*, egg; pl. ova)— A gamete of female animals, produced by the ovary. The oosphere in plants.

Palindrome (Gr. *palindromos*, running back again)—Originally a word, sentence or verse that reads the same from right to left as it does from left to right. In biotechnology, the term is applied to a segment of DNA in which the base-pair sequence reads the same in both directions from a central point of symmetry. *See* the entry for *inverted repeat* for an example of a palindrome. *See also* palindromic sequence.

Palindromic sequence—A segment of duplex DNA whose 5´-to-3´ sequence is identical on each DNA strand. The sequence is the same when one strand is read left-to-right and the other strand is read right-to-left. Typically, recognition sites for type II restriction endonucleases are palindromes. *See* the entry for *inverted repeat* for an example of a palindrome.

Palisade parenchyma—Elongated cells found just beneath the upper epidermis of leaves, and containing many chloroplasts.

Particle radiation—Refers to gamma (g) particles (positively charged) and beta (β) particles (negatively charged), electrons, protons and neutrons. In plant tissue culture, these particles are used to produce mutant cells or organisms.

Parts per million (ppm)—Units of any given substance per one million equivalent units, such as the weight units of solute per million weight units of solution (*i.e.*, 1 ppm = 1 mg/l).

Pathogen (Gr. *pathos*, suffering + *genesis*, beginning)—An organism that causes a disease in another organism.

Pathogen-free—Freedom from disease-causing organisms (bacteria, fungi, viruses, etc.).

Pathotoxin—Very dilute substance synthesized and released by some pathogen, and which interacts with the host metabolism.

Pathovar—In plant pathology, strains of bacteria causing disease in specific plant cultivars.

PBR322—One of the first plasmid vectors widely used; especially used for cloning DNA in *E. coli*. *See* plasmid.

PH—A measure of acidity and alkalinity. Equal to the log of the reciprocal of the hydrogen ion concentration of a solution, expressed in grams per litre. A reading of 7 is neutral (*e.g.*, pure water), whereas below 7 is acid and above 7 is alkaline.

Phosphatase—An enzyme that hydrolyses esters of phosphoric acid, removing a phosphate group from an organic compound.

Phospho-diester bond—A bond in which a phosphate group joins adjacent carbons through ester linkages. A condensation reaction between adjacent nucleotides results in a phospho-diester bond between 3´ and 5´ carbons in DNA and RNA.

Phospholipase A2—An enzyme which degrades type A2 phospholipids.

Phospholipid—A class of lipid molecules in which a phosphate group is linked to glycerol and two fatty acyl groups. A major component of biological membranes. *See* inositol lipid.

Phosphorolysis—The cleavage of a bond by orthophosphate; analogous to hydrolysis referring to cleavage by water.

Polyacrylamide gel electrophoresis (PAGE)—A method for separating nucleic acid or protein molecules according to their molecular size. The molecules migrate through the inert gel matrix under the influence of an electric field. In the case of protein PAGE, detergents such as sodium dodecyl sulphate (SDS) are often added to ensure that all molecules have a uniform charge. Secondary structure can often lead to the anomalous migration of molecules. Therefore it is common to denature protein samples by boiling them prior to PAGE. In the case of nucleic acids, denaturing agents such as formamide, urea or methyl mercuric hydroxide are often incorporated into the gel itself, which may also be run at high temperature. PAGE is used to separate the products of DNA-sequencing reactions and the gels employed are highly denaturing, since molecules differing in size by a single nucleotide must be resolved.

Polyacrylamide gels—Often referred to incorrectly as acrylamide gels. These gels are made by cross-linking acrylamide with *N,N'*-methylene-*bis*-acrylamide. Polyacrylamide gels are used for the electrophoretic separation of proteins, DNA and RNA molecules. Polyacrylamide beads are also used as molecular sieves in gel chromatography, marketed as Bio-gelTM.

Polyadenylation—Post-transcriptional addition of a polyadenylic acid tail to the 3´ end of eukaryotic mRNAs. Also called poly-(A) tailing. The adenine-rich 3´ terminal segments is called a poly (A) tail.

Polyclonal antibody—A serum sample that contains a mixture of immunoglobulin molecules secreted against a specific antigen, each recognizing a different epitope, some antibodies of which bind to different antigenic determinants of one antigen.

Polyembryony—In the ordinary course of events, one embryo is formed in each ovule and is derived from the fertilization of the ovum in the solitary embryo sac. However, two or more embryos could start development even though only one may reach maturity.

Polyethylene glycol (PEG); **carbowax**—A polymer having the general formula: $HOCH_2(CH_2OCH_2)XCH_2OH$ and available in a range of molecular weights from *ca* 1,000 to *ca* 6,000. PEG 4,000 and PEG 6,000 are commonly used to promote cell or protoplast fusion, and to facilitate DNA uptake in the transformation of organisms such as yeast. PEG is also used to concentrate solutions by withdrawing water from them.

Polygene—One of many genes of small effect influencing the development of a quantitative trait; results in continuous variation and in quantitative inheritance. *See* gene.

Polygenic—Controlled by many genes of small effect.

Polylinker—A segment of DNA that contains a number of different restriction endonuclease sites. a.k.a. multiple cloning site (MCS).

Polymer—A compound composed of many identical smaller sub-units; resulting from the process of polymerization.

Polymerase chain reaction (PCR)—A procedure that amplifies a particular DNA sequence. It involves multiple cycles of ***denaturation***, ***annealing*** to oligonucleotide primers, and ***extension*** (polynucleotide synthesis), using a thermostable DNA polymerase, deoxyribonucleotides, and primer sequences in multiple cycles of denaturation-renaturation-DNA synthesis. *See* polymerase.

Polymerase—An enzyme that catalyses the formation of polymeric molecules from monomers. A DNA polymerase synthesizes DNA from deoxynucleoside triphosphates using a complementary DNA strand and a primer. An RNA polymerase synthesizes RNA from monoribonucleoside triphosphates and a complementary DNA strand. *See* poly-(A) polymerase; polymerase chain reaction; RNA polymerase; *taq* polymerase.

Polymerization—Chemical union of two or more molecules of the same kind such as glucose or nucleotides to form a new compound (starch or nucleic acid) having the same elements in the same proportions but a higher molecular weight and different physical properties.

Polymery—The phenomenon whereby a number of genes at different loci (which may be polygenes) can act together to produce a single effect.

Polymorphism—The occurrence of two or more alleles at a locus in a population. a.k.a. genetic polymorphism.

Polynucleotide—A chain of nucleotides in which each nucleotide is linked by a single phospho-diester bond to the next nucleotide in the chain. They can be double- or single-stranded. The term is used to describe DNA or RNA. *See* nucleotide.

Polypeptide—A linear molecule composed of two or more amino acids linked by covalent (peptide) bonds. They are called dipeptides, tripeptides and so forth, according to the number of amino acids present.

Polyploid (Gr. *polys*, many + *ploid*, fold)—Tissue or cells with more than two complete sets of chromosomes, that results from chromosome replication without nuclear division or from union of gametes with different number of chromosome sets, hence triploid ($3x$), tetraploid ($4x$), pentaploid ($5x$), hexaploid ($6x$), heptaploid ($7x$), octoploid ($8x$)).

Polysaccharide (Gr. *polys*, many + Gr. *sakcharon*, sugar) Long-chain molecules, such as starch and cellulose, composed of multiple units of a monosaccharide.

Polysaccharide capsules—Carbohydrate coverings with antigenic specificity that are present on some types of bacteria.

Polyspermy—The entry of several sperm into the egg during fertilization, although only one sperm nucleus actually fuses with the egg nucleus.

Polyvalent vaccine—A recombinant organism into which antigenic determinants have been cloned from a number of different disease-causing organisms, and used as a vaccine. *See* vaccine.

Polyvinylpyrrolidone (PVP) ($(C_6H_9NO)_n$)—An occasional constituent of plant tissue culture isolation media. PVP is of variable molecular weight and has antioxidant properties, so is used to prevent oxidative browning of explanted tissues. PVP is less frequently used as an osmoticum in culture media.

Population density—Number of cells or individuals per unit. The unit could be an area or volume of medium.

Population genetics—The branch of genetics that deals with frequencies of alleles and genotypes in breeding populations. *cf.* quantitative genetics.

Population—A local group of organisms belonging to the same species and interbreeding.

Porcine endogenous retrovirus (PERV)—The provirus (q.v.) of a porcine retrovirus. With the increasing interest in the use of pig organs for xenotransplantation to humans, there has been concern that PERVs could be activated after transplantation, creating an infection in the human recipient.

Positional candidate gene—A gene known to be located in the same region as a DNA marker that has been shown to be linked to a single-locus trait or to a quantitative trait locus (QTL) (q.v.) and whose function suggests that it could be the source of genetic variation in the trait in question.

Positional cloning—The process that commences with searching for markers linked to a particular inherited trait, then using those markers to identify the approximate location of the gene responsible for the trait, and then using various cloning strategies (including chromosome walking, q.v.) to identify, isolate and characterize the gene. Originally, the strategy was called *reverse genetics*, a term that some consider to be a misnomer and misleading.

Position effect—The situation in which a change in phenotype results from the change of the position of a gene or group of genes.

Positive control system—A mechanism in which the regulatory protein(s) is required to turn on gene expression.

Preventive immunization; vaccination—Infection with an antigen to elicit an antibody response that will protect the organism against future infections.

Pribnow box—Consensus sequence near the mRNA start-point of prokaryotic genes.

Primary immune response—The immune response that occurs during the first encounter of a mammal with a given antigen. *cf* secondary immune response.

Primary meristem—Meristem of the shoot or root tip giving rise to the primary plant body.

Primary tissue—A tissue that has differentiated from a primary meristem.

Primary transcript—The RNA molecule produced by transcription prior to any post-transcriptional modifications; also called a pre-mRNA in eukaryotes.

Primer—A short DNA or RNA fragment annealed to a template of single-stranded DNA, providing a 3′ hydroxyl end from which DNA polymerase extends a new DNA strand to produce a duplex molecule.

Primordium A group of cells which gives rise to an organ.

Primosome—A protein-replication complex that catalyses the initiation of synthesis of Okazaki fragments during discontinuous replication of DNA. It involves DNA primase and DNA helicase activities.

Prion—*See* proteinaceous infectious particle.

Probability—Statistics: The frequency of occurrence of an event.

Proband—The individual in a family in whom an inherited trait is first identified.

Probe—For diagnostic tests, the agent that is used to detect the presence of a molecule in a sample. A DNA or RNA sequence labelled or marked with a radioactive isotope or that is used to detect the presence of a complementary sequence by hybridization with a nucleic acid sample.

Probe DNA—A labelled DNA molecule used to detect complementary-sequence nucleic acid molecules by molecular hybridization. To localize the probe DNA sequence and reveal the complementary hybridization sequence, autoradiography or fluorescence is used. *See* multilocus probe; nucleotide.

Promoter—A nucleotide sequence of DNA to which RNA polymerase binds and initiates transcription. It usually lies upstream of (5′ to) a coding sequence. A promoter sequence aligns the RNA polymerase so that transcription will initiate at a specific site. A chemical substance that enhances the transformation of benign cells into cancerous cells. *See* constitutive promoter.

Promoter sequence—A regulatory DNA sequence that initiates the expression of a gene.

Pro-nuclear micro-injection—The initial method of transgenesis in animals, involving injecting many copies of a particular gene into one of the two pro-nuclei of a fertilized ova. Characterised by a very low success rate. As animal cloning (q.v.) becomes more common, pro-nuclear micro-injection will be replaced by micro-injection of cloned genes into a culture of cloned embryos produced by nuclear transfer, which can be tested for expression of the transgene before transfer to recipient females. *See* transgenesis.

Pseudo-gene—A copy (duplicate) of a functional gene that, as a result of mutation following duplication, can no longer function.

PUC—A widely used expression plasmid, containing a galactosidase gene. *See* plasmid.

Pulsed-field gel electrophoresis (PFGE)—A procedure used to separate very large DNA molecules by alternating the direction of electric current in a pulsed manner across a semisolid gel.

Quantitative genetics—The area of genetics concerned with the inheritance of continuously-varying traits. Most practical improvement programs involve the application of quantitative genetics.

Quantitative inheritance—Inheritance of measurable traits (height, weight, colour intensity, etc.) that depend on the cumulative action of many genes.

Quantitative trait—A measurable trait that shows continuous variation; a trait that can not be classified into a few discrete classes.

Quantitative trait locus (QTL)— A locus that affects a quantitative trait. The plural form (quantitative trait loci) is also abbreviated as QTL.

Quantum (L. *quantum*, how much)—An elemental unit of energy. Its energy value is hv, where h, Planck's constant, is 6.62×10^{-27} erg-second and v is the frequency of the vibrations or waves with which the energy is associated. *cf* photon.

Quarantine (It. *quarantina*, from *quaranta,* forty)—Originally, keeping a person or living organism in isolation for a period (originally 40 days) after arrival to allow disease symptoms to appear, if there was any disease present. Now used for regulations restricting the sale or shipment of living organisms, usually to prevent disease or pest invasion of an area.

Quiescent—Quiet, at rest, but not necessarily dormant, and having the potential for resumed activity; can apply to non-meristematic cells.

Random amplified polymorphic DNA (RAPD; pronounced 'rapid')—A technique using single, short (usually 10-mer) synthetic oligonucleotide primers for PCR. The primer, whose sequence has been chosen at random, initiates replication at its complementary sites on the DNA, producing fragments up to about 2 kb long, which can be separated by electrophoresis and stained with ethidium bromide. A primer can exhibit polymorphism between individuals, and polymorphic fragments can be used as markers.

Reading frame—A series of triplets beginning from a specific nucleotide. Each triplet is represented by a single amino acid in the protein synthesized. The reading frame defines which sets of three nucleotides are read as triplets in the DNA, and hence as codons in the corresponding mRNA; this is determined by the initiation codon, AUG. Thus the sequence AUGGCAAAAUUUCCC would read as AUG/GCA/AAA/UUU/CCC/ and not as A/UGC/CAA/AAU/UUC/CC.

Depending on where one begins, each DNA strand contains three different reading frames. *See* open reading frame, overlapping reading frames.

Recognition site—A nucleotide sequence - composed typically of 4, 6 or 8 nucleotides - that is recognized by and to which a restriction endonuclease (restriction enzyme) binds. For type II restriction enzymes (those used in gene-cloning experiments) it is also the sequence within which the enzyme specifically cuts (and their corresponding enzymes methylate) the DNA, *i.e.*, for type II enzymes, the recognition site and the target site are the same sequence. Type I enzymes bind to their recognition site and then cleave the DNA at some more or less random position outside that recognition site. *cf* restriction site.

Recombinant DNA technology—A set of techniques which enable one to manipulate DNA. One of the main techniques is DNA cloning (because it produces an unlimited number of copies of a particular DNA segment), and the result is sometimes called a DNA clone or gene clone (if the segment is a gene), or simply a clone. An organism manipulated using recombinant DNA techniques is called a genetically modified organism (GMO).

Recombinant protein—A protein whose amino acid sequence is encoded by a cloned gene.

Recombinant RNA—A term used to describe RNA molecules joined *in vitro* by T4 RNA ligase.

Recombinant toxin—A single multifunctional toxic protein that has been created by combining the coding regions of various genes.

Recombinant vaccine—A vaccine produced from a cloned gene.

Recombination—The process of crossing over, which occurs during meiosis I. It involves breakage in the same position of each of a pair of non-sister chromatids from homologous chromosomes, followed by joining of non-sister fragments, resulting in a reciprocal exchange of DNA between non-sister chromatids within an homologous pair of chromosomes, **recombination fraction; recombination frequency**. The proportion of gametes that have arisen from recombination between two loci. It is estimated as the number of recombinant individuals among a set of offspring of a particular mating, divided by the total number of offspring from that mating. Represented by the Greek letter theta (θ). Linkage maps are created from estimates of recombination fraction between all pair-wise combinations of loci. *See* map distance.

Replicative form (RF)—The molecular configuration of viral nucleic acid that is the template for replication in the host cell.

Replicon—The portion of a DNA molecule which is replicable from a single origin. Plasmids and the chromosomes of bacteria, phages and other viruses usually have a single origin of replication and, in these cases, the entire DNA molecule constitutes a single replicon. Eukaryotic chromosomes have multiple internal origins and thus contain several replicons. The word is often used in the sense of a DNA molecule capable of independent replication, *e.g.*, 'The shuttle vector pJDB219 is a replicon in both yeast and *E. coli*.'

Replisome—The complete replication apparatus present at a replication fork that carries out the semi-conservative replication of DNA.

Reporter gene—A gene that encodes a product that can readily be assayed. Thus reporter genes are used to determinate whether a particular DNA construct has been successfully introduced into a cell, organ or tissue.

Repressible enzyme—An enzyme whose synthesis is diminished by a regulatory molecule.

Repression—Inhibition of transcription by preventing RNA polymerase from binding to the transcription initiation site: a repressed gene is 'turned off.'

Repressor—A protein which binds to a specific DNA sequence (the operator) upstream from the transcription initiation site of a gene or operon and prevents RNA polymerase from commencing mRNA synthesis. Examples of repressors are the C_1 protein of bacteriophage and the *lac*1 protein of the *lac* operon.

Reproduction—The production of an organism, cell or organelle like itself (self propagation).

Restriction endonuclease [enzyme]—A class of endonucleases that cleaves DNA after recognizing a specific sequence, *e.g.*, *Bam*H1 (5′GGATCC3′), *Eco*RI (5′GAATTC3′), and *Hind*III (5′AAGCTT3′). There are three types of restriction endonuclease enzymes:

Type I: Cuts non-specifically a distance greater than 1,000 bp from its recognition sequence and contains both restriction and methylation activities.

Type II: Cuts at or near a short, and often palindromic (q.v.), recognition sequence. A separate enzyme methylates the same recognition sequence. They may make the cuts in the two DNA strands exactly opposite one another and generate blunt ends, or they may make staggered cuts to generate sticky ends. The type II restriction enzymes are the ones commonly exploited in recombinant DNA technology.

Type III: Cuts 24-26 bp downstream from a short, asymmetrical recognition sequence. Requires ATP and contains both restriction and methylation activities.

Restriction fragment—A fragment of DNA produced by cleaving (digesting, cutting) a DNA molecule with one or more restriction endonucleases.

Restriction fragment length polymorphism (RFLP)—The occurrence of variation in the length of DNA fragments that are produced after cleavage with a type II restriction endonuclease. The differences in DNA lengths are due to the presence or absence of recognition site(s) for that particular restriction enzyme. RFLPs were initially detected using hybridization with DNA probes after separation of digested genomic DNA by gel electrophoresis (Southern analysis). Now they are typically detected by electrophoresis of digested PCR product.

Restriction map—The linear array of restriction endonuclease sites on a DNA molecule. *See* mapping.

Restriction nuclease—A bacterial enzyme that cuts DNA at a specific site.

Restriction site—The specific nucleotide sequence in DNA that is recognized by a type II restriction endonuclease and within which it makes a double-stranded cut. Restriction sites usually comprise four or six base pairs that typically are palindromic (q.v.), *e.g.*, 5´GGCC3´

3´CCGG5´ The two strands may be cut either opposite to one another, to create blunt ends, or in a staggered manner, giving sticky ends, depending on the enzyme involved. *See* restriction endonuclease.

Ribozyme; gene shears—RNA molecule that can catalyse chemical reactions, often cutting other RNAs.

Ribulose—A keto-pentose sugar ($C_5H_{11}O_5$) that is involved in carbon dioxide fixation in photosynthesis.

Ribulose biphosphate (RuBP)—A five-carbon sugar that is combined with carbon dioxide to form a six-carbon intermediate in the first stage of the dark reaction of photosynthesis.

Rinderpest—Cattle plague; a viral infection of cattle, sheep and goats.

Ri plasmid—A class of large conjugative plasmids found in the soil bacterium *Agrobacterium rhizogenes*. Ri plasmids are responsible for hairy root disease of certain plants. A segment of the Ri plasmid is found in the genome of tumour tissue from plants with hairy root disease.

RNase—Ribonuclease. A group of enzymes that catalyse the cleavage of nucleotides in RNA.

Roentgen (symbol: r)—Obsolete unit of ionizing radiation. The SI unit is the sievert (symbol: Sv; 1 Sv ≈ 8.4 r)

Root—The descending axis of a plant, normally below ground, which serves to anchor the plant and to absorb and conduct water and mineral nutrients.

Root apex—The apical meristem of a root; very similar to the shoot apical meristem in that it forms the three meristematic areas: the protoderm (developing into the epidermis); the procambium (which develops into the stele); and the growth meristem (which forms the cortex).

Root cap—A thimblelike mass of cells covering and protecting the apical meristem of a root.

Root culture—The culture of isolated root tips of apical or lateral origin to produce *in vitro* root systems with indeterminate growth habits. Root culture was among the first kinds of plant tissue cultures, and is still largely used in the study of developmental phenomena, and mycorrhizal, symbiotic and plant-parasitic relationships.

Root cutting—Cutting made from sections of roots alone.

Root hairs —Outgrowths from epidermal cell walls of the root specialized for water and nutrient absorption.

Root nodule—A small round mass of cells that is located on the roots of plants and contains nitrogen-fixing bacteria.

Root tuber—Thickened root that stores carbohydrates.

Root zone—The volume of soil or growing medium containing the roots of a plant. In soil science, the depth of the soil profile in which roots are normally found. *See* rhizosphere.

Rootstock—The trunk or root material to which buds or scions are inserted in grafting.

Ruminant animals; ruminants—Animals having a rumen - a large digestive vat in which fibrous plant material is partially broken down by microbial fermentation, prior to digestion in a 'true' stomach (the abomasum). There are also two other stomachs - the reticulum and the omasum. Typical ruminants are cattle and sheep.

Runner—A lateral stem that grows horizontally along the ground surface and gives rise to new plants either from axillary or terminal buds. *cf* stolon.

Rust—A generic descriptor for various plant diseases, especially those caused by a group of parasitic fungi of the phylum *Basidiomycota*, that attack the leaves and stems of crops.

S_1 nuclease—An enzyme that specifically degrades RNA or single-stranded DNA to 5´ mononucleotides. Purified from the filamentous fungus *Aspergillus oryzae*, S_1 nuclease is used in assessing the extent of a hybridization reaction by removing unpaired regions. It is also used to remove the sticky ends of restriction fragments. In S_1 mapping, the coding region of a gene is detected by performing mRNA-DNA hybridization and removing unpaired DNA with S_1 nuclease.

Saccharifaction—Following liquefaction, the hydrolysis of polysaccharides by glucoamylase to maltose and glucose.

Salt tolerance; saline resistance—The ability to withstand a concentration of sodium (Na^+ ion), or of any other salt, in the soil (or in culture), which is damaging or lethal to other plants. Breeding and selection for increased tolerance and resistance in crop plants is of great current interest.

Sap—Fluid content of the xylem and phloem cells of plants. Fluid contents of the vacuole are referred to as cell sap.

Satellite DNA—That portion of the DNA in plant and animal cells consisting of highly repetitious sequences (millions of copies) typically in the range from 5 to 500 bases. Thousands of copies occur tandemly (end-on-end) at each of many sites. It can be isolated from the rest of the DNA by density gradient centrifugation.

Satellite RNA—A small, self-splicing RNA molecule that accompanies several plant viruses, including tobacco ringspot virus. a.k.a. viroid.

Scaffold—The central core structure of condensed chromosomes. The scaffold is composed of non-histone chromosomal proteins.

Scale up—Conversion of a process, such as fermentation of a micro-organism, from a small scale to a larger scale.

Selectable marker—A gene whose expression allows the identification of: A specific trait or gene in an organism. 2. Cells that have been transformed or transfected with a vector containing the marker gene. *e.g.*, β-lactamase, kanr.

Shine-Dalgarno sequence—A conserved sequence in prokaryotic mRNAs that is complementary to a sequence near the 5´ terminus of the 16S ribosomal RNA and is involved in the initiation of translation. *See* ribosomal binding site.

Shoot—A young branch that grows out from the main stock of a tree, or the young main portion of a plant growing above ground.

Shoot differentiation—The development of growing points, leaf primordia and finally shoots from a shoot tip, axial bud, or even a callus surface.

Single-cell line; cell strain—A culture initiated from a single cell, usually from suspension cultures of single cells or small aggregates plated on solidified medium. The latter may incorporate a selective agent, from which tolerant or resistant individual cell lines or cell clones can be selected. *See* selective agent.

Single-cell protein (SCP)—Protein produced by micro-organisms. The dried mass of a pure sample of a protein-rich-micro-organism, which may be used either as feed (for animals) or as a food (for humans).

Site-directed mutagenesis—The introduction of base changes - mutations - into a piece of DNA at a specific site, using recombinant DNA methods.

Site-specific—A term used to describe any process or enzyme which acts at a defined sequence within a DNA or RNA molecule. Type II restriction enzymes are site-specific endonucleases and the recombination systems encoded by some transposons are site-specific, such as is the integration of phage into the *E. coli* chromosome.

Site-specific mutagenesis—A technique to change one or more specific nucleotides within a cloned gene in order to create an altered form of a protein with one or more specific amino acid changes. a.k.a. oligonucleotide-directed mutagenesis; oligonucleotide-directed site-specific mutagenesis.

Stamen (L. *stamen*, the standing-up things or a tuft of thready things) Flower structure made up of an anther (pollen-bearing portion) and a stalk or filament. The stamen is the male part of the flower.

Standard deviation—A statistical measure of variability in a population of individuals or in a set of data; the square root of the variance.

Standard error—A statistical measure of variation in a population of means, used to indicate how well sample estimates represent population parameters.

Stop codon; termination codon—A set of three nulecotides for which there is no corresponding tRNA molecule to insert an amino acid into the polypeptide chain. Protein synthesis is hence terminated and the completed polypeptide released from the ribosome. Three stop codons are found: UAA (ochre), UAG (amber) and UGA (opal). Mutations which generate any of these three codons in a position which normally contains a codon specifying an amino acid are known as *nonsense mutations*. Stop codons can also be called nonsense codons. *See* chain terminator; nonsense mutation; suppressor.

Strain—A group of individuals from a common origin within a species.

Strain isolation—Isolation of any bacterium, animal or plant from the outside world.

Stratification—Treatment of moist seeds at low temperature (+2° - +4°C) to break dormancy.

Stringency—Reaction conditions - notably temperature, salt concentration(s) and pH - that dictate the annealing of single-stranded DNA/DNA, DNA/RNA and RNA/RNA hybrids. At high stringency, duplexes form only between strands with perfect one-to-one complementarity; lower stringency allows annealing between strands with some degree of mismatch between bases.

Stringent plasmid—A plasmid that only replicates along with the main bacterial chromosome and is present as a single copy, or at most several copies, per cell. *See* plasmid.

Stroma—Tissue that forms the framework of an organ.

Structural gene—A DNA sequence that forms the blueprint for the synthesis of a polypeptide.

Structure-functionalism—The scientific tradition that stresses the relationship between a physical structure and its function, *e.g.*, the related disciplines of anatomy and physiology.

Synergids—The two nuclei within the embryo sac at the upper end in the ovule of the flower, which, with the third (the egg), constitute the egg apparatus.

Synkaryon—A nucleus formed by the fusion of nuclei from two different somatic cells during somatic-cell hybridization. *See* hybrid cell.

Synteny—The occurrence of two or more loci on the same chromosome, without regard to the distance between them.

Transformation—1. The uptake and establishment of DNA in a bacterium or yeast cell, in which the introduced DNA often changes the phenotype of the recipient organism.

2. Conversion by various means of animal cells in tissue culture from controlled to uncontrolled cell growth. Typically through infection by a tumour virus or transfection with an oncogene. *See* transformant; transformation efficiency.

Transformation efficiency—The number of cells that take up foreign DNA as a function of the amount of added DNA; expressed as transformants per microgram of added DNA. *See* transformation.

Transformation frequency—The fraction of a cell population that takes up foreign DNA; expressed as the number of transformed cells divided by the total number of cells in a population.

Transforming oncogene—A gene that upon transfection converts a previously immortalized cell to the malignant phenotype. *See* oncogene.

Transgene—A gene from one genome that has been incorporated into the genome of another organism. Often refers to a gene that has been introduced into a multicellular organism.

Transgenesis—The introduction of a gene or genes into animal or plant cells, which leads to the transmission of the input gene (transgene) to successive generations. *cf* transfection.

Transgenic—An organism in which a foreign gene (a transgene) is incorporated into its genome. The transgene is present in both somatic and germ cells, is expressed in one or more tissues, and is inherited by offspring in a Mendelian fashion. *See* transgenic animal; transgenic plant.

Transgressive variation—The appearance in the F_2 (or later) generation of individuals showing more extreme development of a trait than either of the original parents.

Trans heterozygote—A double heterozygote that contains two mutations arranged in the *trans* configuration.

Transient—Of short duration.

Transition—The substitution in DNA or RNA of one purine by another purine, or of one pyrimidine by another pyrimidine.

Transition stage—The integration period of juvenile and reproductive stages of growth.

Transition-state intermediate—In a chemical reaction, an unstable and high-energy configuration assumed by reactants on the way to making products. Enzymes are thought to bind and stabilize the transition state, thus lowering the energy of activation needed to drive the reaction to completion.

Translation—The process of polypeptide synthesis in which the amino acid sequence is determined by mRNA, mediated by tRNA molecules, and carried out on ribosomes.

Translocation (L. *trans*, across + *locare*, to place)—The movement of nutrients or products of metabolism from one location to another. Change in position of a segment of a chromosome to another, non-homologous chromosome.

Transmission electron microscope (TEM)—A microscope which uses an electron beam to obtain images of objects, with a much greater resolving power than a light microscope. *See* electron microscope.

Transplant— noun: A plant grown in a cold frame, greenhouse, tissue culture or indoors for later planting outdoors. To dig up and move a plant to another location.

Transposable genetic element—A DNA element that can move from one location in the genome to another. *See* transposon.

Transposase—An enzyme encoded by a transposon gene and that catalyses the movement of a DNA sequence to a different site in a DNA molecule, by catalysing the excision of the transposon from one site and its insertion into a new chromosomal site.

Transposition—The process whereby a transposon or insertion sequence inserts itself into a new site on the same or another DNA molecule. The exact mechanism is not fully understood and different transposons may transpose by different mechanisms. Transposition in bacteria does not require extensive DNA homology between the transposon and the target DNA. The phenomenon is therefore described as illegitimate recombination.

Transposon—A transposable or movable genetic element. A relatively small DNA segment that has the ability to move (mobile genetic element) from one chromosomal position to another, *e.g.*, Tn 5 is a bacterial transposon that carries the genes for resistance to the antibiotics neomycin and kanamycin and the genetic information for insertion and excision of the transposon.

Transposon tagging—The insertion of a transposable element into or nearby a gene, thereby marking that gene with a known DNA sequence.

Trans test—*See* complementation test.

Transversion—The substitution of a purine for a pyrimidine, or a pyrimidine for a purine.

Tribrid protein—A fusion protein that has three segments, each encoded by parts of different genes.

Trichome (Gr. *trichoma*, a growth of hair)—A short filament of cells.

Tri-hybrid—The offspring from homozygous parents differing in three pairs of genes.

Tri-nucleotide repeats—Tandem repeats of three nucleotides that are present in many genes. In several cases, these trinucleotide repeats have undergone expansions in copy number, and that has resulted in inherited diseases.

Tripartite mating—A process in which conjugation is used to transfer a plasmid vector to a target cell when the plasmid vector is not self-mobilizable. When (1) cells that have a plasmid with conjugative and mobilizing functions are mixed with (2) cells that carry the plasmid vector and (3) target cells, mobilizing plasmids enter the cells with the plasmid vector and mobilize the plasmid vector to enter into the target cells. Following tripartite mating, the target cells with the plasmid vector are separated from the other cell types in the mixture by various selection procedures.

Triplet—A sequence of three nucleotides of DNA which specifies an amino acid. The elucidation of the genetic code involved the binding of charged tRNA species to chemically synthesized ribonucleotide triplets.

Triploid—A cell or organism containing three times the haploid number of chromosomes.

Trisomy (adj: trisomic) (Gr. *treis*, three + *soma*, body)—An otherwise diploid cell or organism that has an extra chromosome of one pair (chromosome formula: 2n + 1). *cf* disomy; monosomy.

Triticale—The species formed by crossing wheat and rye.

Tropism (Gr. *trope*, a turning)—An involuntary plant response to a stimulus, in which a bending, turning or growth occurs, such as phototropism, geotropism or hydrotropism. The response may be positive (towards)—or negative (away from) to the stimulus.

True-to-type—Applied to a plant or propagation source, this term denotes correct cultivar identification and lack of variation in productivity or performance. Verification is determined visually by an expert or through biochemical, serological or other means.

Trypsin—A proteolytic enzyme that hydrolyzes peptide bonds on the carboxyl side of the amino acids arginine and lysine.

Trypsin inhibitors—Substances inactivating the enzyme trypsin, which is needed for digestion of peptides.

Tuber—Food-storing modified roots in plants like potato.

Tubulin—The major protein component of the microtubules of eukaryotic cells.

Tumble tube—A glass tube mainly used *in vitro* to agitate and consequently aerate suspension cultures. The tube, which is commonly attached to a slowly revolving platform, is closed at both ends, with a side-neck opening.

Tumour virus—A virus capable of transforming a cell to a malignant phenotype. *See* virus.

Tunica—An outer one- to four-cell layered region of the apical meristem, where cell division is anticlinal, *i.e.*, perpendicular to the surface. *See* apical meristem.

Turbidostat—An open continuous culture in which a pre-selected biomass density is uniformly maintained by automatic removal of excess cells. The fresh medium flows in response to an increase in the turbidity (usually cell density) of the culture.

Turgid (L. *turgidus*, swollen, inflated)—Swollen, distended; referring to a cell that is firm due to water uptake.

Turgor pressure (L. *turgor*, a swelling)—The pressure within the cell resulting from the absorption of water into the vacuole and the imbibition of water by the protoplasm.

Turion—An underground bud or shoot from which an aerial stem arises. *cf* sucker.

Twins—Two individuals originating from the same zygote.

Ultraviolet light; ultraviolet radiation (UV)—The portion of the electromagnetic spectrum with wavelengths from about 100 to 400 nm; between ionizing radiation (X-rays) and visible light. UV is absorbed by DNA and is highly mutagenic to unicellular organisms and to the epidermal cells of multicellular organisms. UV light is used in tissue culture for its mutagenic and bactericidal properties.

Undefined—A medium or substance added to medium in which not all of the constituents or their concentrations are chemically defined, such as media containing coconut milk, malt extract, casein hydrolysate, fish emulsion or other complex compounds. *cf* organic complex.

Understock—Host plant for a grafted scion, a branch or shoot from another plant; an understock may be a fully grown tree or a stump with a living root system.

Undifferentiated—In a meristematic state or resembling a meristem; lacking the specialized or differential gene expression characteristic of specialized cells.

Unequal crossing over—Crossing over between repeated DNA sequences that have paired out of register, creating duplicated and deficient products.

Unicellular—Describing tissues, organs or organisms consisting of a single cell.

Unisexual—Describing animals and plants possessing either male or female reproductive organs, but not both.

Univalent—An unpaired chromosome at meiosis.

Universal donor cells—Cells that, after introduction into a recipient, will not induce an immune response that leads to their rejection.

Universality—Referring to the genetic code, the codons have the same meaning, with minor exceptions, in virtually all species.

Unorganized growth—*In vitro* formation of tissues with few differentiated cell types and no recognizable structure; many call are unorganized. *cf* organized growth.

Upstream—In molecular biology, the stretch of DNA base pairs that lie in the 5′ direction from the site of initiation of transcription. Usually the first transcribed base is designated +1 and the upstream nucleotides are marked with minus signs, *e.g.*, −1, −10. Also, to the 5′ side of a particular gene or sequence of nucleotides.

Upstream processing—In chemical engineering, those phases of a manufacturing process that precede the biotransformation step. Refers to the preparation of raw materials for a fermentation process. Also called upstream processing.

Uracil—A pyrimidine base found in RNA but not in DNA. In DNA, uracil is replaced by thymine.

Utilization of farm animal genetic resources— The use and development of animal genetic resources for the production of food and agriculture. The use in production systems of AnGRs that already possess high levels of adaptive fitness to the environments concerned, and the deployment of sound genetic principles, will facilitate sustainable development of the AnGRs and the sustainable intensification of the production systems themselves. The wise use of AnGRs is possible without depleting domestic animal diversity. Development of AnGRs includes a broad mix of ongoing activities that must be well planned and executed for success, and compounded over time, hence with high value. It requires careful definition of breeding objectives, and the planning, establishment and maintenance of effective and efficient animal recording and breeding strategies.

Vaccine—A preparation of dead or weakened pathogens, or of derived antigenic determinants, that is used to induce formation of antibodies or immunity against the pathogen.

Vaccinia—The cowpox virus used to vaccinate against smallpox and, experimentally, as a carrier of genes for antigenic determinants cloned from other disease organisms.

Vacuole (L. diminutive of *vacuus*, empty)—A cavity in a plant cell, bounded by a membrane; in which various plant products and by-products are stored.

Vacuum—Vacuum is created through use of a vacuum pump or aspirator pump, to facilitate specific biological preparations, such as inclusions or disinfection of material for *in vitro* culture, etc.

Variable domains—Regions of antibody chains that have different amino acid sequences in different antibody molecules. These regions are responsible for the antigen-binding specificity of the antibody molecule.

Variable expressivity—Variation in the phenotype caused by different alleles of the same gene and/or by the action of other genes and/or by the action of non-genetic factors. *cf* expressivity.

Variable surface glycoprotein (VSG)—One of a battery of antigenic determinants expressed by a micro-organism to elude immune detection.

Variance—In statistics: The sum of the squared deviations, divided by one less than the number of observations. A statistical measure of variation in a population.

Variant—An organism that is genetically different from the wild type organism. a.k.a. mutant.

Variation—Differences between individuals within a population or among populations.

Variegated—Plants having both green and albino tissues. This difference in colour may result from viral infection, nutritional deficiency, or may be under genetic or physiological control.

Variety—A naturally occurring subdivision of a species, with distinct morphological characters and given a Latin name according to the rules of the International Code of Nomenclature. A taxonomic variety is known by the first validly published name applied to it so that nomenclature tends to be stable. *cf* cultivar; pathovar.

Vascular (L. *vasculum*, a small vessel)—Referring to any plant tissue or region consisting of or giving rise to conducting tissue, *e.g.*, bundle, cambium, ray.

Vascular bundle; fascicle—A strand of tissue containing primary xylem and primary phloem (and procambium if present) and frequently enclosed by a bundle sheath of parenchyma or fibres.

Vascular cambium—In biennials and perennials, cambium giving rise to secondary phloem and secondary xylem.

Vascular plants—Plants possessing organized vascular tissues.

Vascular system—A specialized network of vessels for the circulation of fluids throughout the body tissue of an animal. The system of vascular tissue in plants.

Vascular tissue—The tissue that conducts water and nutrients throughout the plant body in higher plants.

Vector (L. *vehere*, to carry)—An organism, usually an insect, that carries and transmits disease-causing organisms. A plasmid or phage that is used to deliver selected foreign DNA for cloning and in gene transfer. *See* Ti plasmid.

Vegetative propagation—Same as asexual or non-sexual propagation.

Vehicle—The host organism used for the replication or expression of a cloned gene or other sequence The term is little used and is often confused with vector. *See* vector.

Velocity density gradient centrifugation—A procedure used to separate macromolecules based on their rate of movement through a density gradient.

Velogenetics—The combined use of marker-assisted selection (q.v.) and embryo technologies such as OPU (q.v.), IVM (q.v.) and IVF (q.v.), in order to increase the rate of genetic improvement in animal populations.

Vermiculite—Coarse aggregate material made from expanded mica having a high cation exchange capacity and high water-holding capacity and used as a rooting medium and a soil additive.

Vernalization—The process by which floral induction in some plants is promoted by exposing the plants to chilling for a certain duration.

Vessel (L. *vasculum*, a small vessel)—A series of xylem elements whose function is to conduct water and nutrients in plants. A container, such as a Petri dish or test tube, used for tissue culture.

Vessel element—A type of cell occurring within the xylem of flowering plants. Many are water-conducting vessels.

Viability—The capability to live and develop normally.

Viability test—Assay of the number or percent of living cells or plants in a population that have been given a specific treatment; such as ascertaining cell viability after cryopreservation treatment, or for seeds.

Viable—Capable of germinating, living, growing and developing.

Vibrio—Comma-shaped bacterium.

Viral vaccines—Vaccines consisting of live viruses rather than dead ones or separated parts of viruses. However, as the virus itself cannot be used, because that would simply give the patient the disease, the virus is genetically engineered so that it elicits the immune response to the viral pathogen without causing the disease itself.

***Vir* genes**—A set of genes on a Ti plasmid that prepare the T-DNA segment for transfer into a plant cell.

Viral coat protein—Protein present in the outer layer of a virus.

Viral oncogene—A gene in the viral genome that contributes to malignancies in vertebrate hosts. *See* oncogene.

Viral pathogen—A disease-causing virus.

Virion—An infectious virus particle. A plant pathogen that consists of a naked RNA molecule of approximately 250-350 nucleotides, whose extensive base pairing results in a nearly correct double helix. *See* satellite RNA.

Viroid—An infectious entity similar to a virus but smaller, consisting only of a strand of nucleic acid without the protein coat characteristic of a virus. *See* satellite RNA.

Virulence (L. *virulentia*, a stench)—The degree of ability of an organism to cause disease. The relative infectiousness of a bacterium or virus, or its ability to overcome the resistance of the host metabolism.

Virulent phage—A phage (virus) that destroys the host (bacterium).

Virus (L. *virus*, a poisonous or slimy liquid)—An infectious particle composed of a protein capsule and a nucleic acid core (DNA or RNA), which is dependent on a host organism for replication. The DNA or a double-stranded DNA copy of an RNA virus genome is integrated into the host chromosome during lysogenic infection or replicated during the cystic cycle. *See* coat protein; DNA; genome; host; nucleic acid; prion; RNA; tumour virus; viroid.

Virus-free—Plant, animal, cell, tissue or meristem which exhibits no viral symptoms or contains no identifiable virus-particles.

Virus-tested—Description of a organism or a cell stock certified as being free of certain specified viruses following recognized procedures of virus diagnosis.

Visible light—The part of the electromagnetic spectrum with wavelengths between 380 nm and 750 nm and perceived by the human eye.

Vitamin B complex—A large group of water soluble vitamins that function as co-enzymes, including thiamine (B_1); riboflavin or vitamin G (B_2); niacin or nicotinic acid (B_3); pantothenic acid (B_5); pyridoxine (B_6); cyanocobalamin (B_{12}); biotin or vitamin H; folic acid or vitamin M (Bc); inositol; choline; and others.

Vitamin C— ascorbic acid.

Vitamin H— biotin;

Vitamins (L. *vita*, life + amine)—Naturally occurring organic substances required by living organisms in relatively small amounts to maintain normal health, and which are added to tissue culture media to enhance growth, usually acting as enzyme co-factors.

Vitrified; water soaked—Cultured tissue having leaves and sometimes stems with a glassy, transparent or wet and often swollen appearance. The process of vitrification is a general term for a variety of physiological disorders that lead to shoot tip and leaf necrosis.

Vivipary—A form of reproduction in animals in which the developing embryo obtains its nourishment directly from the mother via a placenta or by other means.

V_{max} —The maximal rate of an enzyme-catalysed reaction. V_{max} is the product of E_o (the total amount of enzyme) times the value of K_{eat} (the catalytic rate constant).

VNTR—Variable number tandem repeat. A short DNA sequence that is present as tandem repeats and in highly variable copy number.

V regions—Variable regions in antibodies.

Volatilization—The conversion of a solid or liquid into a gas or vapour.

Water potential—Refers to the difference between the activity of water molecules in pure distilled water at atmospheric pressure and 30°C (standard conditions), and the activity of water molecules in any other system. The activity of these water molecules may be greater (positive) or less (negative) than the activity of the water molecules under standard conditions.

Water stress—The condition when plants are unable to absorb enough water to replace that lost by transpiration. The results may be wilting, cessation of growth, or even death of the plant or plant parts.

Wavelength—The distance between two corresponding points on any two consecutive waves. For visible light it is very small and is generally measured in nanometres.

Wax (A.S. *weax*, wax)—Esters of alcohol higher than glycerol, which are insoluble in water and difficult to hydrolyze; wax forms protective waterproof layers on leaves, stems, fruits, animal fur and integuments of insects.

Weed—Simply any plant growing where it is not wanted. In agriculture, used for a plant which has good colonizing capability in a disturbed environment, and can usually compete with a cultivated species therein. Weeds are typically considered as unwanted, economically useless or pest species.

Weediness— In agriculture, the ability of a plant to colonize a disturbed habitat and compete with cultivated species.

Western blot—A technique in which protein is transferred from an electrophoretic gel to a cellulose or nylon support membrane following electrophoresis. A particular protein molecule can then be identified by probing the blot with a radiolabelled antibody which binds only the specific protein to which the antibody was prepared. Useful, for example, for measurement of levels of production of a specific protein in a particular tissue or at particular developmental stage. *cf* blot; northern blot; southern blot.

Wet weight—The gross weight of a product with its full water content, or weight of fully hydrated tissue. *See* fresh weight.

Wetting agent—A substance that improves surface contact by reducing the surface tension of a liquid: *e.g.*, Triton X-10TM added to disinfecting solutions promotes the disinfestation process. *See* detergent; surfactant.

Wild type—An organism as found in nature; the dominant allele usually found in nature, and from which mutations produce other dominants or recessives alleles.

Wilt—Drooping of stems and foliage due to loss of water and decreased turgidity of cells. May be caused by water stress or by disease.

Wilting point (WP)—The moisture content of soil at which plants start to wilt, but not to the extent that they fail to recover when placed in a humid atmosphere. *cf.* permanent wilting point.

Wobble hypothesis—An explanation of how one tRNA may recognize more than one codon. The first two bases of the mRNA codon and anticodon pair properly, but the third base in the anticodon has some wobble that permits it to pair with more than one base.

Xanthophyll (Gr. *xanthos*, yellowish brown + *phyllon*, leaf)—A yellow chloroplast pigment.

X-chromosome—A chromosome associated with sex determination. In most animals, the female has two, and the male has one X chromosome. The opposite occurs in birds, in which the equivalent is the Z chromosome.

Xenia—The immediate effect of pollen on some characters of the endosperm.

Xenobiotic—A chemical compound that is not produced by living organisms; a manufactured chemical compound.

Xenotransplantation—The transplantation of tissue from one species to another species, typically from non-human mammals to humans. This technology has become very important because of a worldwide shortage of human organs for humans requiring a new organ. The most popular non-human species involved in this technology is the pig.

Xerophyte (Gr. *xeros*, dry + *phyton*, a plant)—A plant very resistant to drought or that lives in very dry places.

X-linked—The presence of a gene on the X chromosome. a.k.a. X-linkage.

X-linked disease—A genetic disease caused by an allele at a locus on the X-chromosome. In X-linked recessive conditions, a normal female 'carrier' passes on the allele on her X chromosome to an affected son.

X-ray crystallography—The deduction of crystal structure from analysis of the diffraction pattern of X-rays passing through a pure crystal of a substance.

Xylem (Gr. *xylon*, wood)—A complex tissue specialized for efficient conduction of water and mineral nutrients in solution. Xylem may also function as a supporting tissue, particularly secondary xylem.

Y-chromosome—The partner of the X-chromosome in the male of many animal species.

Yeast artificial chromosome— YAC—Yeast artificial chromosome, used as a vector system for cloning DNA fragments that can be hundreds of kilobases long. Linear cloning vectors constructed from essential elements of yeast chromosomes. They can accommodate foreign DNA inserts of 200 to 500 kb.

Yeast episomal vector (YEp)—A cloning vector for the yeast *Saccaromyces cerevisiae* that uses 2 μm plasmid as origin of replication and is maintained as an extrachromosomal nuclear DNA molecule.

Yeast extract—A mixture of substances from yeast. *cf* organic complex; undefined.

Yeast—A unicellular ascomycete fungus, commonly found as a contaminant in plant tissue culture.

Yeast cloning vectors—The yeasts, and especially *Saccaromyces cerevisiae*, are favourite organisms in which to clone and express DNA. They are eukaryotes, and so can splice out introns, the non-coding sequences in the middle of many eukaryotic genes.

Zinc finger Several types of conserved DNA-binding motifs composed of protein domains folded around a zinc ion; present in several types of eukaryotic transcription factors.

Zone of elongation—The section of the young root or shoot just behind the apical meristem, in which the cells are enlarging and elongating rapidly.

Zoo blot—Hybridization of cloned DNA from one species to DNA from other organisms to determine the extent to which the cloned DNA is evolutionarily conserved.

Zoo FISH Fluorescent *in situ* hybridization of DNA from one species on metaphase chromosomes of another species. Typically, the hybridization is done separately for DNA libraries representing each chromosome. The result is a fascinating picture of the regions of chromosomal homology between species.

Zoospore—A spore that possesses flagella and is therefore motile.

Zygonema (adj. zygotene)—Stage in meiosis during which synapsis occurs; coming after the leptotene stage and before the pachytene stage in the meiotic prophase.

Zygospore (Gr. *zygon*, a yoke + spore)—A thick-walled resistant spore developing from a zygote resulting from the fusion of isogametes.

Zygote (Gr. *zygon*, a yoke)—A diploid cell formed by the fusion of two haploid gametes during fertilization in eukaryotic organisms with sexual reproduction. It is the first cell of the new individual.

Zymogen—Inactive enzyme precursor that after secretion is chemically altered to the active form of the enzyme.

Z-DNA; zig-zag DNA—A form of DNA duplex in which the double helix is wound in a left-hand, instead of a right-hand, manner. DNA adopts the Z configuration when purines and pyrimidines alternate on a single strand, *e.g.*,

5´CGCGCGCG3´ or 5´CACACACACA3´
3´GCGCGCGC5´ 3´GTGTGTGTGT5´

Appendix A

Commonly Used Abbreviations

Ab	Antibody
AIDS	Acquired immune deficiency syndrome
ACP	Acyl carrier protein
ADH	Alcohol dehydrogenase
AMP, ADP, ATP	Adenosine 5'-mono di- and tri-phosphate
APCs	Antigen-presenting cells
cAMP	Cyclic AMP
cDNA	Complementary DNA
COA, COA-SH	Coenzyme A
ConA	Concanavalin A
CR	Complement receptors
DEAE-cellulose	Diethylaminoethyl cellulose
DNA	Deoxyribonucleic acid
DNase	Deoxyribonuclease
DNP	2,4-Dinitrophenol
DTH	Delayed type hypersensitivity
EBV	Epstein-Barr virus
EDTA	Ethylenediaminetetraacetic acid
EPR	Electron paramagnetic resonance
ESR	Electron spin resonance
FAD	Flavin adenine dinucleotide
$FADH_2$	Flavin adenine dinucleotide reduced
Met	N-formyl methionine
FMN	Flavin mononucleotide
$FMNH_2$	Flavin mononucleotide reduced
GDH	Glutamate dehydrogenase
GLC	Gas-liquid chromatography
Glu	Glutamine
cGMP	Cyclic GMP

GS	Glutamine synthetase
HbO_2	Oxyhemoglobin
HEPA	High Efficiency Particulate Air Filter
His	Histidine
HnRNA	Heterogeneous nuclear RNA
Ig	Immunoglobulin
Kb	Kilobase
LAF	Laminar airflow
LDH	Lactate dehydrogenase
Leu	Leucine
MAC	Membrane attack complex
MHC	Major Histocompatibility Complex
NAD+	Nicotinamide adenine dinucleotide
NADH	Nicotinamide adenine dinucleotide (reduced)
NADP+	Nicotinamide adenine dinucleotide phosphate

Index

A

B

C

D

E

F

G

H

I

J

K

L

M

N

O

P

Q

R

S

T

U

V

W

X

Y

Z